The Comparative Reception of Darwinism

The Dan Danciger Publication Series

The Comparative

Reception of

DARWINISM

Edited by THOMAS F. GLICK

UNIVERSITY OF TEXAS PRESS, AUSTIN AND LONDON

QH361
C67
1972

Library of Congress Cataloging in Publication Data

Conference on the Comparative Reception of Darwinism,
    Austin, Tex., 1972.
    The comparative reception of Darwinism.

    (The Dan Danciger publication series)
    "Held . . . under the joint sponsorship of the American
Council of Learned Societies and the University of Texas
at Austin."
    Includes bibliographical references.
    1. Evolution—History—Congresses. I. Glick,
Thomas F., ed. II. American Council of Learned Soci-
eties Devoted to Humanistic Studies. III. Texas.
University at Austin. IV. Title.
QH361.C67   1972   301.15′43′5750162   74–10797

CONTENTS

PREFACE

The Conference on the Comparative Reception of Darwinism was held in Austin, Texas, on April 22 and 23, 1972, under the joint sponsorship of the American Council of Learned Societies and The University of Texas at Austin. The scope of the meeting was adumbrated three years earlier in a presentation to the Spanish National Congress of the History of Medicine, at Valencia,[1] at which time it was apparent not only that there had been no comparative study of the impact of Darwin, but also that many, indeed most, of the national cases had not yet been studied in depth. Darwin himself commented on the differential reception of his ideas in different countries. ("It is curious how nationality influences opinion," he mused, noting the differing receptions accorded his work in France and Germany.) Yet since his time little had been done to examine this phenomenon systematically.

The conference, therefore, was meant to fulfill the double task of presenting a series of case studies on the reception of Darwinism in various European and non-European countries and of subjecting such data to comparative analysis. Accordingly, the interested historians at The University of Texas drew up a provisional framework for discussion, which was circulated to the participants in advance of the preparation of manuscripts. The framework was not intended to be rigid or even to suggest the actual organization of individual papers. It listed a number of rubrics that all participants, it was hoped, might assess whenever relevant in order to provide a basis for comparative analysis. The framework envisioned seven broad categories of discussion:

[1] Thomas F. Glick, "La recepción del darwinismo en España en dimensión comparativa," in *Actas: III Congreso Nacional de Historia de la Medicina (10–12 de Abril de 1969)*, 3 vols. (Valencia: Sociedad Española de Historia de la Medicina, 1971), I, 193–200.

I. The sequence and circumstances of the primary diffusion of the Darwinian corpus (translations, editions, reviews) and the works of major Darwinian biologists and apologists (e.g., reception of *The Origin of Species* and *The Descent of Man*; diffusion of works by Ernst Haeckel, T. H. Huxley, etc.).

II. The sequence of pro- and anti-Darwin arguments in the country, 1859–1885. The terminal date of the discussion must, of necessity, be left to each author. The idea is to cover the main polemical period, assess the situation after the main lines of pro- and anti-Darwin sentiment have crystallized, and discuss the general impact of Darwinism in the immediate postpolemical period.

III. Those factors in the society that encouraged or inhibited the reception of evolutionary ideas:

A. The influence of general philosophical or ideological trends upon the reception of evolutionary ideas.

B. The role of national scientific traditions; quality and methodological orientations of national scientists.

C. The attitude of scientific institutions.

D. Organized religious or political pressures (prohibitions of discussion, censorship of evolutionary books, etc.).

IV. The sociology of Darwinism. What were the social and political backgrounds and linkages of pro- and anti-Darwin forces?

V. Variable institutional, disciplinary, and regional penetration of Darwinism. Was Darwinism received more favorably in some institutions (educational levels, disciplines, regions, cities) than in others?

VI. The impact of Darwinism on subsequent scientific research. In what ways were directions of research altered as a result of the debate over evolution?

VII. The impact of Darwinism on other areas of intellectual endeavor; extension of evolutionary models to the social sciences.

As the early drafts were produced and circulated among the participants before the conference, it became clear that the original framework was too extensively conceived relative to the scope of coverage feasible at such a meeting. Although it was possible to cover the entire spectrum of

Darwinian literature in a small nation like the Netherlands or in those, such as Spain or Mexico, where scientific culture was thin, it was clearly impossible to provide such coverage for the countries of maximum productivity in evolutionary thought. The Russian coverage was divided into two presentations, one for the biological, the other for the social, sciences: a stratagem that might well have been used to advantage in other cases. Moreover, the lack of representation of several key national traditions, especially those of Italy, China, Japan, and orthodox Judaism, was keenly felt. Their exclusion was fortuitous; initial contacts had been made in these, as in some other cases that proved unworkable.

The conference itself was organized into three panels-of-the-whole, each led by a moderator, whose task was to organize discussion along comparative lines. The sessions were The Social Climate of Reception, moderated by Everett Mendelsohn; Darwinism and the Natural Sciences, moderated by David Hull; and Darwinism and the Social Sciences, moderated by Anthony Leeds. A fourth session, on the interrelation of Darwinism and religion, to be led by Harry Paul, was canceled for logistical reasons. Professor Paul's analysis in this volume, however, is an indication of the ground we hoped to cover in such a session.

The chapters of this book are revised versions of the original country papers, in addition to several comparative essays based upon them. Also included is Michele Aldrich's bibliographical essay, which was an integral component of the conference.

The Conference on the Comparative Reception of Darwinism was a collective endeavor, and all who participated contributed more than just an essay. But particular gratitude is owed to the two men who made the Conference possible, Stanley S. Ross, Provost of the University of Texas at Austin, and Frederick Burkhardt, President of the American Council of Learned Societies. My two chairmen, Standish Meacham and Clarence Lasby, were unremittingly encouraging and enthusiastic. Expenses in the preparation of the final manuscript were supplied by the Office of the Graduate Dean of Boston University. Finally, I cite Alex Vucinich and Jon Hodge for all those hours and lunches, as the conference progressed from idea to reality.

T. F. G.

The Comparative Reception of Darwinism

ENGLAND

M. J. S. HODGE

Since Darwin's reception in his own country often makes most sense when related to trends already present, this paper concerns England before, almost as much as England after, 1858.[1] The decisive early developments took place in the first ten years after the announcement of Darwin's theory. Any doubt that common descent—as distinct from natural selection—would prevail ended with George Bentham's presidential address (May 25, 1868) to the Linnean Society and Charles Lyell's new edition, the tenth, of his *Principles of Geology* (first volume, 1867; second volume, 1868). For these two cautious commentators and reluctant converts to descent tacitly agreed that there was no significant English opposition, nothing comparable to Louis Agassiz's defense of immutable species in the United States or the party in France advocating spontaneous generation.[2]

From the late sixties on, then, Darwinism was discussed in two principal contexts. In 1869 the Metaphysical Society began its long years devoted to the fundamental issues raised for science, theology, ethics, and philosophy by higher criticism of the Bible, Comtian positivism, the English Catholic revival, German idealism, thermodynamics, evolution, and nervous physiology. The society, far from enjoying a monopoly on such

[1] I am grateful to Dr. David L. Hull for a chance to consult his edition of the major reviews of the *Origin of Species* (*Darwin and His Critics* [Cambridge: Harvard University Press, 1973]).

[2] *Journal of the Proceedings of the Linnean Society of London* 1868: xciii–xciv; Charles Lyell, *Principles of Geology*, 10th ed., 2 vols. (London: John Murray, 1867–1868), II, 278–283.

subjects, merely gave the most concerted attention to topics any lively mind—like T. H. Huxley, Henry Edward Manning, or William Gladstone—had to confront.[3]

The other striking consequence of a widespread conviction that evolution was established was that natural selection was carefully scrutinized and found far less acceptable. Some of the objections were specific: selection, the critics reiterated, could not have produced complex organs like the vertebrate eye or refined faculties like human speech; but others were quite general: the Darwinian assumptions about heredity and geological time were disputed. Huxley's former student, St. George Mivart, and the future Lord Kelvin's protégé, Fleeming Jenkin, gave the most influential versions of argumentation that was, for the rest of the century, to keep all but a very few biologists in England from joining A. R. Wallace, Joseph Hooker, and, later, E. B. Poulton, in unqualified belief in the adequacy of natural selection.[4]

The early and mid-1860's mark, then, in the English response to Darwin, the crucial first phase that must be considered here. Many events in these years are, however, too well known to need recalling. Moreover, we already have an exemplary study, complete with statistical analysis, of the periodical press, one that fully confirms the impressions we might be likely to gather informally: that, for example, Unitarian and Broad Church publications accepted Darwin more readily than High or Low Church organs did; that many writers who accepted evolution rejected natural selection; that those who admitted one or both frequently made an exception of man; that so-called highbrow periodicals gave more information on Darwinian topics and more support for Darwinian views than did middlebrow periodicals, and these more than lowbrow ones.[5] It has, then, appeared most appropriate, here, to complement Dr. Burkhardt's paper on Darwin's reception in British learned societies by con-

[3] A. W. Brown, *The Metaphysical Society: Victorian Minds in Conflict, 1869–1880* (New York: Columbia University Press, 1947).

[4] P. J. Vorzimmer, *Charles Darwin, the Years of Controversy: The Origin of Species and Its Critics, 1859–1882* (Philadelphia: Temple University Press, 1970); Philip Gilbert Forthergill, *Historical Aspects of Organic Evolution* (London: Hollis and Carter, 1952).

[5] Alvar Ellegård, *Darwin and the General Reader: The Reception of Darwin's Theory of Evolution in the British Periodical Press, 1859–1872* (Göteborg: Elanders Boktryckeri Aktiebolag, 1958).

centrating on those aspects of the English case that either are not so familiar from the standard studies or have been given by those studies an interpretation that seems to need modifying.

HISTORIOGRAPHICAL DISTINCTIONS

When a physicist has to account for the response of a body of water to a falling stone, well-known laws establish what the most useful information will be. The stone's momentum but not its color, the water's depth but not its smell can help explain the waves resulting from the reception of the one by the other. Notoriously, in history we rarely have an equivalent general theory that indicates where in any particular case explanatory light is most likely to lie. In this situation, then, we must be especially careful to distinguish rather than conflate the various questions for which answers are being sought.[6] But just what distinctions does the present case demand? Consider two deliberately idealized narratives: (1) A professional scientist, an officer in the nation's new geological survey, attends in September, 1859, the Aberdeen meeting of the British Association for the Advancement of Science; there, in Sir Charles Lyell's presidential address to Section C, he hears Darwin's forthcoming publication praised as a valuable treatment of the origin-of-species problem. In due course, he buys, reads carefully, and talks over with associates the *Origin of Species*; five months later he is listed by Darwin as a supporter in a letter to Hooker. (2) Following another meeting of the association, at Oxford, a London apprentice overhears his employer commenting with horrified amusement on the implications of a recent cartoon in *Punch* that shows chimpanzees dining in formal attire, their keeper explaining to onlookers that, since an eminent man of science declared them ancestors to his own kind, they have settled for nothing less. The apprentice, invited after work to a secularist lecture given in a nearby Owenite Hall of Science, that evening joins his hosts in hailing a greater weapon than Tom Paine ever wielded, one that assures at last the demise of the clerical hegemony.

The first of these narratives reminds us to distinguish sharply between questions about how the book came to be read by those who read it and

---

[6] I am indebted in this historiographical section to discussions with Dr. Hull and Dr. L. L. Laudan.

questions as to why its argumentation was convincing to some and unconvincing to others. Queries of the former sort are of course rightly answered by citing people's social standing, personal friendships, professional duties, institutional affiliations, private curiosity, and the like. But questions concerning conviction surely require a very different treatment. For what, after all, did the *Origin* have to offer? And what would a professional geologist have brought to his reading of it? The book comprised, as Darwin said, one long argument, compounded naturally from many shorter ones. It is then correct but inadequate to describe the *Origin* as containing a theory—or two—plus so much evidence. For of course Darwin did not simply state his conclusions and then append a list of selected empirical data. Like most publishing theorists, he constructed an argumentative case deliberately designed for a definite readership— we know he even had particular individuals, like Huxley, in mind—people he thought might well come to share not only his beliefs but also his reasons for holding them.

To insist on this may look like begging questions, by taking Darwin's side from the start. But it is not. An earlier version of this paper noted that nothing is achieved by judging Darwin's theories to be true or his argumentation to be essentially strong and then using those judgments to explain why many adopted those theories in the 1860's. To this David Hull has replied that we must indeed invoke some estimate of the strength of Darwin's proposals relative to his generation's empirical information and theoretical presuppositions. But there is no contradiction here; rather, there is an important agreement worth brief explication. In this explication, it is probably best to set aside matters of truth and falsity, not because they are always inconsequential or straightforward—of course they are neither—but because their resolution is not required here. Darwin's theory of natural selection is true; his theory of pangenesis false; but in both cases we account for the theory's initial reception in the same way. For what we have to consider is not the proposal's presumably invariant truth or falsity (or essential rather than relative strength, whatever that might be) but the reasonableness or otherwise of particular people being convinced or unconvinced at a certain time by a particular presentation of it. The apprentice got, via *Punch*, his master, and his work-mates, certain reasons for embracing a palpable misrepresentation of one Darwinian conclusion; for these and other reasons, plainly provided by his social ideology, he came to his new convictions about human origins. The geologist, long, we may suppose, an admirer of Lyell's *Principles* and hitherto quite persuaded by its defense of immutable species, got from his mentor's speech reasons for resolving to give the promised contribution from Darwin—another geologist he would have respected —a fair and thorough hearing—but not, of course, for resolving to believe its conclusions, whatever they turned out to be. Eventually, he did believe them. Why? Because Darwin, in arguments aimed at exactly this kind of reader, gave him reasons good enough to convince without the exercise of any wish or will to believe. To reconstruct the grounds for this conviction we would need, then, only to assume that, while what there are good reasons to believe may change, what it is to believe for good reasons can remain much the same. So the suggestion is entirely sound that Darwin's acceptance in certain quarters is often appropriately explained as being due to his providing demonstrably better solutions to many of the problems he took up—with "better solutions" meaning, not solutions closer to today's orthodoxies, but solutions meeting more fully than any others the ideals of adequate evidence, consistent interpretation, and allowable assumption then prevailing among people concerned with those problems. Bentham and Hooker came to favor much of the Darwinian solution to the origin-of-species problem, not because it was going to meet our standards—if in fact it does—but because it met their own. Certainly, we may find the two sets of standards, ours and theirs, very much alike, but their numerical distinctness is not diminished by this contingent coincidence, and it is that distinctness that makes it possible to go beyond description to explanation in the intellectual history of science. When the time comes, then, that distinctness is all we will need to presume here. But, before getting to questions of conviction, we must explore questions of conveyance.

Conditions for Dissemination, Debate, and Distortion

The means whereby initial discussion of Darwinism spread quickly in England hold few surprises when we recall the general state of the nation. The Industrial Revolution was nearly a century old, and the 1851 census had found more than half the people now living in towns (communities of over five thousand inhabitants, that is). Most striking to contemporary observers was the rapid increase in the numbers, absolute and otherwise, of those in the "middling classes" with the time, money, and ambition to partake of the new railways, adult education, cheap books, and periodicals. The scientific community was already much involved in this increase, perhaps most obviously through the Mechanics Institutes; publishing series, like Murray's Colonial and Home Library, in which Darwin's *Journal of Researches* sold well at seven shillings and sixpence; and the British Association for the Advancement of Science, which, since the 1830's, had been convening each year in a different town or city, but never in London. For the 1859 meeting the president was no less a person than Lyell's friend Prince Albert, an ideal spokesman for the country's scientific interests, especially at a time when the threat from his native Germany's technical triumphs was on many minds. Attendances at the association averaged over two thousand in the 1860's, so that, with local and national publications giving the meetings extensive coverage, some information about the proceedings reached literally hundreds of thousands of homes.[7]

This information differed dramatically, of course, according to the distance from the inner scientific circles. The year 1869 saw the appearance of the tenth thousand of the *Origin of Species*, by then in its fifth edition.[8] Presumably, the book was now owned by most British biologists and by many other scientists as well. To a far wider public, of course, it was known through reviews as a contribution to general literature, for in fields like natural history and geology no sharp separation was as yet made between letters and science. In the quarterlies—the *Edinburgh*, *Westminster*, and *Quarterly* especially—the experts routinely reviewed each other's books much as historians and literary critics still do today. Physics no longer received such a treatment; though James Clerk Maxwell's *Treatise on Electricity and Magnetism* and Lyell's *Principles of*

[7] Ellegård, *Darwin and the General Reader*, Ch. 4, "Darwinism at the British Association."

[8] R. B. Freeman, *The Works of Charles Darwin: An Annotated Bibliographical Handlist* (London: Dawsons of Pall Mall, 1965), p. 44; cf. Darwin's letter to Lyell on January 14, 1860: "I never till to-day realised that it was getting widely distributed; for in a letter . . . a lady . . . says she heard a man enquiring for it at the *Railway Station!!!* at Waterloo Bridge; and the bookseller said that he had none till the new edition was out . . . he had not read it, but had heard it was a very remarkable book!!! . . ." (Francis Darwin, ed., *Life and Letters of Charles Darwin*, 3 vols. [London: John Murray, 1887; New York: Appleton, 1888], II, 266).

Geology fell within the same systematic scientific genre, they were not on a par for critical purposes.[9] But, faced with the *Origin of Species*, reviewers felt no insuperable expository difficulties. The University College, London, physiologist and Unitarian, W. B. Carpenter, reviewed the book both in the middlebrow Unitarian *National Review* and in the specialist *British and Foreign Medico-Chirurgical Review*.[10] He varied the style strikingly, but in both pieces we find the same attention to the Lyellian presuppositions of Darwin's geological and geographical theorizing, the various criteria current in counting species, the recent consensus in favor of fewer, wider specific distinctions, the use of embryonic characters in constructing natural classifications, and support for the stand taken against a mysterious or miraculous origin for species by the ultraliberal Anglican theologian and Savilian Professor of Geometry at Oxford, Baden Powell.[11]

The main trend discernible as one descends from the more highbrow discussions is that emphasis on such abstract and technical issues gives way to an intense preoccupation with the new doctrines' supposed implications for human interests and institutions. Some popular sectarian papers did not bother, apparently, in reporting the British Association's 1863 meeting, to discriminate between Huxley's Darwinian and the ethnologist John Crawfurd's polygenist, special-creationist accounts of man's origin—for both conflicted equally with scripture.[12] At other times the spectacle of eminent experts in vehement disagreement ran religion a good second in arousing readers' curiosity, with the "great brain battle"

9 W. F. Cannon, "Darwin's Vision in *On the Origin of Species*," in *The Art of Victorian Prose*, ed. George Lewis Levine and William Madden (New York: Oxford University Press, 1968), pp. 154–176.

10 [W. B. Carpenter], "Darwin on the Origin of Species," *National Review* 10 (1860): 188–214; idem, "The Theory of Development in Nature," *British and Foreign Medico-Chirurgical Review* 25 (1860): 367–404. This journal reviewed a far wider range of publications than its title might suggest: Henry Thomas Buckle's *History of Civilization in England*, for instance.

11 Powell's long essay, "On the Philosophy of Creation," in his *The Unity of Worlds and of Nature*, 2d ed. (London, 1856), pp. 329–512, is an invaluable analysis of the pre-Darwinian state of the origin-of-species topic. For a recent discussion of Powell, see R. M. Young, "The Impact of Darwin on Conventional Thought," in *The Victorian Crisis of Faith*, ed. Anthony Symondson (London: Society for Promoting Christian Knowledge, 1970), pp. 13–35.

12 See the quotations from the *English Churchman* in Ellegård, *Darwin and the General Reader*, p. 73.

between Huxley and Richard Owen yielding the best copy of all.[13] Again, with people everywhere closely following American events, the mutability of species could be cited against claims, often associated with Southern sympathies, that Negroes were another species separate *ab initio* from whites.[14]

Darwin often complained that competent naturalists failed to grasp what he was saying. Small wonder, then, that in less erudite circles there arose several bizarre misinterpretations. Perhaps the most persistent was the one Hooker corrected Samuel Wilberforce on.[15] Darwin, Hooker explained, was not maintaining that living species like man originated from other living species, for the common ancestors posited by the theory were very likely no longer represented, having long ago become extinct or transformed into new species. One reason for this widespread misunderstanding was the habit of construing any transmutations as movements up or down a universal scale of organic perfection, rather than as branching extensions to a tree of ancestry. Another patent travesty often accompanied this one, when the hardly recondite difference between changes in species and changes in individuals was ignored or deliberately suppressed. Reputable scientists did not help here by often writing "organism" when they obviously meant "species" or "race of organisms," for, by coupling this confusion with a parody of the ideas of Jean Baptiste Lamarck, many uncritical minds ended up convinced that the main issue was whether or not a monkey could with suitable exercise become a man[16]—or at least that was the impression given. For polemical, humorous, or promotional purposes—Darwin's name featured on at least one circus hoarding—a speaker or writer often did best to excite rather than inform. Magazines and novels portraying idle drawing-room chatter amply testify to what we might have suspected anyway: confronted with a deeply disturbing doctrine as to "man's place in nature," people welcomed any witticism belittling its gravity and the pretensions of the sol-

[13] Cyril Bibby, *Thomas Huxley: Scientist, Humanist and Educator* (London: Watts, 1959), pp. 73 ff.

[14] Ellegård, *Darwin and the General Reader*, pp. 301–302.

[15] See the account of the meeting in *The Athenaeum* 2 (July 1860): 64–65.

[16] E. B. Poulton gives several examples in "Theories of Evolution," in his *Essays on Evolution, 1889–1907* (Oxford: Clarendon Press, 1908), pp. 95–119. See also the advance notices of the *Origin* cited by Morse Peckham in the introduction to *The Origin of Species: A Variorum Text* (Philadelphia: University of Pennsylvania Press, 1959).

emn savants responsible for their perplexity.[17] Solemnity was in Victorian England at once a serious and a laughing matter. The new generation of earnest professional biologists, headed by Huxley, Owen, and Hooker, worked hard to end the public reputation of levity and irreverence formerly earned for their calling by people like William Buckland and Darwin's grandfather. The price they paid was joining the new civil servants as targets for the satirical jibes of *Punch* and Gilbert and Sullivan. But this they were happy to do. As several commentators have observed, Bishop Wilberforce's biggest blunder at Oxford was to sabotage his defense of a Christian cosmology as ethically indispensable by tastelessly and flippantly invoking the abhorrent specter of bestial miscegenation in recent human descent. For Huxley and Hooker could seize the opportunity to be "not amused" and to take the higher moral tone. Evident personal sobriety, charity, and sincerity could often discredit hasty opposition to unorthodox opinion, as many careers like Mill's and Darwin's revealed.[18]

THE PROFESSION AND PURSUIT OF SCIENCE

A major source of the trend toward professionalization in the mid-century period was the reforming tradition usually associated with a broadly liberal ideological outlook, a tradition marked by a willingness to change old institutions or devise unprecedented new ones in the name

[17] L. J. Henkin, *Darwinism in the English Novel, 1860–1910* (New York, 1940), provides many instances.

[18] As J. W. Burrow writes, in his introduction to *The Origin of Species*, "The episode epitomized one of the reasons why Darwinism made such headway among the educated classes. Infidelity and materialism had hitherto been generally associated with immorality and lower-class radicalism. As the geologist Hugh Miller remarked a few years earlier, 'it invariably happens that when persons of these walks [of life] become materialists, they become also turbulent subjects and bad men.' It was a deeply rooted assumption. Darwin was later rebuked for publishing *The Descent of Man* 'at a moment when the sky of Paris was red with the incendiary flames of the Commune.' It was of great importance that Darwin and Huxley were gentlemen and family men of complete financial, political and sexual respectability. Huxley turned the tables on opponents like Wilberforce by taking a higher moral line. It was immoral to believe without proof, to refuse, as he said, borrowing the idiom of religion, to 'sit down as a little child, before the fact'" (Charles Darwin, *The Origin of Species*, ed. J. W. Burrow [Baltimore, Maryland, 1968], pp. 41–42). Indeed, as far as respectability went, there were no embarrassing exceptions among the entire Darwinian party. Huxley's rhetoric was consistently indebted to religious morality.

of an appeal to national interests and majority opinion. This willingness often brought with it several characteristic procedures, instruments, and results: a commission of inquiry, a report reducing private prerogatives and substituting instead supervisory boards, and even competitive examinations, all of which increased the demand for competent personnel and for the men and facilities necessary for training them. The significance of such developments is obvious today, but older accounts of Victorian England sometimes neglect them, partly perhaps because these new pundits, as they were appropriately called, derived their income from neither rent, nor profit, nor labor and so find no natural place in classical views of the social consequences of industrialization.[19]

The Geological Survey is an excellent example, as several of Darwin's earliest supporters, such as Joseph B. Jukes, A. C. Ramsay, and Archibald Geikie, were all at one time on its staff. What this connection can explain is not, of course, why they welcomed Darwin's suggestions, but how, particularly under Lyell's and Huxley's influence, they came to ponder them with unusual care and interest. The survey began, typically, when an amateur enthusiasm within an old professional outfit was hived off to form a separate bureaucracy of its own. A Colonel Colby, superintendent early in the century of the army's trigonometrical survey and a keen member of the Geological Society of London, had various local geological mappings published with the Board of Ordnance's charts; so when, in 1835, Lyell, Adam Sedgwick, and Buckland joined together in successfully urging the public advantage of a more systematic survey, the resultant organization began within the military survey. Soon, however, it shed its uniforms and went civilian under the Office of Woods and Forests, so that, within months after Archibald Geikie started with the new organization in 1855, its recruits had to be approved by the new Civil Service Commissioners, who, following an important Order in Council, were to test all candidates for such jobs. Naturally, many suitable men were to come from the Government School of Mines in Jermyn Street, where Huxley held his principal post. Geikie's usual winter stay in London was prolonged in early 1860; for, with Ramsay ill, he had to lecture at the school. Presumably, he read the *Origin* at that time. He remembered

[19] A theme extensively developed in H. J. Perkin's *The Origin of Modern English Society, 1780–1880* (London: Routledge and Kegan Paul, 1969; Toronto: University of Toronto Press, 1969).

years later the deep impression made by the geological chapters; they offered a new revelation, he said, as to how his field must be studied.[20]

The efforts of broadly liberal factions toward reforming Oxford and Cambridge universities helped indirectly to establish Darwinian supporters there, too.[21] In the opinion of many concerned observers, like Lyell, the main needs were to restore to the university and to the professors privileges and duties wrongly usurped by the colleges and the tutors; to reduce ecclesiastical administrative influence; to end religious tests for matriculation, degrees, and fellowships; and to widen the range of subjects for the B.A. examinations, so that coming disciplines like philology, physiology, and modern history would stand equal with the traditionally dominant classics, mathematics, and theology. Such recommendations were sometimes summarized as Germanization.[22] Liberals often made their case for it by urging that the two senior universities were properly national institutions and should be fitted to that end. Judged thus, they were currently deficient; the high expenses and indulgent habits expected of undergraduates kept poorer, more industrious potential entrants away; and, with college instruction in many subjects in the hands of a few young tutors, there was no place for specialized research as the rightful complement to teaching in progressive fields. By the 1850's, such liberal hopes and cavils had taken effect. Parliament issued Royal Commissions

[20] See H. B. Woodward, *History of Geology* (New York and London: Putnam, 1911), and Archibald Geikie, *A Long Life's Work: An Autobiography* (London: Macmillan, 1924). Extracts from an important letter from Jukes to Darwin on February 27, 1860, were recorded by Lyell in his notebooks (*Sir Charles Lyell's Scientific Journals on the Species Question*, ed. L. G. Wilson [New Haven and London: Yale University Press, 1970], p. 361). Likewise a letter from A. C. Ramsay to Darwin, six days earlier, is printed there for the first time (ibid., pp. 355–356).

[21] See J. W. Adamson, *English Education, 1789–1902* (Cambridge: Cambridge University Press, 1930), Ch. 7, "The Universities in the Mid-Century"; and, more fully, Lewis Campbell, *On the Nationalisation of the Old English Universities* (London: Chapman and Hall, 1901); A. I. Tillyard, *A History of University Reform, from 1800 A.D. to the Present Time* (Cambridge: W. Heffer, 1913); and, more recently, W. R. Ward, *Victorian Oxford* (London: Cass, 1965).

[22] Lyell gave a harsh critique of Oxford and Cambridge in his *Travels in North America*, 2 vols. (London: John Murray, 1845), I, 270–316. William Whewell responded remonstratingly, as reported in Lyell's letter to George Ticknor, April 2, 1847, in K. Lyell, ed., *Life, Letters and Journals of Sir Charles Lyell, Bart.*, 2 vols. (London: John Murray, 1881), II, 127. For Germanization, see G. Haines, "Technology and Liberal Education," in *1859: Entering an Age of Crisis*, ed. Philip Appleman, W. A. Madden, and Michael Wolff (Bloomington: Indiana University Press, 1959).

of Inquiry—Lyell testified before the Oxford one—and Parliamentary Acts (Oxford, 1854, and Cambridge, 1856) adopted several of the commissions' Germanizing recommendations. Independently, those governing both universities were granting scientific subjects a larger place in undergraduate and graduate life: witness, most prominently, that very expensive and edifying monument to long efforts to make academic study of nature respectable, the University Museum at Oxford, which still needed its finishing touches when Huxley and Wilberforce clashed in the library there.[23]

Expansion of modern studies increasingly opened colleges, laboratories, and lecture halls to the influence of Scotland, Germany, and London, whence often came the new appointees. University College, London —started in the 1820's by a scientifically minded group of Benthamites and religious dissenters, including Henry P. Brougham and George Grote, who had looked to Berlin and also, it seems, to Jefferson's Virginia for their models—was an important source, and its registrar, W. B. Carpenter, a key person.[24] Michael Foster, a future Darwinian from a Baptist family, trained there in 1849–1852, later substituted for Huxley at the Royal Institution, became in 1870 praelector in physiology at Trinity College, Cambridge, and eventually was that university's first professor of his subject. Earlier, in 1860, George Rolleston, having been an Oxford undergraduate, returned, after medical training at a London teaching hospital with Huxley's revered friend, Sir William Lawrence, to be the first Linacre Professor of Human and Comparative Anatomy at Oxford. Prompted by the debate over primate brains to look into ape anatomy, he sided with Huxley against Owen at the British Association's Cambridge meeting in 1862, having lectured in January on his results at the Royal Institution, where Huxley's future partner in agnosticism, John Tyndall, was now the director.[25]

[23] On the University Museum and its significance, see H. M. Vernon and K. D. Vernon, *A History of the Oxford Museum* (Oxford: Clarendon Press, 1909), and H. W. Acland and John Ruskin, *The Oxford Museum: From Original Edition, 1859, with Additions in 1893* (London: G. Allen, 1893).

[24] Carpenter's various positions in London are described in a "Memorial Sketch" by J. E. Carpenter, in his edition of W. B. Carpenter's *Nature and Man: Essays Scientific and Philosophical* (New York, 1889).

[25] See *Dictionary of National Biography*, s.v. "Foster, Michael," and "Rolleston, George"; George Rolleston, "On the Affinities of the Brain of the Orang Utang,"

The professionalization of science and, as some reformers called it, the nationalization of the older universities, opened up routes and sponsored agents that took Darwinian topics and convictions into several new or newly altered institutions.

SOCIAL THEORIES AND POLITICAL IDEOLOGIES

From time to time one hears a great many concepts current in the last century loosely attributed to Darwin's influence. Of course in most cases we know that it simply was not so. Laissez-faire politics and economics, historicist interpretations of ancient and modern civilizations, educational psychologies that have the child recapitulate the previous mental progress of all mankind, and comparative developmentalist (or, as they are more often, but more misleadingly, called, evolutionist) theories of society—all these, needless to say, flourished in many variants in the century's first half; most, indeed, go well back into the Enlightenment, and some even further. The question arises, then: if these things came too early to owe anything to Darwin, did his acceptance sometimes owe something to them? We may be tempted, not only to presume that it did, but also to justify the presumption by alleging some inherent kinship between these doctrines and Darwin's account of species origins. But obvious analytical considerations show how swift intuitions about affinities can mislead us here. Let us recall briefly what the *Origin* proposed and presupposed. As the reviews noted, the leading presuppositions came from Lyell: an exchange of new species for old has gone on steadily everywhere, ever since the oldest fossiliferous rocks were formed, will surely continue in the future, and is, therefore, going on right now, though too slowly to be noticeable. To this Darwin added that the new species originate in conformity, not to any law of adaptation to conditions nor to any law of progress, but rather to the law of common descent. Within any natural grouping of species there has been an irregularly branching adaptive divergence, in which new species have arisen from pre-existing varieties, the main agent in the conversion of varieties into species being a natural selection of hereditary variants in the struggle for existence.

Natural History Review: A Quarterly Journal of Biological Science 1 (1861): 201–217.

Return now to the economic, political, historiographical, psychological, and anthropological views just mentioned. All their theses, normative and descriptive alike, were solely about man, whether he was considered as one species or many and however he might be deemed to differ from the highest animals. Hence, adherence to one or more of these doctrines was quite consistent with holding any or no theory about the origin of species in general or man in particular. Moreover, if we try to find in Darwin a more extended use of concepts already at work in these doctrines about mankind, we get in almost every case a negative result. First, as has often been noted recently, Darwin had no explanatory use for any general criterion of progress or for any universal laws of development: the only criterion of improvement he recognized was the purely relative one of competitive fitness between contemporary groups with overlapping ranges and vital needs. Second, while he and Wallace did use the populational arithmetic Thomas Malthus made notorious, it was this, not his political economy, that they, like Malthus himself, reckoned to be true of all species.[26] This point, too, is an obvious one, but none the less fatal to that remarkably persistent thesis, pioneered almost jokingly in letters between Karl Marx and Friedrich Engels, that, in their use of Malthus, Darwin and Wallace were extending the laissez-faire capitalist ethos from society to all nature to make a *Weltanschauung* out of the new captain of industry's utopia of progress through unfettered struggle. This thesis can be in no way strengthened by noting what is

[26] This distinction is fully observed in the valuable discussion of Darwin's debt to Malthus in R. M. Young's "Malthus and the Evolutionists: The Common Context of Biological and Social Theory," *Past and Present*, no. 43 (1969): 109–145; but, along with other distinctions, this one is lost sight of in Young's "The Impact of Darwin on Conventional Thought," p. 15: "The debate which we summarize by the idea of evolution also embraced the associationist, utilitarian philosophy of the Philosophic Radicals, Bentham, the Mills, and their followers, who would apply natural laws to men and morality and apply sanctions to induce men to act for the greatest good of the greatest number. The pleasures and pains of utilitarian psychological theory became the rewards and punishments of radical reform movements. In effect, Darwin extended this point of view to the ultimate natural sanctions of survival or extinction. One can also say that Darwinism was an extension of *laissez-faire* economic theory from social science to biology. A similar naturalism was at work in the influence of German historical methodology, which suggested that the Bible should be examined *scientifically* like any other historical document." See also Young's "Evolutionary Biology and Ideology: Then and Now," in *The Biological Revolution*, ed. Watson Fuller (Garden City, N.Y.: Doubleday, 1972), pp. 241–282.

true: that Malthus argued from any species's tendency to exhaust its resources to the conclusion that charity toward paupers will eventually
bring even more misery, so that the Poor Law violates the greatest-happi-
ness principle. Perhaps it will be objected that the points made here are
misplaced, on the grounds that Darwin's public did, in fact, see him as
implicitly endorsing Malthusian attacks on state aid to the impoverished
or Ricardian defenses of free enterprise or even the average man's concern to "get on" in a competitive society. However, there appears to be
no solid evidence from the 1860's for this generalization, and the most
pertinent social ideology at that time was something very different, something related to Darwinian theories in another way altogether.[27]

Herbert Spencer and A. R. Wallace are probably the best examples
here, for though they parted ideological company much later on, when
Wallace came out for socialism and against Spencer's individualism, their
earlier concordance was almost total.[28] What had impressed Wallace, in
Spencer's *Social Statics* of 1851, was the sweeping proposal that, through
the improving adaptation of men to "the social state," any society inevitably advances toward a future when each citizen's moral sense of others' rights will be as strong as that of his own, so that all need for governmental restraints on individual liberties will disappear. Intent, above
all, on making ethics scientific, Spencer hoped to promote realization of
what he still construed, deistically, as a Divine Idea, the greatest happi-

---

[27] Indeed, if any association with laissez-faire theories, via Malthus, did ever play a
significant part in early English debates, it would seem to have been to Darwin's dis-
advantage. Thus Ellegård writes in his conclusion: "It is noticeable that the arguments
specifically designed to support Natural Selection were sparse indeed. For instance,
though both Darwin and Wallace explicitly acknowledged their debt to Malthus, the
resemblance between Natural Selection theory and the economic theory of *laisser-faire*
[*sic*] was not often advertised on the Darwinian side: instead, opponents used it to
discredit Darwinism. The Manchester School was not popular in the 1860's" (*Darwin
and the General Reader*, p. 334). However, Ellegård gives no direct evidence for this
claim. To a letter to Lyell on May 4, 1860, Darwin appended a postscript: "I have noted
in a Manchester newspaper a rather good squib, showing that I have proved 'might is
right' and therefore that Napoleon is right and that every cheating tradesman is also
right" (quoted from facsimile of the original letter in J. C. Greene, *The Death of
Adam: Evolution and Its Impact on Western Thought* [Ames: Iowa State University
Press, 1959], p. 308).

[28] See especially Ch. 34, "Land Nationalization to Socialism, and the Friends They
Brought Me," in A. R. Wallace's *My Life: A Record of Events and Opinions*, 2 vols.
(London: Chapman and Hall, 1905; republished in facsimile, 1969).

ness. He hoped to do this by showing that the essence of progress, in nature or society, is individuation with differentiation and mutual adaptation among the members, whether these be animal cells or human beings. F. W. J. von Schelling (as rendered by Samuel Coleridge) and later Karl von Baer provided his program with the requisite general account of differentiation; Lamarck's theory of the inheritance of acquired characters and later the Darwinians' principle of natural selection supplied suitably general theories of adaptation. We can see, then, how Wallace, long an ardent follower of Robert Owen's secularist philosophy and cooperative commercial ventures, would find consilience, not conflict, between Spencer's championing of the individual—especially the man of business—against the state and the loyalty to the union movement that Wallace shared with his older brother (a journeyman carpenter). We can see, too, how he could recommend natural selection as both explaining the present diversity among human races and making more credible than ever a distant but terrestrial millennium that would see that diversity vanish, with everyman becoming, if not an angel, at least an aristocrat. In a declaration extravagant even by the standards of the rather cranky Anthropological Society, he predicted that natural selection, having eliminated all but the most "intellectual and moral" race, would then improve the adaptation of these men "to the exigencies of the social state" and so produce a "paradise" in which "perfect freedom of action will be maintained," with the "well balanced moral faculties" permitting no transgression "on the equal freedom of others" and "compulsory government" replaced by "voluntary associations for all beneficial public purposes."[29] Unfortunately, as Wallace later recognized, any relation between fitness and happiness would be, at best, a contingent correlation. And, reviewing the *Descent of Man*, he admitted that, if Darwin was right, selection would sometimes increase the competitive survival chances of a tribe without further enhancing its "moral sense."[30]

[29] A. R. Wallace, "The Origin of Human Races and the Antiquity of Man Deduced from the Theory of 'Natural Selection,'" *Journal of the Anthropological Society of London* 2 (1864): clviii–clxxxvii. A verbatim report of the lively discussion following the paper is included. On the society, see J. W. Burrow, "Evolution and Anthropology in the 1860's: The Anthropological Society of London, 1863–71," *Victorian Studies* 7 (1963): 139–154.

[30] A. R. Wallace, Review of *The Descent of Man*, *The Academy* 2 (March 15, 1871): 178.

However, even to make such factual claims about the natural origins of the moral sense was to take sides, not on what our values should be, but on how our evaluations are to be understood. Late in the eighteenth century, English writers like William Paley had commonly paired a utilitarian ethic with a conventional Christian cosmology. But by the 1850's William Whewell and Adam Sedgwick, though admirers of Paley's natural theology, had come out against utility and the moral sense, urging instead an intuitional ethics indebted to Kant and akin to a much older Cambridge Platonism. By means of a Platonic theology, they joined this easily enough with a providential progressionist geology of which we will hear more shortly. By contrast, of course, Spencer had tried to go beyond abstract utilitarian principles by combining his completely naturalistic cosmological account of progression with a detailed determination of the prospects for greater happiness in diverse social systems and circumstances. There was no mistaking how irreconcilable these two mid-century programs were: the one Christian and conservative, the other secular and radical. Needless to remark, it had not been Darwin's original intention to contribute to either, but the idea of descent through natural selection found a ready use in one and not in the other.

Revealed and Natural Theology

A paradox may now be resolvable. Darwin offered no cosmology at all, only explicitly limited nomological claims about a restricted set of events, the origins of the extinct and extant organic species that had lived on this planet since some unspecified time before the most ancient fossils then known were formed. But discussion repeatedly escalated to the cosmological limit. Why? Part of the answer lies in Spencer's (and Robert Chambers's) earlier decisions to seek in cosmology their understanding of human morals. Another part lies in what positivists were, patronizingly but reasonably, to call the anthropomorphic theology and anthropocentric geology of those who, like Whewell and Sedgwick, opposed Lyell's teachings.

Sedgwick's *Discourse on the Studies of Cambridge University* (5th ed., 1850) and Whewell's *Plurality of Worlds* (2d ed., 1854) defended the likelihood that the earth, having started as an uninhabitable chaos, had had a physically and organically progressive history that in some sense led up to and culminated in the appearance of man, the "crown of creation" in the Apocryphal phrase; for God, after arranging for the earth to be successively improved, then created on it higher and higher types of creatures adapted to the improved conditions.[31] Once understood, awareness of this thesis saves us from two errors. The first would be to follow Huxley in making Moses responsible for Sedgwick's or Whewell's hostility to Darwinism: an error, for Whewell was in the 1850's as adamant as anyone in the belief that the Mosaic cosmogony was entirely superseded, even in its chronology for the most recent arrival, man. The other error would be to think that anyone was likely at this time to be a Platonist first and a Christian second. The Divine Ideas were meant to account, not only for the providential plan in the world, but also for the ability of the human intellect (a unique image of God) to understand that plan and his own privileged place in it. The difficulties between this cosmology and Darwinism cannot be reduced, then, to conflicts over the argument from design and miracles, though these were of course obvious cruces.

Having conceded twenty years before that geology could be no direct help to Biblical theology—in, say, evidencing the Noachian flood—Sedgwick had been all the keener to insist on his discipline's indirect and analogical service to Christianity through natural theology.[32] Not only was paleontology to enlarge our knowledge of adaptive design; it was also to provide an analogy in the book of nature to the progressive revelation recorded in scripture. An embryo's eyes fit it in the womb to its future adult needs "and so demonstrate provident intelligence," said Sedgwick, continuing unequivocally: "Should any one deny conclusions such as these, I can only reply, that his mind is differently constituted from my own, and that we have no common ground on which to build a reasonable argument."[33] Years later he wrote in his copy of Richard Owen's discourse *On the Nature of Limbs*: ". . . analogy one creation prelude to another—so also—one revelation prelude to another—it is progressive."[34]

Such passages show us why calmly reasonable argument was simply

[31] William Whewell, *Of the Plurality of Worlds: An Essay. Also a Dialogue on the Same Subject*, 2d ed. (London: J. W. Parker, 1853), pp. 198–199.

[32] See, especially, Adam Sedgwick, *A Discourse on the Studies of the University* (Cambridge and London: J. W. Parker, 1833), pp. 104–106.

[33] Ibid., pp. 20–21.

[34] Richard Owen, *On the Nature of Limbs* (London, 1849). Sedgwick's signed copy of this work is now in the library of the University of Toronto.

not to be expected from many even of the most liberal scientist-theologians. But, more than that, they indicate the significance of a strange move, common among people explicitly in sympathy with Darwin's case for common descent. In this move one construed descent as "evolution" as embryogeny writ large, so that the providentialist commitments to a progress, including that toward man, all planned and designed beforehand, could be preserved. Ironically enough, the most striking instance of this move was made by Lyell himself. His notebooks show him, in the early 1860's, fully aware that three positions—his old antiprogressionist doctrine of the *Principles of Geology*, Darwin's new antiprogressionist teaching in the *Origin of Species*, and any version of a planned development through progressive descent—were quite distinct and irreconcilable.[35] And no less emphatically, largely in order to save the privileged status of man, with his unique moral and intellectual relation to a benevolent, omniscient God, Lyell (to Darwin's candid dismay) embraced the third option, complete with the corollary that there must be, in addition to what he deemed the purely eliminative power of natural selection, some truly productive, creative natural agents capable of reliably executing the ordained evolution leading to the highest plants, animals, and man.[36] The most succinct statement of this common position is probably that given by Carpenter, in an essay on "Darwinism in England" written during a Mediterranean cruise and first published "in a little Valetta journal named *Il Barth*."[37] After making the usual point that the variation that selection accumulates has yet to be traced to a cause, Carpenter goes on:

Consequently we must look to *forces* acting either *within* or *without* the organism, as the real agents in producing whatever developmental variations it may take on. Of the action of such forces, we at present know scarcely anything; and Mr. Darwin has not given us much help towards the solution of the problem. But this much seems to me clear: that just as there is at the present time a determinate capacity for a certain fixed kind of development in

[35] Lyell, *Sir Charles Lyell's Scientific Journals*, pp. 336–338, 382–383, 406–409, 418–423; idem, *Principles of Geology*, II, 485–494.

[36] See note 35. On Lyell, descent, and man, see, further, an article devoted to this by Michael Bartholomew, "Lyell and Evolution: An Account of Lyell's Response to the Prospect of an Evolutionary Ancestry for Man," *British Journal for the History of Science* 6 (1973): 261–303.

[37] Carpenter, *Nature and Man*, p. 105.

each germ, in virtue of which one evolves itself into a zoophyte, and another into a man, so must the primordial germs have been endowed each with its determinate capacity for a particular course of development; in virtue of which it has evolved the whole succession of forms that has ultimately proceeded from it. That the "accidents" of Natural Selection should have *produced* that orderly succession, is to my mind inconceivable; I cannot but believe that its evolution was part of the original Creative Design; and that the operation of Natural Selection has been simply to limit the survivorship, among the entire range of forms that have thus successively come into existence, to those which were suited to maintain that existence at each period.[38]

For Carpenter, "evolution," if intelligible at all, must have proceeded according to some order that was in principle specifiable beforehand. Natural selection is not an explanatorily sufficient "force," because it does not operate according to an established law and order in its effects. These demands for the intelligibility of "evolution" prevented people like Carpenter from sharing one of Darwin's deepest assumptions: that, precisely because in descent there is no evidence of such an intelligible order, natural selection might provide a suitable explanation. Darwin had assumed that the very adaptive divergence entailed by the positing of common ancestries itself discredited any appeal to determinate capacities analogous to what seem evidenced by embryos. Which of the actual and possible descendant species did the ancestor have a "determinate capacity" to grow into? All equally, must be the answer; but then the problem becomes the one natural selection was to solve: how can indefinitely many species descend from just one, with all kept adapted throughout the entire time to endlessly shifting conditions? The analogy between growth and descent (so strongly hinted at by the word *evolution*) was so attractive metaphysically that it went unnoticed how completely Darwin's theorizing presupposed a denial of it.[39]

---

[38] Ibid., p. 110. Ellegård has rightly stressed the metaphysical gap between this position and a whole-hearted acceptance of Darwin's own position (*Darwin and the General Reader*, p. 31).

[39] The significance of the lack of this analogy in Darwinian theory is analyzed in my paper, "The Universal Gestation of Nature: Chambers' *Vestiges* and *Explanations*," *Journal of the History of Biology* 5 (1972): 127–151. On the instructive history of the word "evolution," see a paper forthcoming on that subject, by P. J. Bowler, in the *Journal of the History of Ideas*.

PHYSICS, METHODOLOGY, AND ONTOLOGY

On one common mid-century interpretation, Newtonian physics met just those demands for intelligibility that Carpenter found unsatisfied by natural selection. A commitment to Newton and to God could, then, lead very directly to a rejection of Darwin. William Hopkins, famed Cambridge mathematics coach, physical geologist, and evident reader of Whewell's philosophy of science, presents the best example of this in his long review in *Fraser's Magazine*.[40] Citing as instances the Keplerian laws of planetary motion and the optical laws of refraction and polarization, he distinguishes a class of "laws of phenomena." By contrast, the inverse square law of gravitation and the laws of wave propagation he calls laws of "physical causes." A "proof" of a hypothesis concerning a physical cause accordingly consists, "first, in the accuracy with which we can determine its necessary consequences," and, second, in the degree of accord "between those consequences and the existing observed phenomena." Gravitational theory excels on both counts; the "undulatory theory" of light also does respectably; but, says Hopkins, other physical theories, like those proposed for heat, magnetism, and electricity, are mostly far inferior.[41]

To his distinction of two kinds of laws and his criteria of "proof," Hopkins adds the stipulation that "explanation," as opposed to "mere description," must provide a deduction from proven laws of physical causes. This requirement determines his whole appraisal of the *Origin*. For, of course, the only candidate for a physical cause that he finds there is natural selection; but no explanation is achieved by this, because there is no acceptable proof that, in the initial conditions obtaining, this "force" has necessitated any of its supposed effects. Moreover, since the law of common descent is, considered on its own, only a "descriptive" law, it can have, Hopkins insists, no explanatory power to solve problems in biogeography and comparative anatomy. It was, manifestly, his ontology as much as his methodology that sustained Hopkins's dissatisfaction. Explanatory causes had to be for him either Newtonian forces or some other agents that, like them, lawfully mediate between God and matter.

---

[40] William Hopkins, "Physical Theories of the Phenomena of Life," *Fraser's Magazine* 61 (June 1860): 739–752; 62 (July 1860): 74–90.
[41] Ibid. 61 (June 1860): 740.

In denying that the doctrine of successive creations is—and so could be found wanting as—a "physical theory," he implicitly indicates what a satisfactory theory would have to provide. No more, he says, "than that of final causes" does the doctrine of creations "pretend to be a *physical theory*." It does not pretend to "define the nature of any physical cause" responsible for "certain extraordinary phenomena of nature." Rather, it expresses disbelief "that such phenomena are, like ordinary phenomena, due solely to the action of ordinary natural causes," together with belief in "some higher order of causation in nature" than that "usually recognised in secondary causes," whether this higher causation is to be sought in "properties originally impressed on matter by the Creator," or in "a more immediate emanation from the Divine mind."[42]

COMPARATIVE ANATOMY, BIOGEOGRAPHY, AND BIOLOGY

Some of what has been discussed so far may appear to have been rather peripheral to the real business, so to speak, of the initial scientific assimilation of Darwin's proposals in England. Indeed Darwin and his principal associates, Hooker and Huxley, often did discount four major kinds among early English reactions: those from long-standing opponents of Lyell, such as Sedgwick, Owen, and Whewell, who were predictably hostile for the reasons just sketched; those from anyone like Hopkins, who, as a physicist, measured theoretical innovations in biology against the arguably irrelevant standards of Newtonian science as he understood it; those from unambitious, conservative naturalists who, however competent in systematic biology, dismissed all theorizing about the origin of species as inconclusive, divisive speculation; and finally those from popular, religious, or literary critics who, whether favorable like G. H. Lewes or defiant like Wilberforce, could claim—laymen's interests or first-class degrees in mathematics notwithstanding—no regular membership in the researching community. This was, moreover, a reasonable attitude. But what made it so was the contemporary situation in various areas of biological inquiry, areas where descent or natural selection plainly promised to supply genuine solutions to difficulties already recognized as pressing. The *Origin* reveals other possible choices, but

42 Ibid. 62 (July 1860): 84.

two cases will be discussed here that show how even that mundane scientific business was often conducted along lines explicable only when referred to themes developed above.

The first concerns comparative anatomy. As Rolleston noted in 1861, it was then acknowledged, "even in England," that normal organisms possess "many structures" for which "no teleological explanation will suffice"; and, he says, theories to explain them have proliferated as nowhere else and never before. Of the four he summarized, two, "adherence to type" and "genealogical, yet modified, transmission," that is, descent, concern us here.[43]

Adherence to type had recently become in England an explanatory rather than a merely descriptive claim, mainly through Owen's efforts. The *explicandum* was quite uncontroversial. Running throughout many divisions of the animal and plant kingdom were structural resemblances answering to no functional similarities. To take the favorite example, not only does the pentadactyl structure in the monkey's grasping hand, the mole's shoveling paw, and the sparrow's flying wing resist any single teleological explanation—its presence in the penguin's swimming flipper, where its details make no contribution to functional proficiency, resists any such explanation at all. To summarize this conclusion by saying that a common plan or a unity of type transcends a diversity in ends offended no one's ontological inhibitions. However, Owen had followed Continental authors in explicitly Platonic interpretations. Many diversely adapted species can be made from "the same materials" because in each a different, active, specific *idea* controls development, while their common organizational features, found especially in serial parts, are due to an antagonistic "polarizing force." Thus any vertebrate body comprises various modifications of a standard structural unit—best exemplified perhaps by fish gill-arches—the "Archetype," the most general *idea* found in the group. From here, of course, Owen could move empirically to the possibility that, to give the most famous case, vertebrate skull components were homologous with ordinary vertebrae. Not so empirically, he

---

[43] George Rolleston, "On Correlations of Growth, with a Special Example from the Anatomy of a Porpoise," *Natural History Review: A Quarterly Journal of Biological Science* 1 (1861): 484–485.

could appeal to God's power and wisdom in adapting a few common plans to many diverse ends.[44]

To these moves Huxley had in the 1850's responded in his usual deflationary empiricist way. As a descriptive abstraction in our minds, a common plan was methodologically acceptable, indeed indispensable. But each application must be tried and that to the vertebrate skull rejected, he maintained in a long paper of 1858.[45] Moreover, physiology and embryology gave no support for any special adaptive and general polarizing forces mediating between the divine mind and ordinary animal matter.

To Huxley, then, Darwin offered an explanatory program perfectly contrasting with Owen's. Empirical common ancestors were to replace transcendent common archetypes. The replacement may appear neither subtle nor novel, but in fact it was both. No one prior to Darwin had put mutable species into the service of *common* descent, nor had earlier explanations for divergence assumed that it was always adaptive. Certainly, Huxley only admitted natural selection as a provisionally legitimate hypothesis, but, just as common ancestry was a known cause of similarities, so hereditary variation and potential geometrical increase were, in wild species, known powers capable, as domestic selection showed, of producing limited adaptive divergence. To make these assumptions explanatori-

[44] Richard Owen, *On the Archetype and Homologies of the Vertebrate Skeleton* (London, 1848), p. 172. See Huxley's much later comments on these passages in his essay, "Owen's Position in the History of Anatomical Science," in *The Life of Richard Owen*, ed. Richard Owen, 2 vols. (London: John Murray, 1894; facsimile reprint: Gregg, 1970), II, 273–332.

[45] T. H. Huxley, "On the Theory of the Vertebrate Skull," *The Scientific Memoirs of Thomas Henry Huxley*, ed. Michael Foster and E. R. Lankester, 4 vols. (New York: D. Appleton, 1892–1898), I, 538–606. This paper was a Croonian Lecture given before the Royal Society, June 17, 1858. It appeared in the *Proceedings of the Royal Society* 9 (1857–1859): 381–457, and in *Annals and Magazine of Natural History* 3 (1859): 414–439. See also idem, "On the Common Plan of Animal Forms," Abstract of a Friday Evening Discourse at the Royal Institution, May 12, 1854, in *Scientific Memoirs*, I, 281–283, and *Proceedings of the Royal Institution of Great Britain* 1 (1851–1854): 444–446. Spencer gave a thorough critique of Owen's views, in the *British and Foreign Medico-Chirurgical Review* (October 1858), which he included as an appendix to *Principles of Biology* (2 vols. [New York: D. Appleton, 1897], II, 517–535). On Owen and Darwinism, generally, see R. M. MacLeod, "Evolutionism and Richard Owen, 1830–1868: An Episode in Darwin's Century," *Isis* 56 (1965): 259–280.

ly useful, one extrapolated from the many races or varieties of a species to the many species of a genus, order, or class, an extrapolation sanctioned, for Huxley, by its analogy with the whole Lyellian program in geology and paleontology, where horizontal extrapolations from the "present system of nature" were consistently preferred over vertical escalations to causes, natural or otherwise, more powerful than any now in action.

Common ancestors, being as fully adapted to their way of life as any descendants would be to theirs, might well be as specialized. Moreover they, or their close relatives, might persist long after descendant groups had sprung from them. Huxley welcomed *Archaeopteryx*, in 1862, and certain ancient reptiles, in 1868, as extinct links with an earlier descent of birds from quadrupeds.[46] His 1862 anniversary address to the Geological Society, in assembling the evidence against Owen's claims for a progress within the known fossiliferous rocks from "embryonic" or "generalized" to specialized types in any group, argued throughout for the concluding endorsement of Darwinism: "Contrariwise, any admissible hypothesis of progressive modification must be compatible with persistence without progression through indefinite periods. And should such an hypothesis eventually be proved to be true, in the only way in which it can be demonstrated, viz., by observation and experiment upon the existing forms of life, the conclusion will inevitably present itself, that Palaeozoic, Mesozoic, and Cainozoic faunae and florae, taken together, bear somewhat the same proportion to the whole series of living beings which have occupied this globe, as the existing fauna and flora do to them."[47]

While Huxley had good explanatory reasons for favoring ancestors as an alternative to archetypes, his reasons for welcoming natural selection were more negative: there was, in the early 1860's, nothing to rule it out as a suitable complement to an irregular branching descent, but it had no overwhelming advantages. We may ask, then, as our second case, whether there were people with good reasons for going farther than this, to the

[46] T. H. Huxley, "The Coming of Age of 'The Origin of Species,'" in *Darwiniana* (London: Macmillan, 1893), pp. 227–243.

[47] "Anniversary Address to the Geological Society," delivered February 21, 1862, first published in *Quarterly Journal of the Geological Society of London* 18 (1862): xl–liv; reprinted in Huxley, *Scientific Memoirs*, II, 512–529.

claim that natural selection was not only acceptable but compellingly so? There were; they were those who, like Hooker, Henry W. Bates, and Wallace, primarily concerned themselves with geographical—including what we would call ecological—questions about species and varieties. Again a brief sketch of earlier developments is necessary. In the 1840's, Hooker and Edward Forbes began deliberately to exploit the new explanatory avenues opened up for biogeography in Lyell's *Principles of Geology*.[48] By identifying the changes in climate and land and sea distribution that had taken place within the lifetime of species still extant, they reconstructed the successive migrations from different sources leading to the present phytogeographical situation in Britain and in New Zealand. Such reconstructions invoked various generalizations about species colonizations, particularly the rule that, contrary to what teleological considerations would imply, species of types previously unrepresented in an area would—witness today those animals and plants carried far and wide by man—often expand at the expense of native types. Any theory as to how new species arise had to explain, then, how these would be, not only adapted to their immediate original location, but also competitively superior to the aboriginals in lands they reached later after long migration. This requirement Darwin naturally claimed to meet. If the first placental mammal species arose by the selective improvement of some marsupials outside Australia, and if these placentals were then diversified into descendant species adapted to various ways of life, including those pursued by Australian marsupials, then this would explain why, on eventually reaching that continent, the placental mammals would triumph there. Generalizing, the origin of species by natural selection provided a constant cause for those successive colonization waves, back and forth across

---

[48] See the extensive review by J. D. Hooker in his "Presidential Address to Section E (Geography)," in *Report of the Fifty-first Meeting of the British Association for the Advancement of Science* (London, 1882), pp. 727–738. On Hooker's and Forbes's debts to Lyell's *Principles*, see, especially, the "Introductory Essay" to Hooker's *Flora Novae Zelandiae, Part I: Flowering Plants* (London, 1852), and Asa Gray's discussion and reprinting of much of it, in *American Journal of Science and the Arts*, 2d ser. 17 (1854): 241–252, 334–350; Edward Forbes, "On the Connexion between the Distribution of the Existing Fauna and Flora of the British Isles, and the Geological Changes Which Have Affected Their Area, Especially during the Epoch of the Northern Drift," *Memoirs of the Geological Survey of Great Britain* 1 (1846): 336–432; and Lyell and Forbes's correspondence following this, in Lyell, *Life, Letters and Journals*, II, 106–113.

the earth's endlessly shifting migratory routes and barriers, that geology increasingly evidenced.[49]

Successful colonizations by species of alien types were, then, frequently recognized as both an anomaly on the theory of independent creations and an uncontested fact in Darwin's favor. Lyell's friend Charles Bunbury recorded that on January 29, 1867, he had "a walk with Charles Lyell and the Malletts." Charles Lyell said that

nothing contributed more to shake his belief in the old doctrine (which he formerly held) of the independent creation of species, than the facts of which so many have lately been recorded, relating to the rapid naturalization of certain plants in countries newly colonized by Europeans. He remarked that these introduced plants many of which have spread to an enormous extent and with surprising rapidity in the Australian colonies, New Zealand and parts of S. America, belong in many cases to families entirely wanting in the indigenous floras of the countries in which they have thus settled themselves, and hardly ever to families prevailing in or characteristic of those indigenous floras. When one sees, he said, a particular genus or order of plants abounding very much in a particular country and exhibiting there a great variety of specific forms, one is naturally inclined to suppose (on the "independent creation" hypothesis) that there are particular local conditions in that country, which render it peculiarly suitable and favorable to that family of plants. But when we see an introduced stranger which has no affinity to that prevalent family, intruding itself in its place, overpowering and superseding it, this explanation becomes less satisfactory, and one is led to search rather for some law of descent with variation, to explain the multiplicity of nearly allied forms in a particular region.[50]

Archipelagos offered the best way to test such generalizations because they presented the phenomena in miniature. The testing strategy is nowhere better exemplified than in Wallace's long paper (1864) on the Malayan Papilionidae or swallow-tailed butterflies.[51] Having distin-

---

[49] Charles Darwin, *On the Origin of Species* (London: John Murray, 1859), pp. 336–338, 404–406.

[50] Mrs. Henry Lyell, ed., *The Life of Sir Charles J. F. Bunbury, Bart.*, 2 vols. (London: John Murray, 1906), II, 216–217. Cf. Charles Lyell, *The Antiquity of Man*, 2d ed. (London, 1863), pp. 422–423.

[51] References here are to the revised version of the introductory portion that Wallace published in his *Contributions to the Theory of Natural Selection* (London and New York: Macmillan, 1870), pp. 130–200, as "The Malayan Papilionidae or Swallow-tailed Butterflies, as Illustrative of the Theory of Natural Selection."

guished six categories of variation or differences—running from individual variants through varieties and subspecies to species—as successive
stages in the origin of species, he gives cases where the varieties of some
species and the species of some genus are all marked, on a particular
island, by some distinctive character, peculiarly tailed wings in the most
striking instance. Such a case furnishes, he argues, "a strong corroborative testimony" to the "origin of species by successive small variations;
for we have here slight varieties, local races, and undoubted species, all
modified in exactly the same manner, indicating plainly a common cause
producing identical results."[52] Wallace admits that here, as often, no local advantage for the modification is plainly apparent. However, in other
cases, the variety or the species "mimics" a distasteful "model" not found
elsewhere in the group's range. Just as selection of favored variants, rather than a direct Lamarckian effect of external conditions, seems a reasonable cause for these mimetic adaptations, so generally one will do best to
look for local selective benefits arising from the presence or absence of
certain predators or food sources or the like.

In ending his paper Wallace urges the importance of studying "what
may be termed the external physiology of a small group of animals, in-
habiting a limited district." Darwin, he says, has drawn to this study
"some small share of that research which had been almost exclusively
devoted to internal structure and physiology." Wallace points out that
"The nature of species, the laws of variation, the mysterious influence of
locality on both form and colour, the phenomena of dimorphism and of
mimicry, the modifying influence of sex, the general laws of geographical
distribution, and the interpretation of past changes of the earth's surface,
have all been more or less fully illustrated by the very limited group of
the Malayan Papilionidae; while, at the same time, the deductions drawn
therefrom have been shown to be supported by analogous facts, occurring
in other and often widely-separated groups of animals."[53]

Wallace's manifesto was not just a programmatic piety. A major requirement for the success of Darwinism in England was that a *tertium
quid* be recognized, something theoretical like the best physiology, but,
like the old antitheoretical natural history, concerned with species rather

---

[52] Ibid., p. 174.
[53] Ibid., pp. 199–200.

than individuals. George Bentham, discussing, in 1863, the new term
"biology"—not without a wistful recollection of his uncle's "biontology"
—acknowledged that this Darwinian aim was already fulfilled, less than
four years after the *Origin* had first appeared: "The science of life relates
to the life of the species and to that of the individual. The life of the
species includes its origin, increase, dispersion, migrations, diminution,
and final extinction. This is touching on delicate ground, which I could
have wished to have avoided; but the subject has acquired that degree of
importance, that no biological investigations can now be considered satisfactory which do not apply directly or indirectly to the great questions in
agitation."[54]

Other things being equal, one would conclude a paper on this subject
by listing the items it has proved necessary to take into account in making
sense out of Darwin's reception in England. I hope however that I have
already shown why any such listing must be inappropriate. It would presuppose that we might weigh in the same balance as commensurable
"factors" the Lyellian tradition in geology, say, and the movement for
university reform. But the questions those items helped to answer were
too unalike for this assumption to do anything but diminish our understanding of why things went as they did.

[54] George Bentham, "Address as President on May 25, 1863," in *Journal of the Proceedings of the Linnean Society of London* 1863: xii.

ENGLAND AND SCOTLAND
The Learned Societies

FREDERICK BURKHARDT

As is well known, Darwin's theory of the origin of species by means of natural selection was, together with Alfred Russel Wallace's paper advancing the same thesis, first announced to the scientific world at a meeting of the Linnean Society on July 1, 1858. There was no discussion. Twenty-eight years later J. D. Hooker, who with Sir Charles Lyell had communicated the papers to the society, wrote to Darwin's son Francis: "The interest excited was intense, but the subject was too novel and too ominous for the old school to enter the lists, before armouring. After the meeting it was talked over with bated breath: Lyell's approval and perhaps in a small way mine, as his lieutenant in the affair, rather overawed the Fellows, who would otherwise have flown out against the doctrine. We had, too, the vantage ground of being familiar with the authors and their theme."[1]

Darwin wrote to Hooker on July 5, 1858, to thank him and, apparently in response to a question of Hooker's, said: "I can easily prepare an abstract of my whole work, but I can hardly see how it can be made scientific for a Journal without giving facts, which would be impossible. Indeed [even] a mere abstract cannot be very short. Could you give me any idea how many pages of the Journal could probably be spared for me? ... If the Referees were to reject it as not entirely scientific, I could perhaps publish it as a pamphlet."[2]

[1] Francis Darwin, ed., *The Life and Letters of Charles Darwin*, 3d ed., 3 vols. (London: John Murray, 1887), II, 126.
[2] Ibid., pp. 126–127.

The abstract, upon which Darwin immediately set to work, grew very large, and his decision to publish the volume with John Murray was no doubt due mainly to its length, although he had been offered the possibility of publishing a series of articles in the *Journal* of the Linnean Society. Whether he felt that the volume's length had also enabled him to make it "scientific" is not clear. But his anxiety on that score and his apparent agreement with the standards of the Linnean Society seem to indicate that he had doubts that the form in which he was now presenting his theory—as an abstract of a still unpublished larger work (which Darwin himself said seemed "a queer plan")[3]—would be acceptable to the professional scientific community.

In any case, the *Origin* made the theory a public issue rather than one restricted to a professional scientific body. Thereafter it was discussed and debated by scientists, clergymen, and literary men in magazines and newspapers, in lecture halls in and out of universities,[4] but, on the whole, not in the learned societies. An examination of the details of the reception given to Darwin by the learned societies, as revealed in their proceedings and transactions, gives us fairly clear evidence that this was a matter of considered policy and was, at least in part, a consequence of the prevailing professional standards of science and scientific method.

THE ROYAL SOCIETY OF LONDON

Between 1859 and 1870 no paper in either the *Proceedings* or the *Philosophical Transactions* of the Royal Society had as its subject a discussion of Darwin's theory of the origin of species. Darwin's name is scarcely even mentioned, and explicit references to his views are equally rare in the scientific papers.

There are, however, some signs of an awareness that a revolutionary

---

[3] Ibid., p. 132.

[4] It should be noted that the British Association for the Advancement of Science and the Royal Institution are not included in this analysis. Although both were highly regarded scientific organizations, the BAAS had as its primary mission the advancement of public understanding of science, and the Royal Institution was mainly a research institution, which sponsored meetings and lectures for a semipopular audience. For an excellent exposition of Darwin's reception by British organizations and publications with a general audience, see Alvar Ellegård, *Darwin and the General Reader: The Reception of Darwin's Theory of Evolution in the British Periodical Press, 1859–1872* (Göteborg: Elanders Boktryckeri Aktiebolag, 1958).

theory was in the air. In 1860, Dr. W. B. Carpenter was bringing to a
close a four-part report on his major work on the microscopic fossil
forms, the Foraminifera. During the printing of the paper in the *Transactions* he added a concluding summary, including remarks "on some of
the higher questions which have recently been brought prominently under the consideration of the scientific world." He wrote: ". . . it appears
to me to be a justifiable inference . . . that the wide range of forms which
this group contains is more likely to have come into existence as a result
of modifications successively occurring in the course of descent from a
small number of original types, than by the vast number of originally distinct creations which on the ordinary hypothesis would be required to
account for it."[5]

Carpenter never mentions Darwin, though he is obviously aware that
he is producing evidence in confirmation of Darwin's theory of the trans-
mutability and descent of species.

The Foraminifera were a rather special order of organism for the systematist. As Carpenter says, "[there is] no other group of natural object
in which such a ready comparison of great numbers of individuals can be
made. . . . thousands, even tens of thousands of them can be contained in
a pillbox."[6] And again: "No other group affords anything like the same
evidence on the one hand of the derivation of a multitude of distinguishable forms from a few primitive types and on the other of the continuity
of those types through a vast succession of geological epochs."[7]

Carpenter expressed himself more explicitly in favor of Darwin in his
two critiques of the *Origin* in *The National Review* and in the *British
and Foreign Medical and Chirurgical Review*. In the former he said,
"We are disposed to believe that Mr. Darwin and Mr. Wallace have assigned a *vera causa* for that diversification of original types of structure
which has brought into existence vast multitudes of species, sub-species
and varieties."[8]

The next reference to Darwin in the Royal Society comes in 1864, not
in a paper but in the citation on the occasion of the award of the Copley

---

[5] William Carpenter, "Researches on Foraminifera," *Philosophical Transactions of the
Royal Society of London* 150 (1860): 569–570.
[6] Ibid., p. 569.
[7] Ibid., p. 583.
[8] *The National Review* 10 (1860): 209.

Medal to Darwin—generally considered the highest scientific honor of the time. Darwin had been passed over for this award in 1863, but his friends on the council were determined that this should not happen again. Thomas Henry Huxley, writing to Darwin about the meeting of the council, said: "Many of us were somewhat doubtful of the result. . . . But the affair was settled by a splendid majority. . . . It would have been an indelible reproach to the Royal Society not to have given it to you, and a good many of us had no notion of being made to share that ignominy."[9]

The opposition was not, however, completely overcome. The president of the Royal Society, Gen. Edward Sabine, was strongly antagonistic to Darwin. In his citation he stated that the award of the Copley Medal had been founded on Darwin's researches in geology, zoology, and physiological botany. After detailed descriptions of this work, the statement goes on:

In his most recent work *On the Origin of Species*, although opinions may be divided or undecided with respect to its merits in some respects, all will allow that it contains a mass of observation bearing upon the habits, structure, affinities, and distribution of animals, perhaps unrivalled for interest, minuteness, and patience of observation. Some among us may perhaps incline to accept the theory indicated by the title of this work, while others may perhaps incline to refuse, or at least to remit it to a future time, when increased knowledge shall afford stronger grounds for its ultimate acceptance or rejection. Speaking generally and collectively, we have not included it in our award.[10]

As it was read at the Royal Society meeting, the wording of the last sentence was, ". . . we have expressly omitted it from the grounds of our award." Sabine made the change after Huxley objected and asked that the minutes of the council meeting be read to prove that no such exclusion had been discussed or decided upon.[11]

The citation goes on:

This on the one hand; on the other hand, I believe that, both collectively and individually, we agree in regarding every real bona fide inquiry into the truths

---

[9] Leonard Huxley, ed., *Life and Letters of Thomas Henry Huxley*, 2 vols. (New York: D. Appleton, 1901), I, 275.

[10] *Proceedings of the Royal Society of London* 13 (1863–1864): 508.

[11] Huxley, *Life and Letters*, I, 276.

of nature as in itself essentially legitimate; and we also know that in the history of science it has happened more than once that hypotheses or theories, which have afterwards been found true or untrue, being entertained by men of powerful minds, have stimulated them to explore new paths of research, from which, to whatever issue they may ultimately have conducted, the explorer has meanwhile brought back rich and fresh spoils of knowledge.[12]

In 1868 another reference to Darwin appears in the citation awarding the Royal Medal to Alfred Russel Wallace: "Another remarkable essay 'On the Tendency of Varieties to Depart Indefinitely from the Original Types' . . . contains an excellent statement of the doctrine of Natural Selection which the author . . . had developed independently of Mr. Darwin; and apart from its intrinsic merits, this paper will always possess an especial interest in the history of science as having been the immediate cause of the publication of the *Origin of Species*."[13]

So much for the direct and more or less explicit references to Darwin. Their paucity is the more remarkable since, in the *Proceedings* and the *Transactions* throughout the decade, we find a considerable number of papers on subjects directly relevant to the Darwinian theory, which must have been listened to with that theory uppermost in mind. In May, 1859, for example, a paper by Joseph Prestwich, F.R.S., was read: "On the Occurrence of Flint Implements Associated with the Remains of Animals of Extinct Species . . ."; it discussed the evidence for the coexistence of man and some now extinct mammals.[14] Such coexistence was entirely consistent with the Darwinian hypothesis and aroused intense interest, in that, if proven, it would place in jeopardy the belief in the special creation of man in geologically recent times. Revived by new evidence, the subject was pursued throughout the sixties by geologists and paleontologists—among them Lyell, Hugh Falconer, John Lubbock, and George Busk.

Another question of great scientific as well as public concern was that of man's relationship to the apes. Sir Richard Owen, the leading comparative anatomist of the time, had pronounced on the question by stating flatly at the British Association for the Advancement of Science meet-

[12] *Proceedings of the Royal Society of London* 13 (1863–1864): 508.
[13] Ibid. 17 (1868–1869): 148.
[14] *Philosophical Transactions of the Royal Society of London* 150 (1860): 279–317.

ing of 1860 that an "impassable gulf" existed between man and the ape in that man's brain contained a posterior lobe that was lacking in the apes. This assertion was just as flatly denied at the same session by Huxley, and the controversy was the implicit subject of many papers in several learned societies during the following years. Thus, when W. H. Flower read a paper at the Royal Society in January, 1862, "On the Posterior Lobes of the Cerebrum of the Quadrumana,"[15] submitting evidence that the posterior lobes exist in all of the Quadrumana, he was taking Huxley's side in what was in effect a debate about Darwin's theory, though Darwin was never mentioned.

Flower returned to the theme in 1865, replying to Owen's criticism of his paper.[16] In 1867 a major paper on the subject was presented by St. George Mivart, "On the Appendicular Skeleton of the Primates."[17] The conclusion reached, as stated in the abstract, was that "Man differs less from the Higher Apes than do certain primates below him from each other, and he, *thus judged*, evidently takes his place amongst the members of the sub-order anthropoidea."[18]

Another much discussed topic, which was regarded as providing a major difficulty for the Darwinian theory, was the lack of intermediate forms—the "missing links" in the fossil record. Thus, the discovery of the *Archaeopteryx* fossil in the lithographic limestone of Bavaria aroused great interest, revealing as it did characteristics of both reptile and bird. In 1863 Sir Richard Owen read a paper to the Royal Society on the specimen.[19] Owen makes no reference whatever to its bearing on the Darwinian question, but confines himself to a description in the vein of the professional systematist and paleontologist of the time and without hesitation classifies it as a bird. In 1868 Huxley made some "Remarks upon the *Archaeopteryx Lithographica*" intended to correct some errors in Owen's description (a frequent practice of Huxley's and one in which he

---

[15] Ibid. 152 (1862): 185–201.

[16] *Proceedings of the Royal Society of London* 14 (1865): 134.

[17] *Philosophical Transactions of the Royal Society of London* 157 (1867): 299–429.

[18] *Proceedings of the Royal Society of London* 15 (1866–1867): 320. The significance of the italics, which are Mivart's, is that Mivart, though pro-Darwinian at this time, was also a pious Catholic, who argued that the gulf between man and the rest of animal creation was real, but spiritual and not to be accounted for on natural grounds.

[19] *Philosophical Transactions of the Royal Society of London* 153 (1863): 33–47.

took particular pleasure), but again no mention is made of the bearing of the *Archaeopteryx* on the Darwinian theory.[20]

Huxley's paper was read on January 30, 1868. It is interesting that on February 7, possibly stimulated by writing the Royal Society paper, Huxley gave one of the popular Friday Lectures at the Royal Institution on the subject of "The Animals Which Are Nearly Intermediate between Birds and Reptiles."[21] "We, who believe in Evolution," says Huxley, "reply that these gaps were once non-existent, that the connecting forms existed in previous epochs of the world's history, but that they have died out." But, when asked to produce these extinct forms, the evolutionist can only reply that they are not forthcoming because of the imperfection of the geological record. The best one can do now is to ask of the geological record: (*a*) Are any fossil birds more reptilian than any of those now living? (*b*) Are any fossil reptiles more birdlike than living species? And Huxley endeavored to show that both questions must be answered in the affirmative, giving the *Archaeopteryx* as his example of a bird exhibiting a closer approximation to reptilian structure than any modern bird. He then concluded: "I think I have shown cause for the assertion that the facts of Palaeontology, so far as Birds and Reptiles are concerned, are not opposed to the doctrine of Evolution, but on the contrary are quite such as that doctrine would lead us to expect; for they enable us to form a conception of the manner in which Birds may have been evolved from Reptiles, and thereby justify us in maintaining the superiority of the hypothesis, that birds have been so originated, to all hypotheses which are devoid of an equivalent basis of fact."[22]

A dramatic contrast is offered by these two articles, by the same man on the same subject, written almost at the same time. The one written for a learned society is factual, descriptive, devoid of reference to theoretical considerations. The other, read before a mixed audience of scientists and laymen, is a sustained pro-Darwinian argument based on the relation of the data furnished by the fossils to the evolutionary hypothesis. It is reasonable to suppose that the latter type of article was considered inappropriate for the learned body.

[20] *Proceedings of the Royal Society of London* 16 (1867–1868): 243–248.

[21] T. H. Huxley, *The Scientific Memoirs of Thomas Henry Huxley*, ed. Sir Michael Foster and E. Ray Lankester, 4 vols. (London: Macmillan, 1901), III, 303–313.

[22] Ibid., p. 313.

The Geological Society of London

In a letter to Wallace in May, 1860, Darwin, agreeing that the chapter in the *Origin* on the imperfection of the geological record is "the weakest of all," continued, "but yet I am pleased to find that there are almost more geological converts than of pursuers of other branches of natural science. . . . I think geologists are more easily converted than simple naturalists because more accustomed to reasoning."[23]

One would not, however, come to this sense of the hospitality of geologists to Darwin from the papers read at the meetings of the Geological Society. The pattern of the Royal Society prevails here too—no papers read during the decade discuss the Darwinian theory, and direct references to it are relatively infrequent.

On the other hand, we find some evidence that it was a generally accepted practice to refrain from the consideration of theoretical issues in the papers presented to the society. In February, 1860, for example, President John Phillips in his anniversary address commented that "communications at our meetings so rarely touch on theoretical subjects that, but for the discussions which ensue, it might seem as if we had ceased to doubt, or ceased to hope and aspire after higher generalizations."[24] Phillips believed that the advisability of this practice of avoiding speculation was demonstrated by the advance of science:

Geology . . . has fairly taken her place among the Inductive Sciences; and by acting in the spirit which has won our emancipation from the tyranny of hypothesis, we have established firmly the authority of real laws and phenomena. From time to time, indeed, the historical aspect of geology reveals itself, in efforts more or less successful, to tear aside the veil which hides the origin of things, and to deduce not only the modern features of the land and sea from ancient physical revolutions, but the actual forms of life on the globe from earlier types modified by some assumed law of variation operating through

[23] Darwin, *Life and Letters*, II, 309.

[24] *Quarterly Journal of the Geological Society of London* 16 (1860): xxxii. No record of the discussions of the society appears in the *Journal* until 1869. From other sources we know such discussions often involved important debates, such as the famous "Devonian Battle" of 1839. See Archibald Geikie, *Life of Sir Roderick Murchison*, 2 vols. (London: John Murray, 1875), I, 265. Speaking of the early years of the society, Geikie comments: "There was a freshness about the young science, and men still fought about broad principles, intelligible and interesting to most listeners. The inevitable days of sub-division and detail had not yet come" (ibid., p. 195).

unlimited time. Let us not, while wandering in this dark labyrinth of cosmogony, lose our hold on the slender thread which may bring us back to the light of true philosophy.

He then referred to "the practical character of the papers presented at our meetings" as indicating that the Geological Society needed no such warning and, after summarizing a number of them, concluded: "What a contrast is offered by this mass of efforts all tending to enlarge knowledge and fortify the basis of theory, with the heap of crude conjectures which formerly took the place now firmly held by the true, however imperfect, natural history of the earth which we have founded on sections of the strata and classifications of organic remains!"[25]

That the practice remained in force throughout the decade is shown by Huxley's anniversary address of 1869, in which he suggested that the avoidance of speculation by the society might have been an error:

. . . you will find, if you look back into our records, that our revered fathers in geology plumed themselves a good deal upon the practical sense and wisdom of this proceeding. As a temporary measure, I do not presume to challenge its wisdom; but in all organized bodies temporary changes are apt to produce permanent effects; and as time has slipped by, altering all the conditions which may have made such mortification of the scientific flesh desirable, I think the effect of the stream of cold water which has steadily flowed over geological speculation within these walls, has been of doubtful beneficence.[26]

Huxley's observations are certainly borne out by the papers read at the meetings, but the strictures against theory apparently did not apply to the anniversary addresses of the presidents over the years. Adam Sedgwick, for example, who was president in 1831 when the first volume of Lyell's *Principles of Geology* was the talk of all the profession, delivered a forceful criticism of the uniformitarian theory that formed the basis of that

---

[25] *Quarterly Journal of the Geological Society of London* 16 (1860): xxxi. Phillips was, as his remarks suggest, an anti-Darwinian. In his Rede Lecture at Cambridge in May, 1860, on "The Succession of Life on the Earth," he concluded that "from the facts laid down . . . the Darwin theory of natural selection and the development of life from one original type, has no foundation to rest upon" (quoted from a summary in the *Cambridge Herald and Huntingdonshire Gazette*, May 19, 1860).

[26] Huxley, *Scientific Memoirs*, III, 416; also in *Quarterly Journal of the Geological Society of London* 25 (1869): xxviii–liii.

work.[27] Again, in 1856, W. J. Hamilton took the occasion of the anniversary address to attack Baden Powell's *Essay on the Philosophy of Creation*.[28]

Thus, one would expect that the references to Darwin would appear mainly in the presidential addresses of the society, and this is indeed the case.

In February, 1859, seven months after the Darwin-Wallace papers had been read at the Linnean Society, the Geological Society awarded its Wollaston Medal to Darwin for "labours which combine the rarest acquirements as a naturalist with the qualities of a first-class geologist."[29] No mention was made of the views set forth in the Linnean paper, nor did Lyell, who received the medal on behalf of the ailing Darwin, make any reference to them. The award and citation do, however, provide strong evidence of the high standing Darwin held among the scientists of the time. It is sometimes forgotten what a surprise and shock it must have been to have a naturalist of Darwin's reputation suddenly come out with heretical views on the species question.

Phillips, in the 1860 anniversary address already referred to, discussed the species problem and its relation to geology, seeking to minimize the importance of the theoretical issues to the practice of the science. None of the hypotheses, he said, was "wholly without a foundation of fact, though none of them can be held to penetrate more than a small way in the mystery of the origin of species." After referring to the views of Lamarck, to the *Vestiges of the Natural History of Creation*, and to Darwin's "more practical view of the derivation of some specific forms of one period from others of earlier date by descent with modification," he continued, "We may accept all this, and yet consistently retain the conviction that the changes which are possible by such causes are circumscribed within the many essential types of structure which appear to be a part of the plans of creation."[30]

In 1861, Leonard Horner, presenting, as had become customary in the society, a summary of the important developments of the year, referred

---

[27] *Proceedings of the Geological Society of London* 1 (1826–1833): 302–312.
[28] *Quarterly Journal of the Geological Society of London* 12 (1856): cxiv–cxv.
[29] Ibid. 15 (1859): xxiii.
[30] Ibid. 16 (1860): xlviii.

to the *Origin*, praising Darwin's acute mind, his twenty years of devotion to his subject, his numerous ingenious experiments, vast mass of facts, and calmly stated conclusions, finished by quoting from Huxley's review of the *Origin* in the *Westminster Review* of April, 1860: "We have ventured to point out that it does not, as yet, satisfy all [the] requirements [of scientific logic]; but we do not hesitate to assert that it is as superior to any preceding or contemporary hypothesis, in the extent of observational and experimental basis on which it rests, in its rigorously scientific method, and in its power of explaining biological phenomena, as was the hypothesis of Copernicus to the speculations of Ptolemy."[31]

This quotation is the most positive and favorable statement about the *Origin* made before any of the learned societies during this period. Coming on the formal occasion of the anniversary meeting, from the presidential chair, such a statement must have been carefully considered. Huxley, the secretary of the society, undoubtedly gave Horner permission to reveal his authorship of the review and may indeed have instigated the use of the unusually favorable notice, since by this time he had made up his mind to become Darwin's champion. His famous rhetorical victory over Bishop Samuel Wilberforce at the Oxford meeting of the British Association and his direct challenge of Owen on the difference, or lack of it, between the brains of apes and men had by this time made him *the* defender in the eyes of the public and the learned world. Huxley relished combat and lost no opportunity to make the case for Darwin. But, though Huxley himself presented more than a half-dozen papers on paleontological subjects to the society between 1859 and 1863, he never departed from the conventional scientific format of factual description to argue the merits of Darwin's theory. The presidential chair was apparently the only appropriate place for comments on theoretical matters.

Huxley delivered the anniversary address in 1862 *vice* Horner, who was ill. It was devoted to the subject of paleontology and sought to demonstrate that much more needed to be known about the geological record before that science could be regarded as providing secure evidence of the development of life forms. Huxley's objectives were, first, to shake Owen's theory that life forms had been created in generalized types, which had then developed progressively into more complex and varied

31 Ibid. 17 (1861): xxxix.

forms, and, second, to defend Darwin against those who claimed that the paleontological record refuted his theory of descent.[32]

In 1863 and 1864 there were only a few references to Darwin—all of them in the anniversary addresses of A. C. Ramsay, an early convert to the Darwinian views. None was more than a passing reference expressing the opinion that Darwin's theory was consistent with Ramsay's view of the relationship between breaks in the succession of certain British strata and the changes in the species therein.[33]

Of the papers read at the society's meetings during the decade, only one explicitly offered its findings as providing corroborative evidence for Darwin's views on geographical distribution and variations of species. The paper was "A Description of the Echinodermata from the Strata of the Southeastern Coast of Arabia," by P. Martin Duncan, secretary (later president) of the society. He found Darwin's theory applying in the strongest manner to the assemblage of species in southeastern Arabia.[34]

There were of course many papers that were relevant, in one or another aspect, to the evolutionary hypothesis. In the 1860's several articles appeared on findings in caves containing flint instruments and fossil remains of extinct animals. All followed the usual pattern of reticence on controversial or speculative subjects except one by Falconer in 1865 on the "Asserted Occurrence of Human Bones in the Ancient Fluviatile Deposits of the Nile and Ganges." In discussing the history of man in these regions, he argued that Georges Cuvier was wrong about the advent of man being comparatively late and wrong in his method of relying for his evidence on the records and traditions of ancient people. "Archeological ethnology now transforms the subject," he said.[35]

Another interesting series of papers began in November, 1864, with one "On the Occurrences of Organic Remains in the Laurentian Rocks

---

[32] Ibid. 18 (1862): xl–liv. In a letter to Hooker while he was at work on this address, Huxley wrote: "Darwin is everywhere met with 'Oh, this is opposed to palaeontology or that is opposed to palaeontology' and I mean to turn round and ask, 'Now Messieurs les Palaeontologues, what the devil *do* you really know?'" (Huxley, *Life and Letters*, I, 220).

[33] *Quarterly Journal of the Geological Society of London* 19 (1863): xxix–lii; 20 (1864): xxxiii–lx.

[34] Ibid. 21 (1865): 360–361.

[35] Ibid., p. 383.

of Canada," by Sir W. E. Logan, accompanied by papers on the structure
of these remains by J. W. Dawson, principal of McGill University in
Canada, Dr. W. B. Carpenter, the leading expert on Foraminifera, and
T. Sterry Hunt, F.R.S.[36] The significance of these remains, named
Eozoön canadense by Dawson, was that, if they were indeed organic, they
were the first evidence of living forms in the Pre-Cambrian strata. On
the Darwinian view, one should expect to find such evidence of life,
though, if Carpenter was right in identifying it as a gigantic foraminifer,
a very large gap was opened between it and succeeding forms of life.
Darwin, on Carpenter's authority, added *Eozoön* in the fourth edition of
the *Origin* as evidence that pushed back the origins of life. The antievolutionists, and Dawson was among them, maintained that *Eozoön*
revealed the grandest of all gaps in the record: there was no evidence
that *Eozoön* was defeated in a struggle for existence; it was merely
"superseded" by other species.

Two Irish mineralogists, William King and Thomas Rowney, vigorously denied that the specimens of *Eozoön* were organic, though they
had once believed them to be so. The argument, carried on not only in
the Geological Society but also in the Royal Society, The Royal Irish
Academy, and learned societies in Boston and Canada, reached a degree
of bitterness and invective that is most uncommon in the pages of the
learned journals. As the historian of the *Eozoön* controversy has said, "it
provides an example of the manner in which every aspect of nineteenth
century paleontology was scrutinized for its bearing on evolution."[37] The
significance of *Eozoön*, as he says, was "seldom stated by the disputants,"
but it very clearly accounts for the interest, and at least partially for the
heat, of the argument. Indeed, considering the general avoidance of controversy in the papers read at the learned societies, the *Eozoön* dispute is
quite exceptional in having taken so many pages of their proceedings.

Huxley, in his address of 1870, like most Darwinians, sided with Carpenter and found the importance of *Eozoön* to be that "the ascertained
duration of life upon the globe was nearly doubled at a stroke."[38] Look-

[36] Ibid., pp. 45–71.

[37] Charles F. O'Brien, "Eozoön Canadense, 'The Dawn Animal of Canada,'" *Isis*
61 (2), no. 207 (Summer 1970): 206.

[38] Huxley, *Scientific Memoirs*, III, 527; also in *Quarterly Journal of the Geological
Society of London* 26 (1870): xliv.

ing back at the controversy, one can understand the interest of the Darwinians, since evidence for Pre-Cambrian life was necessary to the argument for evolution. The *Eozoön* specimens were concrete data, matters of "fact," needing description and classification. For this reason the first *Eozoön* papers may have seemed noncontroversial descriptions, albeit with considerable implications for the evolutionary case. Once begun, the argument as to fact and interpretation could hardly be suspended or abandoned as "unscientific."

Two other papers clearly relevant to Darwin's theory were read by Huxley in 1869. They are good examples of his adherence to the ruling practice of presenting data without discussing their bearing on evolutionary theory, even though in his eyes they clearly provided supportive evidence. Both papers dealt with dinosaurs—their affinity with birds and their proper classification.[39] In a letter to Ernst Haeckel while at work on them, Huxley reveals that "in scientific work the main thing just now about which I am engaged is a revision of the Dinosauria with an eye to the 'Descendenz Theorie.' The road from Reptiles to Birds is by way of Dinosauria to the Ratitae. The bird 'phylum' was struthious and wings grew out of rudimentary forelimbs. You see that among other things I have been reading Ernst Haeckel's *Morphologie*."[40]

The theme of Huxley's anniversary address of 1869 was occasioned by the very serious doubt cast on the plausibility of the Darwinian doctrine by the arguments of the leading physicist of the day, Sir William Thomson (later Lord Kelvin). Thomson calculated that the earth as a cooling body could not, on the principles of thermodynamics, have supported life for the many millions of years that the uniformitarian geologists and evolutionists required if their theories were to stand up. "It is quite certain," he said, "that a great mistake has been made—that British popular geology at the present time is in direct opposition to the principles of Natural Philosophy."[41]

Darwin found this one of the most worrisome obstacles to his hypothesis, and, at the time, there was scarcely any answer to Thomson's calculations. Huxley could only say that "Biology takes her time from

[39] *Quarterly Journal of the Geological Society of London* 26 (1870): 12–31, 32–50.
[40] Huxley, *Life and Letters*, I, 325.
[41] Quoted by Huxley, *Scientific Memoirs*, III, 408; also in *Quarterly Journal of the Geological Society of London* 25 (1869): xxxviii.

geology. The only reason for believing in the slow rate of change in living forms is the fact that they persist through a series of deposits which geology informs us have taken a long while to make."[42] This is little more than saying, "There must be something wrong with these calculations." Huxley ends his argument by asking, "But is the earth nothing but a cooling mass . . . ? And has its cooling been uniform? An affirmative answer to both these questions seems to be necessary to the validity of the calculations on which Sir W. Thomson lays so much stress."[43]

The questions had to wait until Curie's discovery of radium and its properties, when a clear negative answer to both could be made.

From the point of view of the history of science, the dilemma is an interesting one. Thomson considered it deplorable that geologists ignored the principles of physics and that they resented his intrusion into their field. Most geologists and biologists, however, could scarcely conceive that the uniformitarian assumptions that worked so well in organizing and explaining their data could be false. In the evolution controversy, Thomson's arguments strengthened the older generation of geologists, who believed in evolution by successive creations in each strata, for a shortened time span made a process guided by intelligent design more plausible. The Darwinians, however, could only stay with their hypothesis in the obdurate hope that Thomson's conclusions were not as obviously true as they seemed. Darwin's view in the sixth edition of the *Origin* is the best possible, considering the state of knowledge at that time: "This objection . . . [is] probably one of the gravest yet advanced. I can only say, firstly that we do not know at what rate species change as measured by years, and secondly that many philosophers are not as yet willing to admit that we know enough of the constitution of the universe and of the interior of our globe to speculate with safety on its past duration."[44]

Though the problem of geological time remained, the meaning of the geological record was becoming steadily more clear. During the 1860's so much paleontological evidence had accumulated that was clearly con-

---

[42] Huxley, *Scientific Memoirs*, III, 425–426.

[43] Ibid., p. 426.

[44] Charles Darwin, *The Origin of Species*, 6th ed. (New York: Modern Library, 1936), p. 357.

sistent with, if not, indeed, confirmatory of, the general evolutionary theory that Huxley devoted his anniversary address of 1870 to a sustained argument on this theme:

It is now no part of recognized geological doctrine that the species of one formation all died out and were replaced by a brand-new set in the next formation. On the contrary, it is generally, if not universally agreed that the succession of life has been the result of a slow and gradual replacement of species by species, and that all appearances of abruptness of change are due to breaks in the series of deposits, or other changes in physical conditions. The continuity of living forms has been unbroken from the earliest times to the present day.[45]

THE LINNEAN SOCIETY

On May 24, 1859, President Thomas Bell summarized the developments of the past year and, in a passage that was to become famous, observed that "it has not, indeed, been marked by any of the striking discoveries which at once revolutionize, so to speak, the department of science on which they bear; it is only at remote intervals that we can reasonably expect any sudden and brilliant innovation which shall produce a marked and permanent impress on the character of any branch of knowledge, or confer a lasting and important service to mankind."[46]

The "unfortunate Bell," as Gavin de Beer has quite properly called him, was Professor of Zoology at King's College, London, and, though not one of the more original men in his field, he did a great deal to advance the cause of science by his effective organizational and administrative work. He was, of course, by no means alone in his failure to recognize the revolutionary implications of the Darwin-Wallace papers.

In his next anniversary address Bell again ignored the Darwinian theory, and this time it was not because he thought it was unimportant. The *Origin* had been out for six months and had been reviewed at length in several of the leading intellectual journals. Most of the reviews were negative, but it was abundantly clear that the work could not be dismissed. Bell's omission, by this time, was a considered one, and the reason is stated quite clearly by George Bentham, his successor as president, in his address of 1862. Referring to Darwin's work on orchids as a

[45] Huxley, *Scientific Memoirs*, III, 526–527; also in *Quarterly Journal of the Geological Society of London* 26 (1870): xliii.

[46] *Journal of the Proceedings of the Linnean Society* 4 (1860): viii–ix.

good field for further work, Bentham continues: "I do not refer to those speculations on the origin of species, which have excited so much controversy; for the discussion of that question, when considered only with reference to the comparative plausibility of opposite hypotheses, is beyond the province of our Society. Attempts to bring it forward at our meetings were very judiciously checked by my predecessor in this Chair and I certainly should be sorry to see our time taken up by theoretical arguments not accompanied by the disclosure of new facts or observations."[47]

Bentham, a botanist, was sympathetic to Darwin and was gradually convinced to give up the immutability of species. But he never yielded on the policy that Bell had set. In 1863 he reaffirmed it: "Although . . . we cannot allow the time of our own meetings, nor the pages of our publication to be devoted to the abstract discussion of any such theories, yet we should give every encouragement to the search after facts on the one side or the other irrespective of what we may deem extravagant in the results which might be deduced from them."[48]

The strictures did not, however, apply to the anniversary address, and Bentham devoted almost every one of his to reports on the current status of the evolution controversy, including reviews of books bearing on the problem,[49] though he refrained from discussing the points on which he would agree or disagree with Darwin as "scarcely within the legitimate scope of our society to enter."[50] He did, however, make a judgment about the scientific status of the theory: "Mr. Darwin has shown how specific changes *may* take place and . . . has endeavored to show how they do take place. His is not therefore a theory capable of proof, but 'an unimpeachable example of a legitimate hypothesis' requiring verification, as defined by J. S. Mill in his excellent chapter on Hypothesis"[51]

---

[47] Ibid. 6 (1861–1862): lxxxi.

[48] Ibid. 7 (1862–1863): xiii.

[49] Indeed, his addresses provide an excellent record of the development of biological science during the years he was in office.

[50] *Journal of the Proceedings of the Linnean Society* 7 (1862–1863): xvi.

[51] Ibid., p. xv. The quotation is from Mill's *System of Logic*, 5th ed., vol. 2, and refers to Darwin.

Asserting that it is the special province of the society to collect facts bearing upon any biological hypothesis, Bentham continued:

With regard to the process of verification or refutation of the Darwinian hypothesis by actual investigation, there has not been time yet for much progress. Mr. Darwin has not yet published those detailed evidences of his propositions which might give to fresh observers a fair starting point, and it is only from naturalists who have long given up their minds to similar subjects that we can for some time expect anything important for or against his views. What has as yet been published of any merit, as far as I am aware, is more or less in their favour.[52]

The society had indeed been hospitable to papers presenting facts purporting to confirm the new species doctrine. In November, 1859, Wallace sent in a paper on the "Zoological Geography of the Malay Peninsula" —an important extension of his views on the geographical distribution of species. As a logical inference from his earlier paper, he derived the generalization that in all cases where we find an island with "species mostly identical or closely allied to the adjacent country, we are forced to the conclusion that a geologically recent disruption has taken place."[53]

In June, 1860, J. D. Hooker read a paper on "Outlines of the Distribution of Arctic Plants." Hooker had found Darwin's views "a jam pot"[54] and had put them to work in his own botanical investigations. Just before the publication of the *Origin*, Hooker had brought out his *Introductory Essay to the Flora of Tasmania*, in which he had applied the Darwinian theories of mutability and descent as a working hypothesis to the problems of geographical distribution of plants in Tasmania. In his paper on Arctic plants he continues the work of supporting Darwin with his own data and observations. It is replete with confirmatory statements like: "It appears to me difficult to account for these facts unless we admit Mr. Darwin's hypotheses" and "Mr. Darwin shows how aptly such an explanation meets the difficulty," or, again, "Mr. Darwin's hypothesis accounts for many varieties of one plant being found in various alpine and arctic regions of the globe" and "It appears therefore to be no

[52] *Journal of the Proceedings of the Linnean Society* 7 (1862–1863): xix.
[53] Ibid. 5 (1860–1861): 183.
[54] Darwin, *Life and Letters*, II, 139.

slight confirmation of the general truth of Mr. Darwin's hypothesis that . . . ," etc.[55]

Earlier that same month Sir John Lubbock had read a paper, "On Some Oceanic Entomostraca collected by Capt. Toynbee,"[56] in which he found enough intermediate forms to cause him to suggest that the gaps between species "remain only in our knowledge and not necessarily in nature. . . . How worthless then is the argument against the mutability of species which depends upon the supposed absence of 'links'!"[57] Lubbock also found that the Entomostraca presented remarkable examples of a generalization Darwin had made in "his admirable work" on the difference in secondary sexual characteristics between species.

On November 21, 1861, two very important papers were contributed, one by Darwin, "On the Two Forms, or Dimorphic Condition, in the Species of *Primula*," and the other by Henry W. Bates, on "Contributions to an Insect Fauna of the Amazon Valley."[58]

Darwin's paper was the first of a series on experiments he was conducting on plant reproduction and hybridism. In all of them he stressed findings that suggested that sterility was not, as the critics of Darwinism argued, a distinguishing criterion of a species. Huxley too assumed that true species are fertile only with their own kind and pointed out that, until varieties were produced under domestication that were fertile only with one another, the logical foundation of the theory would be insecure. Darwin, though agreeing that "here is a great hiatus in the argument," thought Huxley had overrated it: "Do oblige me by reading the latter half of my Primula paper in the *Linnean Journal*, for it leads me to suspect that sterility will hereafter have to be largely viewed as an acquired or *selected* character—a view which I wish I had had facts to maintain in the *Origin*."[59]

The paper by Bates was to become a classic in the literature of evolution. Though its observations and descriptions were in the standard Lin-

[55] *Transactions of the Linnean Society* 23 (1860): 253–254.
[56] Ibid., pp. 173–191.
[57] Ibid., p. 174.
[58] *Journal of the Proceedings of the Linnean Society* 6 (1861–1862): Botany, pp. 77–96 (Darwin); Zoology, pp. 73–77 (Bates—abstract only). The full Bates paper appears in *Transactions of the Linnean Society* 23: 495–566.
[59] Letter, Charles Darwin to T. H. Huxley, January 14, 1862, in Darwin, *Life and Letters*, II, 384.

nean style, the particular kind of evidence it provided for the Darwinian hypothesis was both original and dramatic. It dealt with the phenomenon of mimicry—in this case, the incredibly minute imitation of one species of butterfly by another—which, Bates contends, the hypothesis of natural selection explains more rationally than any other.[60]

The *Proceedings of the Linnean Society* for the year 1864 include four papers by John Scott of the Royal Botanical Gardens of Edinburgh, communicated by Darwin. They are accounts of experiments on sterility and reproduction in plant species made at Darwin's suggestion and are presented by Scott as corroborating and extending the paper on the *Primulae*.[61]

In 1864 and subsequent years, Wallace, Bates, and Darwin presented papers reporting their observations and investigations in factual terms, with low-keyed but clear references to the evolution question. In 1868, a paper "On Some Remarkable Mimetic Analogies among African Butterflies" by Roland Trimen was read, carrying forward Bates's and Wallace's observations of this phenomenon.[62] The paper concludes:

Such results as these are of the deepest significance. Inexplicable as they must ever remain when regarded on the theory of the independent creation of all organic beings precisely as we now behold them, they become clearly intelligible when viewed as the natural consequences of the innate variability of species, and the preservation and development by inheritance, through all time and under all changes of surrounding conditions, of every successive variation advantageous to the organism originating it. In the infinitely complicated "struggle for life" any advantage, however slight, inevitably has its effect, and the individuals possessing it will not only hold their ground to the exclusion of less fortunate competitors, but will transmit the precious quality to some at least of their descendants.[63]

---

[60] *Journal of the Proceedings of the Linnean Society* 6 (1861–1862): Zoology, p. 77. Darwin was so pleased with Bates's article that, fearing it might be overlooked, he wrote an anonymous and highly laudatory review of it in the *Natural History Review* 3 (1863): 219–224. Darwin was not content with describing the data and the explanation Bates gave of mimicry but also showed how difficult and unsatisfactory a creationist explanation would have to be.

[61] *Journal of the Proceedings of the Linnean Society* 8 (1863–1864): Botany, pp. 78–126, 162–167, 169–196, 197–206.

[62] *Transactions of the Linnean Society* 26: 497–522.

[63] Ibid., p. 520.

Throughout the decade almost no paper was presented to the society that produced any evidence against the general Darwinian theory or its related hypotheses. The only extensive paper that purported to present such negative evidence came from an old opponent of Darwin, Andrew Murray, an entomologist. In December, 1868, he read a paper on "Geographical Relations of the Chief Coleopterous Faunae," which does not mention Darwin by name but, after discussing various theories of land and water distribution to account for the distribution of insects, argues the anti-Darwinian thesis that "actual continuity of soil and non-interruption of barriers" is the only explanation.[64]

The overwhelming majority of papers published by the society during the decade continued in the traditional line of presenting descriptions of new species and genera. Except for the mimetic phenomena, the taxonomical problems raised by the specimens provided very little in the way of systematic evidence pro or contra on the Darwinian question. But, as the specimens poured in from surveys and voyages all over the globe, a problem of major interest became that of explaining the geographical distribution of species, and the data appeared consistently more explicable from the Darwinian standpoint. It is therefore not surprising that many of the strongest early Darwinians were those who studied species and varieties as they occurred in nature: field naturalists like Wallace, Bates, H. B. Tristram,[65] Trimen, and those with extensive field experience like Hooker, Huxley, and Darwin himself.

Darwin had expected the *Origin* to produce a great revolution and simplification in taxonomy, but almost no change occurred during the period under review nor, for that matter, is the situation much different today. What did happen was that the discipline steadily lost status during the rest of the nineteenth century, and experimental biology, vigorously promoted by Huxley and other evolutionists, gained intellectual and professional ascendancy.

[64] *Journal of the Linnean Society* 11 (1873): Zoology, pp. 1–89.

[65] The Reverend H. B. Tristram was an ornithologist who published the first article presenting evidence in corroboration of the Darwin-Wallace theory: "Writing with a series of about 100 Larks of various species from the Sahara before me, I cannot help feeling convinced of the truth of the views set forth by Messrs. Darwin and Wallace in their communications to the Linnean Society" ("The Ornithology of Northern Africa," *Ibis* 1, no. 4 [October 1859]: 429). Tristram later changed his view and became an opponent of Darwinism on theological grounds.

The contributions to the "species problem" that the papers of the Linnean Society made during this period were not in the area of its major interest, taxonomy, but in the allied ones of geographical distribution and plant physiology. The Fellows of the society listened, for the most part, to descriptions of new species and genera and only occasionally to papers presenting data of more or less probative significance to the speculative theory they were enjoined from discussing. But, as if to make up for this restrictive discipline, they were provided almost every year after 1861 with a solid discourse on Darwinism from the privileged position of the president, George Bentham.

THE ZOOLOGICAL SOCIETY OF LONDON

The Zoological Society did not have the prestige of the three learned bodies discussed so far. Nonetheless, its papers were not conspicuously inferior to those of the "major" societies, and, indeed, the Zoological was a forum for many of the same contributors—Huxley, Owen, Wallace, Bates, Flower, and others familiar to the *Proceedings* of the Royal and Linnean societies.

The reception to Darwin was also very much the same, though the Zoological Society has the distinction of having published the only paper devoted to an analysis and critique of the Darwinian thesis in theoretical terms (see below). The other references to the new hypothesis were all in connection with discussions of some new species or other data. Although they were relatively infrequent, some of these references were of particular interest because they were negative and critical in an outspoken way that did not occur in the papers read at the Royal, Geological, or Linnean societies.

The papers of Dr. John Edward Gray, for example, provided excellent statements of the reaction to Darwin of a leading professional systematist. Gray was vice-president of the Zoological, president of the Entomological, and a fellow of the Royal Society. As Keeper of the Zoological Specimens of the British Museum, he was the most distinguished classifier in the country and an obdurate and formidable defender of species as real entities.

In a paper presented in 1863, he wrote:

My firm opinion, founded on forty years' experience, and after having had through my hands perhaps more specimens of animals of different classes than most living zoologists, if not more than any other, is that species are permanent; indeed they appear to me to be the only groups of individuals that seem to be well-defined and separated from other groups by a distinct and unvarying character. . . .

It is no doubt true, as Mr. Darwin observed . . . that the "origin or derivation of species from gradual change, however produced, does appear to connect large classes of facts"—that is to say, if such a derivation could be proved; but, unfortunately during all my experience and after most careful search (for the origin of species has always been a most interesting subject of my contemplation) I have never found the slightest evidence for the support of such a theory, or the least modification of any species leading to such an opinion.[66]

Gray makes much the same case in several other articles.[67] The fixity of his conviction is revealed most clearly, however, in his easy disposal of Bates's Darwinian explanation of mimicry in butterflies. In a paper on the "Revision of the Species of Trionychidae Found in Asia and Africa," he quotes from Louis Agassiz the dictum that "under surprisingly similar external appearances marked structural peculiarities amounting to generic differences are hidden," and he continues: "I believe that such cases are much more common than has hitherto been suspected, and it is on such superficial resemblances that Mr. Bates' observations and theories respecting the Brazilian Butterflies are founded—notions which will vanish into the air when the insects are more carefully examined by a systematic entomologist."[68] This was an unkind cut at the amateur naturalist, Bates, and missed the point so completely that it is difficult to see it as anything but hopeless bias. But Gray, according to his taxonomical lights, saw only distinct species—the difference was "real," and resem-

[66] "Revision of the Species of Lemuroid Animals," *Proceedings of the Zoological Society of London* 1863: 134–135.

[67] In "Description of a New Species of Cuscus (C. Ornatus)," for example: "Some of the converts to the theory of the mutation of species may think that this animal is an instance in point; but such a hypothesis derives no support from the observations I have made" (*Proceedings of the Scientific Meetings of the Zoological Society of London* 1860: 2). And again, in "Observations on the Box Tortoises": "Yet on [such] doubts and lack of knowledge do the theorists wish to support the theory that species pass gradually into each other. . . . Never was a theory more baseless so far as our knowledge is concerned" (ibid. 1863: 174).

[68] *Proceedings of the Scientific Meetings of the Zoological Society of London* 1864: 77.

blance was "superficial." Bates knew well enough that his butterflies were different species. His problem was to explain the astonishing resemblance, which was also "real," and natural selection alone did this for him. Each, from the other's point of view, was begging the question, but it was not so clear then as it is now which was "scientific." As it turned out, "Old Dom. Gray," as Huxley called him, was right to hold on to the distinctness and continuity of specific characters, and Darwin was right about the "origin" of species through natural selection, but the solution to the contradiction was not to come until after Gregor Mendel had been rediscovered in 1900 and the science of genetics based on his findings had advanced another quarter of a century.

Another distinguished opponent of Darwinism, Professor Richard Owen, entered the lists in 1863 with a paper "On the Aye-aye."[69] He had read a paper on the same subject the previous summer at the British Association meeting, but it was then more explicitly entitled "On the Characters of the Aye-aye, as a Test of the Lamarckian and Darwinian Hypothesis of the Transmutation and Origin of Species." His argument rests upon the implications of the remarkable combination of features the lemurlike animal has for getting at wood-boring larvae: ears to hear the larvae, teeth to gouge, digits to probe and extract, feet to grasp the limb, and tail to balance itself. "Thus we have not only obvious, direct and perfect adaptations of particular functions . . . but we discern a correlation of these several modifications with each other, and with modifications of the nervous system and sense organs . . . the whole determining a compound mechanism to the perfect performance of a particular kind of work."[70]

Our only guide to an explanation of this phenomenon, says Owen, is that, by analogy with our own intellectual activities, we are led to a conception of a cause of this correlation and adaptation—an intellect "in a transcendentally higher degree."

Owen then discusses Georges Buffon, Lamarck, and Darwin, and the aye-aye is used as a test of their views of the origin of species. In his eyes, all fail to explain how the structure of the animal could have been developed. Natural selection fares no better than the Lamarckian use-

---

[69] *Transactions of the Zoological Society of London* 5 (1866): 33–97.
[70] Ibid., p. 88.

and-disuse hypothesis: "We know of no changes in progress in the Island of Madagascar, necessitating a special quest of wood-boring larvae by small quadrupeds of the Lemurine or Sciurine types of organization. Birds, fruits, and insects abound there in ordinary proportions; and the different forms of Lemuridae co-exist with their several minor modifications."[71]

From the argument as a whole it is quite clear that Owen believed that teleology was essential to biological science and that, though he was willing to concede the development of species through continuous "secondary" causes, he thought these nevertheless in their broad development and course required a guiding mind or divine volition. Natural selection as advanced by Darwin could not provide a satisfactory explanation of the origin of species, although Owen seems to have thought that it might have a role in their extinction.

Darwin was not impressed by Owen's criticism. The argument from the aye-aye's structure was after all no different from that based on the correlations of the woodpecker and of the human eye, which he had faced in the *Origin*.

A major point at issue was whether teleology was to be given scientific status in biology. Darwin, Huxley, and their followers regarded "final causes" as sterile, but such prestigious figures as Owen, Sedgwick, Sir John F. W. Herschel, and William Thomson continued to proclaim the need for design as an explanatory principle in biology.[72]

Teleology was also the issue in the one theoretical discussion of Darwinism published by the society. It appeared in a paper read in January, 1861, by E. Vansittart Neale, who, though a fellow of the Zoological Society, had no scientific training; he was a lawyer, Christian Socialist, and leader in the cooperative and workers' association movements. His paper was called "On Typical Selection as a Means of Removing the Difficulties Attending the Doctrine of the Origin of Species by Natural Selection."[73] The strength of Darwin's view, said Neale, lay in "the sat-

---

[71] Ibid., p. 95.

[72] In his presidential address at the British Association meetings of 1871, Thomson invoked Herschel's view to this effect (*Report, British Association for the Advancement of Science* 1871: cv).

[73] *Proceedings of the Scientific Meetings of the Zoological Society of London* 1861: 1–11.

isfactory explanation afforded . . . of the analogies and differences observed in the various forms of living beings which have been or actually are tenants of our globe." Its weakness lay in its exclusive emphasis on the principle of change, which produces varieties. There is also a principle of permanence, which produces species. The varieties that man can produce in pigeons, say, remain distinct only so long as man prevents their interbreeding. In nature, species interbreed with their own kind. The offspring of hybrids are sterile. This is the principle of permanence. Neale commented: "Darwin has no answer to 'Whence came this principle of permanence?'"

He then proposed a reconciliation between the Darwinian supporters of change and those who supported permanence or fixity by means of a "conception of the typical character pervading all organic life." Finally, to account for the permanence of types, he said, "we must resort to the hypothesis of an intelligent action as the only intelligible one. Accordingly it is to an intelligent choice exercised upon the infinity of possible variations capable of arising in different organisms, through the laws belonging to their natures, that I would attribute the formation of species by what I venture to call Typical Selection." In short, according to Neale's view, there is an intelligence or "ordering will," which pervades nature, selects individuals with a "disposition to vary in the desired direction," and sees to it that these individuals become more prolific. Typical selection, he maintained, explains some things better than Darwin's theory—for example, the great differences between a sparrow and a condor: no human breeder could produce such a change, even given a great deal of time. Typical selection also explains the lack of intermediate types in the geological record. This was a great problem for Darwin, but not for Neale, since, under typical selection, "advance is more direct."

The "reconciliation" Neale proposed may well have been the reason the article was accepted for publication despite its exclusively theoretical contents and its nonprofessional authorship, for by 1861 the Darwinian controversy had become extremely vehement and was dividing the scientific community into hostile camps. The possibility of an intellectually respectable reconciliation between evolution and creationism would have had great appeal.

Darwin wrote his reaction to Neale's argument at the end of his copy of the paper: "In the plainest and coarsest language the author makes God a great breeder of animals, . . . who selects and works like an Improver of Short Horns or a Pigeon Fancier."[74]

Other papers presented during the decade may be briefly mentioned. In 1860, Alfred Newton, a young pro-Darwinian ornithologist, to illustrate his paper "On Some Hybrid Ducks"[75] exhibited the skins of birds he had bred to refute the common view that hybrids between distinct species are *inter se* infertile. Then, after describing cases in which hybrid offspring differed remarkably from parents, he says, "It is not . . . difficult to see what use may be made of this singular circumstance by those who advocate the views of Mr. Darwin, but into any consideration of the question I forebear to enter, contenting myself merely by noticing the fact."[76]

In 1861 Huxley (then vice-president of the society) submitted "On the Brain of *Ateles Paniscus*" as pertinent to "the controversy which has arisen respecting the nature and the extent of the differences in cerebral structure between Man and the Apes [and which] gives an especial value to all new facts."[77]

In 1862, W. K. Parker, in his paper "On the Osteology of Genera *Pterocles*," said, "It is . . . almost impossible for the most devout believer in separate creations to keep the idea of 'ancestral relationship' altogether out of his mind when considering such birds . . . at any rate dogmatism on either side, on a subject so far beyond the reach of our feeble faculties and limited knowledge, has in it something of profanity."[78]

In that year too, four articles on new species by Wallace appeared, but all are purely descriptive.

In 1863 Bates presented two papers—one on butterflies in Panama and the other on insects in Madagascar—with some important hypothesizing on geographical distribution suggested by the specimens. The papers are good examples of how a naturalist working on this type and

---

[74] In Darwin's collection of reviews, now in the Cambridge University Library, Neale's paper is no. 45.
[75] *Proceedings of the Scientific Meetings of the Zoological Society of London* 1860: 336–339.
[76] Ibid., p. 338.
[77] Ibid. 1861: 247.
[78] Ibid. 1862: 257.

range of data with the evolutionary concept in mind found it fruitful in producing new meanings and suggestions for further research.[79]

Between 1864 and 1865 there appeared a number of articles by pro-Darwinians (Wallace, Huxley, Flower, Anton Dohrn, William Bernhard Tegetmeier, et al.), but they are all of a factual, descriptive kind, in no way to be distinguished from the normal run of taxonomical papers that appeared during the same period. In 1868 Huxley presented a paper on gallinaceous birds, in which he applied evolutionary theory explicitly to his classifications: "From the point of view of the Evolution theory, all the Gallo-columbine birds must be regarded as descendants of a single primitive stock; and the relations of the different groups should be capable of representation by a genealogical tree, or *phylum* as Haeckel calls it in his remarkable *Generelle Morphologie*. Such a *phylum* can only be put forward with confidence when a tolerably complete knowledge of the development and of the palaeontological history of a group has been obtained."[80] But there was an air of assurance that this knowledge would be forthcoming.

By 1869 the Darwinian theory was clearly in the ascendant in the learned societies—accepted, not as demonstrated truth, but as a view for which there was now enough factual data to permit its being used with confidence as a working hypothesis. This applied especially to the general theory of descent. Natural selection, as providing the "cause" of evolution, was still very much criticized and indeed was to lose ground in the later decades of the nineteenth century because of difficulties that could not be solved on the basis of the genetic knowledge of the time.

The intellectual revolution was nevertheless in full swing. New areas of research were opened, older findings were re-evaluated, and many new experiments were suggested by the great hypothesis. Paleontological and embryological evidence accumulated, and vast quantities of new specimens and their geographical and geological distribution were studied. Not only were almost all of the new data consistent with the Darwinian view, but the various lines of evidence also converged to provide steadily more convincing circumstantial proof of the continuous development of species throughout the history of life on the planet.

[79] Ibid. 1863.

[80] "On the Classification and Distribution of the *Alectoromorphae* and *Heteromorphae*," ibid. 1863: 312.

THE ENTOMOLOGICAL SOCIETY OF LONDON

In May, 1867, Darwin wrote to Ernst Haeckel: "The belief in the descent theory is slowly spreading in England, even amongst those who can give no reason for their belief. No body of men were at first so much opposed to my views as the members of the London Entomological Society, but now I am assured that, with the exception of two or three old men, all the members concur with me to a certain extent."[81]

Entomologists were a mixed bag, from a professional scientific point of view, ranging from the butterfly chasers with their cabinets to naturalists like Bates and Wallace and systematists like T. V. Wollaston and F. P. Pascoe. But by 1867 the Council of the Entomological Society was made up predominantly of fellows of the Linnean Society, with Sir John Lubbock as president, and the society had acquired standing in the learned world.

In relation to the Darwinian controversy, entomology had in many respects an advantage over other branches of zoology, in that the collection and study of the Insecta provided plentiful and accessible data both for theorizing on the species question and for verification and experimentation.[82]

During the 1860's, as one might expect from the conduct of the learned societies so far examined, the Entomological Society published no paper directly concerned with a discussion of the evolutionary hypothesis. As in the other learned societies, there are occasional references to Darwin or to the *Origin* in the papers and in the anniversary addresses of the presidents. But, unlike those of the other London societies, the *Proceedings of the Entomological Society* include a record of the discussions of papers by the members. These reveal an informality and directness that is refreshing after the normal taxonomic style of the papers, and in them we find some interesting and frank interchanges on Darwin and the species problem.

The earliest reference to Darwin is in a paper by Henry W. Bates in 1860 on South American butterflies. Referring to an argument Darwin

---

81 Darwin, *Life and Letters*, III, 69.

82 Wallace, for example, gave as an important reason that he and Darwin reached the same theory the fact that both had been ardent beetle hunters. See *The Darwin-Wallace Celebration Held on Thursday, 1st July 1908 by the Linnean Society of London* (London, 1908), pp. 8–9.

had made for the extension of glaciers to the equatorial regions, he says that "the present distribution of *Papilio* does not support the hypothesis of such a degree of refrigeration in the equatorial zone of America, or at least does not countenance the supposition of any considerable amount of extinction."[83]

The *Proceedings* for January 5, 1863, report that a paper was read "On the Geographical Distribution of European *Rhopalocera.*"[84] The author, W. F. Kirby, having "expressed himself as an adherent to the Darwinian theory of the origin and development of species," makes an interesting suggestion for further research: "If the geologists can tell us how long Corsica has been separated from the mainland we shall have some more valuable data by which to calculate the length of time that the formation of a species requires, for a large number of Corsican insects are already good species and many more have become subspecies, which is the last step towards the formation of a new species." In a statement that marked him as an atypical member of the learned fraternity of the time, Kirby said: "In conclusion, I do not think anyone present, however much he may differ from the opinions expressed in this paper, will regret my bringing prominently forward Mr. Darwin's theory of the Origin of Species, for it is the great scientific question of the day and ought to be freely discussed here as well as in other societies; and my own opinion is, that in a great measure, by the study of geographical distribution in the widest sense the theory must stand or fall."[85]

At the February 1, 1864, meeting President F. P. Pascoe exhibited some new species of butterfly, one of which resembled another and might be taken as "another case of 'mimetic resemblance' but, if so, the stronger insect was here imitating the weaker, which seemed not quite consistent with the view that the imitated form was copied with a view to the protection of the imitating." Wallace answered that the instance was quite consistent with the theory: "An insect might be very weak in structure and yet be a proper subject for mimicry; many insects of weak structure were extremely abundant, were in fact dominant species; such species no doubt possessed some protection against their enemies with

---

[83] *Transactions of the Entomological Society of London,* n.s. 5 (1858–1861): 352. (The *Transactions* also include the Proceedings of the Society.)

[84] Ibid., 3d ser. 1 (1862–1864): 481–492.

[85] Ibid., p. 491.

which we were unacquainted and of which other species of stronger structure were deprived."[86]

The next few years contain only a few references to Darwin. For the most part it was Bates and Wallace who used the meetings to provide examples that fit the Darwinian hypothesis. In one, a spider imitating a flower in order to trap butterflies was discussed by Bates;[87] in another, "Mr. Darwin's beautiful explanation of the apterous condition of many *Coleoptera* of usually winged genera in Madeira" was applied to Celebes butterflies by Wallace.[88]

At one meeting, Professor J. O. Westwood launched into an extended criticism of the Batesian theory of mimicry, using a Brazilian butterfly as an exhibit. Bates replied to Westwood that his specimen was a mere monstrosity, which had no bearing whatever on the question of the origin of species.[89] "The Darwinian theory," he said, "dealt only with variations that were propagated and not with monstrosities the peculiarities of which were not transmitted to their descendants." He then went on to attack the absurdities to which Westwood would be led if he consistently applied his creationist theory in explaining the numerous shades of variations found in one and the same locality. Bates then once again asserted that the Darwinian theory provided the only explanation of mimicry.[90]

In 1867, Wallace requested the assistance of the members in helping him clear up a problem raised by Darwin. The latter had arrived at the conclusion that brilliant coloring was as a rule due to sexual selection, but the bright hues of many larvae—especially of butterflies—could not, being sexless, owe their colors to sexual selection. Wallace, believing that "nothing in nature is without its cause and believing in the principle of natural selection or the preservation of the fittest," suggested that the color must be in some way useful and that it might be that, as a rule, brilliantly colored larvae were distasteful to birds. "It was on this point that he wished to collect observations and statistics and he should be glad if any who kept birds, particularly indigenous birds, would make

86 Ibid. 2 (1864–1866): Proceedings, p. 14.
87 Ibid., Proceedings, p. 29.
88 Ibid. 4 (1865–1868): 308.
89 Ibid. 5: xxxvii.
90 Ibid., p. xxxix.

experiments with different larvae to ascertain which were eaten and which rejected."[91]

At the March 18 meeting a paper was read "On Species and Varieties," by Captain Thomas Hutton, F.G.S., on the subject of breeding together as a test of distinct species. The paper was never published in the *Transactions* of the society. The abstract in the *Proceedings* concludes with what may be an explanation: "The remainder of the paper was a criticism of the Darwinian theory of Natural Selection, the writer's views being principally enforced by arguments beyond the province of the Entomological Society."[92]

In 1868 Bates was elected president of the society. During that year several sessions were devoted to discussions of questions put to the entomologists by Darwin, whose health did not permit his coming to meetings. He wanted detailed information of the numerical proportion of the sexes of insects in nature: were there, for example, any species in which the females were in excess over the males? Mr. H. T. Stainton thought this could be ascertained only by breeding. Observations in the field, he observed, were often unreliable and at variance with controlled breeding experiments.

Darwin was also interested in differences in colors in male and female butterflies and moths and in the distinction between sexual and protective coloring in insects. "Mr. M'Lachlan said Mr. Darwin had recently put two queries to him. Do male dragon-flies fight with one another? And, do many or several males follow one female? He confessed his inability to answer with certainty either of these apparently simple questions."[93]

In his presidential address of 1869,[94] Bates expatiated on the great importance of evidence from geographical distribution, which he felt had greater strength than the evidence of modification derived from domestication, on which Darwin had placed such emphasis in the first place:

In distribution we see the operation of changes independent of artificial inter-

91 Ibid., p. lxxxi.
92 Ibid., pp. lxxxii–lxxxiv.
93 Ibid. 1868: x, xiv.
94 Ibid., pp. lv–lxix.

ference and once admitting that species do slowly and intermittingly extend their areas of dissemination and that certain local forms are modifications of their sister-forms the whole process of the formation of species by natural means lies straight way open to our investigation, the steps of modification being capable of proof, by logical induction after the premises just mentioned are granted. . . . I have been surprised to find how defective are most of our collections in suites of specimens, and our books in recorded facts of this nature. Few Entomologists lay themselves out to collect series of specimens illustrative of this subject. I can assure them from experience that they would find it most interesting to do so.[95]

Later in 1869, two papers were read reporting on the experiments suggested by Wallace. One, "On Insects and Insectivorous Birds and Especially on the Relation between the Colour and the Edibility of *Lepidoptera* and Their Larvae," by J. Jenner Weir, F.L.S., concluded: "To sum up, I have quite satisfied myself that insectivorous birds, as a general rule, the Cuckoo perhaps being an exception, refuse to eat hairy larvae, spinous larvae, and all those whose colours are very gay, and which rarely or only accidentally conceal themselves. On the other hand they eat with great relish all smooth skinned larvae of a green or dull-brown colour which are nearly always nocturnal in their habits, or mimic the colour or appearance of the plant they frequent."[96]

The other, "Remarks upon Certain Caterpillars, etc., Which are Unpalatable to Their Enemies," by A. G. Butler, F.L.S., contained observations on the feeding habits of lizards, frogs, and spiders, which ". . . ate voraciously. But some caterpillars and moths would be seized only to be dropped in disgust." Mr. Butler concludes, "Thus in three instances I have proved the same caterpillars to be distasteful to insect persecutors; surely such evidence may be looked upon as conclusive of the fact, that some species have an advantage over others, and therefore are more likely to survive in the great struggle for existence."[97]

In his presidential address of January, 1870, Bates returned to the theme of geographical distribution, especially in relation to the fauna and flora of oceanic islands. T. V. Wollaston and Andrew Murray had maintained that there must have been connecting land between them and

[95] Ibid., p. lxviii.
[96] Ibid. 1869: 21–26.
[97] Ibid., pp. 27–29.

the continents from which they derived their species. Bates sided with Lyell and Darwin in accounting for dissemination by winds and currents. He called attention to Darwin's "important generalization that the chance of permanent establishment of migrants as species in a locality is in inverse proportion to the degree in which the locality is already well-peopled with similar forms. On this view a land newly emerged from the sea, or with a stock of species diminished by extinction, would in the course of time be appropriated by the waifs and strays which are brought to its shores."[98]

Looking back over the proceedings of the Entomological Society, it appears that the Darwinian thesis was discussed more frequently and openly there than in any of the other societies. This is partly, but only partly, to be explained by the fact that insect specimens, being far more numerous than mammals or fossils, provided a wider range of data for testing and thus for discussing the theory. It is likely, however, that the major reason was that the Entomological Society had less formal and strict standards for what was "appropriate" in scientific discussions. Judging from the number and kind of references, one might say of the five societies that their readiness to discuss Darwin varied inversely with their scientific status and prestige. Such a generalization gets some support, as we shall see, from the reception of the provincial societies.

LEARNED SOCIETIES OUTSIDE LONDON

During the 1820's a large number of societies were established in the provinces of Britain. Commonly called Literary and Philosophical Societies, they originally set out to be learned bodies, but as the years passed most of them became less scientific in the professional sense.

The career of the Leeds Philosophical and Literary Society is typical. In 1869 the society celebrated its fiftieth anniversary as "the oldest scientific institution in Yorkshire and one of the most flourishing in the provinces." In reviewing the half-century, the "Annual Report" makes the observation that originally "it was doubtless thought that many of the papers would be concerned with the natural history of the neighborhood —the local geology, flora and fauna of the district." These expectations were not fulfilled. "The meetings of the Society, instead of being com-

---

[98] Ibid. 1870: xxxvi–xxxvii.

posed mainly of students of the sciences, consist mainly of a miscellane-
ous audience with little interest in technical papers, whilst societies have
been established devoted to each particular branch of Natural History
or Philosophy. The result has been to leave to such societies as ours rather
the popularization and dissemination of knowledge than its local investigation."[99]

In pursuit of this function, fellows of the London societies were often
invited to give lectures. Owen, Wallace, Stainton, Huxley, and many
other names familiar to the London scientific bodies appear on the lecture lists of Leeds and other societies, and many of the subjects in the
1860's were, as one might expect, related to Darwinism: Professor H. D.
Rogers spoke "On Reported Traces of Primeval Man" during the session
of 1860–1861; Sir Richard Owen gave four lectures on natural history
and paleontology in 1862; Professor George Rolleston of Oxford lectured on the "Distribution of Species" in 1866; H. T. Stainton spoke on
"The Habits and Transformation of Insects" in 1866–1867; and R. Gor-
don Latham, M.D., F.R.S., lectured on "The Ratio of the Brain and Intellect in Man and Brutes" in the same year.[100]

In keeping with their more popular audiences and more elastic conception of what was "philosophical" or scientific, the provincial societies
tended to treat the *Origin* as a recent development about which their
membership would want to be informed. On May 7, 1860, at the Cambridge Philosophical Society, perhaps the most formal and academic of
the provincial societies, the Reverend Adam Sedgwick read a paper "On
the Succession of Organic Forms . . . and on Certain Theories Which
Profess to Account for the Origin of New Species"[101] (see below).

In December, 1860, at the Literary and Philosophical Society of Manchester, the Reverend W. N. Molesworth, M.A., read a paper "On the

[99] "Annual Report," *Proceedings of the Leeds Philosophical and Literary Society*
1869–1870: 4.

[100] Summaries of the "Annual Reports" and a complete list of lectures appear in E.
Kitson Clark's *The History of 100 Years of Life of the Leeds Philosophical and Literary
Society* (Leeds: Jowett and Sowry, 1924).

[101] *Proceedings of the Cambridge Philosophical Society* I (1843–1865): 223. Only
the title is included, and it is interesting that the paper was not reproduced in the
Transactions of the society. A summary appeared in a local newspaper on May 19,
1860 (see below, note 117).

Origin of Species," which asked that the theory be "considered in that spirit of philosophical calmness with which it had been proposed by the Author."[102]

At Liverpool, in October, 1860, the president, the Reverend H. H. Higgins, spoke "On Darwin's Theory of the Origin of Species" and commented on various reviews. Higgins was generally sympathetic and inclined to defend Darwin against what he considered unfair attacks. His address started a debate in the society on the merits of Agassiz's idealistic view of species as opposed to Darwin's.[103] The debate continued intermittently until October, 1862, when Higgins in his valedictory address referred to the progress made in zoology and botany by "the naturalists who, by Mr. Darwin's work on the *Origin of Species* have been taught and have already extensively learned, the true meaning of natural history—that it is the science not merely of the distinctions between living forms, but of comparative life. Much as I disagree with the conclusions of Mr. Darwin I regard the publication of his book as an epoch in natural history."[104]

In Scotland, the prestigious Royal Society of Edinburgh acknowledged the arrival of the Darwinian theory in 1860 with an extensive review by Andrew Murray.[105] After a good concise summary of the *Origin*, Murray marshaled many of the arguments being made against the theory at the time, such as the difference between domesticated and natural varieties and species, the absence of transitional forms, and the difficulties of the geological evidence. Murray conceded that Darwin's facts were fatal to the view of the independent creation of each individual species but held that the difficulties of the fossil record were better solved by the theories of progressive development promulgated by Lorenz Oken and Agassiz, because, he said, "*their* theory allowed us to retain our belief in the great argument on which the whole of Natural Theology is based."[106]

[102] *Proceedings of the Literary and Philosophical Society of Manchester* 2 (1860–1861): 24.
[103] *Proceedings of the Literary and Philosophical Society of Liverpool* 15 (1860): 42.
[104] Ibid., p. 8.
[105] *Proceedings of the Royal Society of Edinburgh* 4 (1857–1862): 274–291.
[106] Ibid., p. 290.

Other than this review, however, no paper during the next decade concerned itself directly with a discussion of Darwin's theory. But, as in the London societies, we find the anniversary addresses less reticent. The Duke of Argyll in December, 1860, devoted his address to a critique of the *Origin*:

I think it is impossible not to consider the publication of Mr. Darwin's work on the *Origin of Species* as an event in the history of scientific speculation. The influence which such theories have had in stimulating and directing the progress of actual discovery, entitles them, when they come from distinguished men, and when they rest on any large amount of careful observation, to the marked attention of such societies as this. It cannot be denied that Mr. Darwin's book claims our respect on both these grounds. It may be true, as I think it is, that all the facts he has brought together . . . bear a very small proportion to the purely speculative conclusions which go to make up his theory on the *Origin of Species*. Yet probably there is no other man now living who could have made such a rich collection.[107]

After giving his objections to the theory, most of which had been advanced by Murray and others, His Grace ended by saying: "The conclusions arrived at by Mr. Darwin are essentially but another form of the old theory of development." In 1864 the duke returned to the theme, but this time his object was to demonstrate that Darwin had not offered any "law" or explanation inconsistent with a "Creative Purposive Intelligence."[108] Using the hummingbird as a critical example, he found natural selection inadequate to account for the origin and spread of its species. "On the other hand," he said, "if I am asked if each separate species has had a separate creation, not born—but separately made—I must answer I do not believe it." He then concluded that the facts suggested the operation of some creative law in fulfillment of creative purpose.[109]

The Royal Society of Edinburgh was also the forum in which Sir William Thomson presented his arguments against uniformitarian geology in a paper "On the Secular Cooling of the Earth," read in April,

---

[107] Ibid., p. 371.
[108] Ibid. 5 (1862–1866): 265–292.
[109] Ibid., p. 277, passim.

1862,[110] and in "The 'Doctrine of Uniformity' in Geology Briefly Refuted," read in December, 1865.[111]

In the latter, Thomson, with a single paragraph of text and a page of mathematical calculations, demolished the possibility of the enormous time span needed by the Darwinians for the development of the organisms now living on the planet. Although he did not refer to Darwin or to natural selection, the implication and force of the paper were clear. As noted earlier, the problem raised by Thomson was severe and was the occasion of a rebuttal by Huxley at the Geological Society in 1869.

In 1865, despite its apparently cool reception to his theory, the Royal Society of Edinburgh—greatly to Darwin's surprise and pleasure—elected him to honorary membership.

The Royal Physical Society of Edinburgh, though its papers show a greater concern with natural history than those of the Royal Society, did not review the *Origin*, nor did Murray, the president, refer to Darwin in his anniversary address of 1859, which consisted of a summary of progress in entomology during the preceding three years. The later presidential addresses, however, are of more interest. In 1861, T. Strethill Wright, M.D., made it clear in reviewing the year's work of the society that avoidance of discussion of Darwinism was no accident:

Such has been the result of the past session. Good steady work has been done, and patiently recorded. We are men of work, not of talk. We have given forth no voice on the grand hypothetical questions which are now troubling the commonwealth of Natural Science. We have been singularly apathetic as to whether or not the stock of our first parent struggled upwards through innumerable adversities from a monad to a man. I fear, indeed, that we are prejudiced people and would rather leave the question as we found it settled many a year ago at our mother's side. . . . But we have been jotting down hard little facts—rough diamonds, which by-and-bye we may see taken up and ground and polished and set by other hands . . . These small facts are the foundations of adamant on which the vast inverted pyramids of science are balanced.[112]

In a similar vein, James M'Bain, the president in November, 1863, began his address by announcing that he was departing from "almost a

---

[110] Ibid. 4 (1862): 610–611.   [111] Ibid. 5 (1862–1866): 512.
[112] *Proceedings of the Royal Physical Society of Edinburgh* 2 (1861): 298.

fundamental principle with our Society—namely to observe and record facts rather than indulge in theory and speculation."[113] But M'Bain then went on to defend philosophical speculation as supplying a stimulus and giving "a direct aim to inquiry and investigation on the science it bears upon."[114]

Over in Glasgow, at the Philosophical Society meetings during this period, there was, so far as the record shows, an almost total absence of discussion of the Darwinian theory. In 1865 there is a note that the vice-president "exhibited the skull of a Gorilla. He directed attention to . . . the points of resemblance between this and the human skull and the much more important points of distinction between them."[115]

In 1866, John Young, a professor of natural history at Glasgow, read a paper on "The Scientific Premonitions of the Ancients" in which he, apparently a supporter of Darwin, dealt severely with those who sought to lessen Darwin's achievement by calling it merely a restatement of Lamarck or of Restif de la Bretonne or who maintained he was anticipated by the Greeks. "The progress of any important discovery has been well traced as rousing, first, opposition to the doctrine; next, denial of its originality; a last stage might be added—assertion of its previous universal acceptance. It is not true—it is not new—everyone knew it before; so it slides into the common creed and in a short time its date of discovery is known only to the learned."[116]

As for the Greeks, Young found them "incompetent, from defect of knowledge, to construct a theory of the origin of species."

Of all the reviews and addresses in these societies, the paper read by Sedgwick at the Cambridge Philosophical Society in 1860 is the most significant for the light it sheds on the conception of science of the time. Coming from the grand old professor of geology at Cambridge, it provides a fine example of the conservative scientific case against Darwin. The summary published in the local newspaper, which was prepared by Sedgwick himself, begins with a synopsis of the whole series of fossil-bearing rocks and then gives a summary of the Darwinian theory, which

[113] Ibid. 3 (1862–1866): 111.
[114] Ibid., p. 124.
[115] *Proceedings of the Philosophical Society of Glasgow* 6 (1865–1868): 132.
[116] Ibid., p. 163.

presumes to account for the succession of organic forms in these formations. Sedgwick's first criticism then follows: "Darwin's theory is not
inductive—is not based on a series of acknowledged facts, leading to a
general conclusion evolved, logically, out of the facts, and of course including them."

The proper duty of science is to study secondary causes, the only ones
"submitted to our senses" as the stepping stones to higher physical
truths. To advance beyond "these elements of inductive physical truth,"
however, and "to begin to speculate about the origin of species" is to
deal with "matters out of the province of observation and experiment."

Sedgwick then argues that we can observe all through organic nature
"a mutual adaptation of parts, fitted for the condition of each individual being; and a wonderful adaptation of its organs to all the complicated conditions of the surrounding world. These things we can see and
comprehend; and they prove with the force of demonstration that there
is, exterior to, and far above the mere phenomena of nature, a great
prescient and overruling cause."

After some examples from the geological record which test the Darwinian hypothesis, such as the great changes in flora and fauna between
formations and the absence of "links" in the fossil record, he finds that
"The hypothesis is not supported by a good inductive argument derived
from living nature." The report ends with a characteristic Sedgwickian
peroration:

The author concluded by expressing thankfulness that he had been trained in
the stern inductive truths of the Newtonian Philosophy. He expressed his hope
that this scheme of teaching would long keep its ground in this University,
and that the sons of our *Alma Mater* might long continue to seek truth with
honest zeal, and follow her lead through whatever track she might guide them.
He warned them never to forget their guide, by the seductions of hypotheses;
but to generalize inductively from facts well established; and ever, as they advanced, to test their first generalizations, and to expand them, (and sometimes
it might be to contract them,) so as to bring them into true logical accordance
with their knowledge of the laws of the material world. Pursued in this spirit
of caution and severe induction, we have no results to be afraid of.[117]

[117] *The Cambridge Herald and Huntingdonshire Gazette*, May 19, 1860. See also
Sedgwick's long letter to Darwin, making much the same points regarding final causes
versus natural selection (Darwin, *Life and Letters*, II, 247–250).

Conclusions

The reception of the learned societies to Darwin in the 1860's may be summarized as follows. There was a general avoidance of the discussion of his theory on the grounds that speculation was not the province of a scientific body. Scientific papers submitted to the societies could refer to the Darwinian hypothesis, provided they presented factual data either in support or in opposition to it. Neither the *Origin* nor any later work was reviewed at any meeting of a major London society, though in the provinces and in Scotland reviews of the *Origin*, mainly critical, were sometimes presented. Most of the discussion of Darwinism took place in the annual addresses of the presidents of those societies that followed the custom of having the chair summarize important developments in the field. Of the scientific papers in the biological sciences, the vast majority were descriptions of new species or catalogues of collections of specimens. The papers that made reference to Darwin's theory for the most part presented data which supported it, though some scientists of great prestige, like Owen and John Edward Gray, produced what they considered to be negative evidence. As the 1860's progressed, the *Proceedings* of the various societies reveal a fairly steady and rapid growth in the acceptance of the general evolutionary doctrine, i.e., that the species now on earth were descended through continuously operating natural causes from earlier existing forms. The major unsettled scientific question was whether natural selection was the explanation of how species originated, and this was to remain unsettled throughout the rest of the century and well into the present one.

In accounting for the reception of the learned societies, the evidence from their records is abundant that it was largely due to the conception of sound scientific method that was implicit in their rejection of "speculation." This conception, which Sedgwick's paper makes explicit, presented the straight course of science as the patient accumulation and organization of facts and the cautious building of generalizations upon them, which finally culminated in "laws" of nature. Darwin had gone beyond the facts, though he had accumulated a vast store of them. Unlike the conventional scientist of the time, who collected and described without theorizing (or thought he did), Darwin appeared to have observed and collected with theory and hypothesis in mind, which led to the criticism that he assumed what was to be proven.[118]

This generally accepted conception of science dictated the descriptive, factual form and content of the papers that were submitted to and published by the learned societies. It is clear from the records that the supporters of Darwin also adhered to this policy; even so aggressive a defender as Huxley did not depart from the orthodox format in the many papers he submitted to four of the five London societies.

The accepted standards of investigation described above appeared to be particularly appropriate to the scientific fields most affected by the species problem—paleontology, botany, zoology—in all of which description and organization—that is to say, taxonomy—were considered the most important work to be done and that which had, until the Darwinian revolution, the highest prestige. It was, moreover, generally assumed that the life sciences were different from the physical sciences in that, for the time being at least, they must remain primarily descriptive and classificatory.

There was another difference between the two kinds of science as they were viewed by many scientists in the sixties and seventies: namely, that in biological science teleology was considered not merely legitimate but essential to explain the phenomena. The organic world, unlike the inorganic, because of its interrelated functionings and adaptations, was, according to this view, observably purposive in its nature. Although this concept may be regarded simply as a vestige of creationism, it was not so regarded by those who held it—references to creative law, prescient intelligence, and purposive language in descriptions and explanations were not generally considered unscientific. Purposive adaptation and function seemed to a Sedgwick or an Owen part of the data in biological phenomena.

So long as this view, or ambiguity concerning it, exercised any considerable influence among biological scientists, it must have acted as a deterrent in the reception of Darwinism in the learned societies.

---

[118] For example, Sedgwick, in the *Cambridge Herald*: ". . . its [Darwin's hypothesis's] strength is said to lie in the facts supplied by Geology. But when we bring these facts to the test, we find no strength in them, except that which is gained by a virtual

A third reason for scientific reticence and caution may well have been a considered view that controversy in a learned body was destructive or at least unproductive. Thomas Bell seems to have had this in mind when he banned Darwin's theory as a subject of discussion in the Linnean Society. And there is no doubt that the views advanced in the *Origin* produced almost immediately upon their publication a much more intense reaction of hostility and fear than the theories of Lamarck or even the *Vestiges of Creation*, and calm professional discussion would have been almost impossible. In such circumstances it was no doubt convenient to have at hand the precedent of excluding the discussion of speculative and theoretical subjects.

---

assumption of the hypothesis. No good theory was ever built out of such assumptions— by a series of hypothetical reasonings from the unknown to the known."

ENGLAND

Bibliographical Essay

M. J. S. HODGE

Francis Darwin's monumental *Life and Letters of Charles Darwin* provided the first extensive information about the reception and spread of his father's doctrines. Though he exercised the customary right to chop and change things, the mass of letters and notes printed there and in the sequel, *More Letters of Charles Darwin*, remains an invaluable source. T. H. Huxley contributed to the *Life and Letters* a short essay, "On the Reception of the 'Origin of Species,' " but it is sketchy and often misleading, the main principle of interpretation being a simplistic and faulty contrast between a naturalistic outlook traced through Lyell to Descartes and various theistic prejudices ascribed to Whewell.[1]

It is only recently that historians have gone beyond Francis Darwin and Huxley in either their documentary sources or their interpretive assumptions, but a number of older publications remain, needless to say, indispensable. Besides the familiar volumes of correspondence, biography, and autobiography pertaining to Charles Lyell, Huxley, Joseph Hooker, Herbert Spencer, Adam Sedgwick, Richard Owen, and A. R. Wallace, three books by Edward Bagnall Poulton included previously unpublished letters and careful analysis of Darwin's influence: *Essays on Evolution, 1889–1907, Charles Darwin and the Origin of Species*, and *Charles Darwin and the Theory of Natural Selection*. A few years

---

[1] Francis Darwin, ed., *Life and Letters of Charles Darwin*, 3 vols. (London: John Murray, 1887; American ed., New York: Appleton, 1888); Francis Darwin and A. C. Seward, eds., *More Letters of Charles Darwin*, 2 vols. (New York: Appleton, 1903); T. H. Huxley, "On the Reception of the 'Origin of Species,' " in *Life and Letters*, II, 179–204 (American ed.: I, 533–558).

ago the complete correspondence between Darwin and John Stevens Henslow appeared, edited by Nora Barlow. A document perhaps unmatched in the entire history of science in the insight it gives into a major scientist's change of mind on a fundamental theoretical issue is *Sir Charles Lyell's Scientific Journals on the Species Question*, edited by Leonard G. Wilson.[2] These journals, which run from November, 1855, to December, 1861, in addition to revealing Lyell's own deliberations on a wide variety of scientific and metaphysical matters, frequently cast light on the opinions of others in and beyond Darwinian circles.

Discussion of the periodical literature must begin, naturally, with Alvar Ellegård's remarkable monograph, *Darwin and the General Reader*.[3] Although Ellegård's primary aim was to establish the range and roots of public opinion through analysis of press coverage of Darwin, he provides excellent accounts of several major topics, such as miracles, missing links, and the positions for and against Darwin taken at the British Association. On some subjects, Ellegård's interpretations are probably destined to be modified—his claim, for instance, that John Stuart Mill's philosophy of science was able, as William Whewell's was not, to provide a rationale for accepting Darwinism, and his tendency to find religious commitments behind almost all opposition to descent and natural selection—but even on these topics his detailed documentation and much of his succinct commentary will surely stand.

The numerous more general studies provoked and promoted by the Darwin centennial in 1959 contain no comprehensive treatments of Darwin's English reception, but they often quote from previously overlooked or unpublished sources bearing on this topic. There are, for example, few more instructive comparisons (or better correctives to any unqualified generalization about liberals and Darwinism) than that be-

---

[2] Edward Bagnall Poulton, *Essays on Evolution, 1889–1907* (Oxford: The Clarendon Press, 1908); idem, *Charles Darwin and the Origin of Species* (London: Longmans, 1909); idem, *Charles Darwin and the Theory of Natural Selection* (London: Cassell, 1896); Nora Barlow, ed., *Darwin and Henslow: The Growth of an Idea; Letters 1831–1860* (London: Murray, 1967); Leonard G. Wilson, ed., *Sir Charles Lyell's Scientific Journals on the Species Question* (New Haven and London: Yale University Press, 1970).

[3] Alvar Ellegård, *Darwin and the General Reader: The Reception of Darwin's Theory of Evolution in the British Periodical Press, 1859–1872* (Göteborg: Elanders Boktryckeri Aktiebolag, 1958).

tween the responses to common descent made by those two ironically estranged Dubliners, Archbishop Richard Whately and his former protégé at Oriel, the future cardinal, John Newman. See Whately's letter of February 19, 1860, to Sedgwick in Gertrude Himmelfarb's *Darwin and the Darwinian Revolution* and Newman's remarks in a notebook of 1863, quoted in John Lyon's "Immediate Reactions to Darwin."[4] The radical defects of Himmelfarb's understanding of Darwinian science are too manifest to need emphasis here. Her chapters on "The Darwinian Party" and on "The Anti-Darwinian Party" do bring out, though, three important points: Darwin's major reputation as a scientist prior to the *Origin*; the enormous solidarity, camaraderie, and prestige enjoyed by the triumvirate of Lyell, Hooker, and Huxley; and, by contrast, the failure of Darwin's opponents to achieve the same esprit de corps, coordinated strategy, or public respect. The contrast would make an excellent case-study in the role personalities and institutions can play in getting a fair hearing for an unorthodox innovation. Owen, for example, by implicit sniping at Sedgwick in his review and by failing to keep secret his collaboration with Bishop Samuel Wilberforce, and Whately, by publishing without permission an unguardedly impetuous letter of Sedgwick's in the *Spectator*, gave anti-Darwinian efforts a reputation for inept conspiracy. Huxley, on the other hand, seized a splendid opportunity to involve Darwinian theory in a general movement to upgrade British biology, especially relative to German efforts, when he took over, as editor, the *Natural History Review* in 1861. For four years this journal published original articles and reviews by an impressive array of people, all of whom, though not positively embracing Darwinian doctrine, wanted to see biology become responsibly theoretical and systematically empirical in ways the old natural history had rarely been. By 1864, the editors included such names as W. B. Carpenter, John Lubbock, P. C. Sclater, George Busk, and Wyville Thomson, while George Bentham, Hooker, Darwin, Wallace, Henry W. Bates, Lubbock, George Rolleston, Huxley, G. H. Lewes, Carpenter, Sclater, Hugh Falconer,

---

[4] Gertrude Himmelfarb, *Darwin and the Darwinian Revolution* (Garden City: Doubleday, 1959; edition cited is paperback, New York: Norton, 1968), pp. 270–271; John Lyon, "Immediate Reactions to Darwin: The English Catholic Press's First Reviews of the 'Origin of the Species' [*sic*]," *Church History* 41 (1972): 78, n. 2. Lyon's article is a bibliographically helpful but analytically naïve discussion of its subject.

Edward Blyth, and many others less famous had written especially for it. For the state of the science in those years no source is more instructive.

It has long been recognized that the history of the *Origin of Species* as a book throws much light on the Darwinian debates. Morse Peckham's introduction and his text, *The Origin of Species: A Variorum Text*, initiated serious study of this history, while Peter J. Vorzimmer's *Charles Darwin: The Years of Controversy* gives a detailed discussion of Darwin's response to those who challenged his assumptions about the causes of variability, role of isolation, frequency of atavism, and the like.[5] Vorzimmer shows how, by 1872, Darwin's case for natural selection had become seriously weakened by inconsistency in his responses to critics and by the empirical counterevidence they had accumulated. However, the implication of this, for the book's initial reception, is mainly, I think, to confirm that until the mid-sixties no one had given a good reason for not provisionally admitting natural selection as a legitimate hypothesis, in the sense claimed by Bentham, Huxley, and Henry Fawcett, as well as Mill, whose account of hypotheses they all invoked. Vorzimmer, somewhat surprisingly, nowhere considers the geochronology issue. This is, however, thoroughly treated by J. D. Burchfield in his dissertation, "The Age of the Earth: The Theories and Influence of Lord Kelvin."[6] Burchfield discusses fully the irrefutable force of Kelvin's arguments; the willingness of some geologists, like Lyell, to discount established thermodynamic principles; and the difficulty of showing, in the case of any particular scientist, that chronological considerations alone were decisive in diminishing the appeal of natural selection.

Our understanding of Darwin's relation to contemporary moral and social theory has gained immeasurably from several recent studies, especially, John Wyon Burrow's masterly *Evolution and Society*, the paperback impression of which carries an important new preface; George W. Stocking's *Race, Culture and Evolution*; and Antony G. Flew's *Evolutionary Ethics*. Flew rightly praises an exemplary article by D. Raphael,

---

[5] Charles Darwin, *The Origin of Species: A Variorum Text*, ed. Morse Peckham (Philadelphia: University of Pennsylvania Press, 1959); Peter J. Vorzimmer, *Charles Darwin: The Years of Controversy; The Origin of Species and Its Critics, 1859–1882* (Philadelphia: Temple University Press, 1970).

[6] J. D. Burchfield, "The Age of the Earth: The Theories and Influence of Lord Kelvin" (Ph.D. dissertation, Johns Hopkins University, 1969).

"Darwinism and Ethics," in *A Century of Darwin*, edited by S. A. Barnett. An older paper, more purely historical, is still useful: Kenneth F. Ganz, "The Beginnings of Darwinian Ethics." Herbert Spencer is now, very properly, being seen, once more, as a well-respected and instructive critic (especially in his essays) of almost everything going on in his day. J. D. Y. Peel's *Herbert Spencer: The Evolution of a Sociologist* is a lucid and penetrating study. The section on Spencer in Robert M. Young's *Mind, Brain, and Adaptation in the Nineteenth Century* is also valuable.[7]

Darwinism's place in the general history of British "rationalism" and "free thought" remains largely undocumented. But considerable information is to be found in two recently reprinted classics—John MacKinnon Robertson, *A History of Freethought in the Nineteenth Century*, and Alfred William Benn, *The History of English Rationalism in the Nineteenth Century*—and in Warren Sylvester Smith's *The London Heretics, 1870–1914*. The relation of Darwinian theory to the contemporary philosophies of science of John Herschel, Mill, and Whewell raises complex and controversial matters, as clearly shown by David Hull in his incisive paper, "Charles Darwin and Nineteenth-Century Philosophies of Science," and in the introduction to his important collection, *Darwin and His Critics*.[8]

In tracing responses to Darwin in the learned societies, one gets little

7 John Wyon Burrow, *Evolution and Society: A Study in Victorian Social Theory* (Cambridge: Cambridge University Press, 1966; paperback ed., 1970); George W. Stocking, Jr., *Race, Culture and Evolution: Essays in the History of Anthropology* (New York: Free Press, 1968; paperback ed., 1971); Antony G. Flew, *Evolutionary Ethics* (New York: St. Martin, 1968); D. Raphael, "Darwinism and Ethics," in *A Century of Darwin*, ed. Samuel A. Barnett (London: Heinemann, 1958); Kenneth F. Ganz, "The Beginnings of Darwinian Ethics," *University of Texas Publication: Studies in English*, no. 3939 (1939): 180–209; J. D. Y. Peel, *Herbert Spencer: The Evolution of a Sociologist* (New York: Basic Books, 1971); Robert M. Young, *Mind, Brain, and Adaptation in the Nineteenth Century* (Oxford: Clarendon Press, 1970).

8 John MacKinnon Robertson, *A History of Freethought in the Nineteenth Century*, 2 vols. (London: Watts, 1929; reprint ed., New York: Humanities, 1969); Alfred William Benn, *The History of English Rationalism in the Nineteenth Century*, 2 vols. (London: Longmans, 1906; reprint ed., New York: Russell and Russell, 1962); Warren Sylvester Smith, *The London Heretics, 1870–1914* (London: Constable, 1967); David Hull, "Charles Darwin and Nineteenth-Century Philosophies of Science," in *Foundations of Scientific Method: The Nineteenth Century*, ed. Ronald N. Giere and Richard S. Westfall (Bloomington and London: University of Indiana Press, 1973); idem, *Darwin and His Critics* (Cambridge: Harvard University Press, 1973), pp. 1–77.

help, as Frederick Burkhardt confirms, from the various histories of them that have appeared from time to time. Two articles do, however, enlarge our knowledge of the Linnean Society's reception of the 1858 paper: Barbara G. Beddall's "Wallace, Darwin, and the Theory of Natural Selection" and J. W. T. Moody's "The Reading of the Darwin and Wallace Papers."[9]

Of the many cooperative studies of the Victorian period, one deals with Darwin's impact in some detail: Philip Appleman, W. A. Madden, and M. Wolff, eds., *1859: Entering an Age of Crisis*, a volume that shows, once and for all, that Darwinism was only one among many profound challenges to conventional mid-century opinions. Two standard works should also be mentioned: William Laurence Burn's *The Age of Equipoise: A Study of the Mid-Victorian Generation* and George Kitson Clark's *The Making of Victorian England*; both offer authoritative alternatives to several popular misconceptions. Finally, the puzzling failure of social unrest to match intellectual upheavals is considered in T. R. Tholfsen's "The Intellectual Origins of Mid-Victorian Stability."[10]

[9] Barbara G. Beddall, "Wallace, Darwin, and the Theory of Natural Selection: A Study in the Development of Ideas and Attitudes," *Journal of the History of Biology* 1 (1968): 290–320; J. W. T. Moody, "The Reading of the Darwin and Wallace Papers: An Historical 'Non-Event,'" *Journal of the Society for the Bibliography of Natural History* 5 (1971): 474–476.

[10] Philip Appleman, W. A. Madden, and M. Wolff, eds., *1859: Entering an Age of Crisis* (Bloomington: University of Indiana Press, 1959); William Laurence Burn, *The Age of Equipoise: A Study of the Mid-Victorian Generation* (London: Allen and Unwin, 1964); George Kitson Clark, *The Making of Victorian England* (London: Methuen, 1962); T. R. Tholfsen, "The Intellectual Origins of Mid-Victorian Stability," *Political Science Quarterly* 86 (1971): 57–91.

GERMANY

WILLIAM M. MONTGOMERY

Darwin's theory of evolution arrived in Germany in 1860 with the first translation of the *Origin of Species*. This event occasioned immediate debate, not only among biologists, but also among clergymen, philosophers, and scientifically educated laymen. During the 1860's and 1870's a flood of books and articles appeared in Germany that touched on Darwin's theory in one way or another. It would be impossible to describe the entire scope of the German response to Darwinism within the limits of this paper. Therefore, I shall confine my discussion to the German biologists. They were professionally concerned with the question of evolution, they understood it better than anyone else, and their reactions were at least as widely varied as those of any other group. For these reasons the biologists deserve preferred attention. In presenting their story, I want first to give a short description of the foreign sources of information that they drew on most heavily—the evolutionary books that were translated into German. Next I intend to describe the participants in the scientific debate in order to suggest some of the personal factors that affected their judgments. And, finally, I shall devote the bulk of the paper to the scientific arguments for and against Darwinism. German scientists made important contributions to evolutionary theory, contributions that helped determine what Darwinism came to be.

One measure of the success of evolutionary theory in Germany is the large number of translations of foreign authors. Even before the *Origin*

NOTE. This paper is based on my dissertation, "Evolution and Darwinism in German Biology, 1800–1883." The research has been carried out in Germany under an NDEA Title IV grant awarded by the University of Texas at Austin.

appeared, there had been two separate translations of Robert Chambers's *Vestiges of the Natural History of Creation*. This work seems to have influenced Hermann Schaaffhausen and Bernhard Cotta and probably encouraged other German evolutionists before Darwin did.[1] One of Chambers's translators, Carl Vogt, although unpersuaded by Chambers at the time, later became an outspoken advocate of Darwinism. Beginning in 1860, the *Origin* itself appeared in five separate German editions. The first two of these were prepared by the paleontologist Heinrich Bronn, who, like Vogt, was not convinced by his author. Bronn also completed a German edition of Darwin's book on orchids before he died in 1862. Thereafter, Darwin's books were translated by Julius Viktor Carus, a convinced follower, who kept the German public supplied with each new book as it appeared. In 1875 Darwin's German publisher, the Schweizerbart firm of Stuttgart, inaugurated an edition of his collected works. This task was thus accomplished in Germany even before it was in England. In addition to Darwin's books, Carus also translated T. H. Huxley's *Man's Place in Nature* in 1863. Its message of human evolution was startling; next to the *Origin*, no foreign work on evolution made as deep an impression on German readers. Many of Huxley's other books were rendered into German by various individuals. Of particular importance for evolution were *A Manual of the Anatomy of Vertebrate Animals* and *A Manual of the Anatomy of Invertebrate Animals*. Adolf Bernhard Meyer translated the Darwin and Wallace essays of 1858 as well as three books by Wallace: *The Malay Archipelago* (1869), *Contributions to the Theory of Natural Selection* (1870), and *The Geographic Distribution of Animals* (1876). And the materialist evolutionist Ludwig Büchner translated Charles Lyell's *The Antiquity of Man*, which appeared in three German editions between 1864 and 1874. Of course, antievolutionary writers were published too. Louis Agassiz's *Essay on Classification* came out in German, as did an edition of *The Plan of Creation*, a series of posthumous lectures. Armand de Quatrefages's *The Human Species* was translated, but St. George Mivart's *Genesis of Species* was not. One important foreign anti-Darwinist did not require translation; the Baltic Russian Karl Ernst von Baer wrote in

---

[1] Hermann Schaaffhausen, *Anthropologische Studien* (Bonn: Adolf Marcus, 1885), p. 146; see also Bernhard von Cotta, *Briefe über Alexander von Humboldt's Kosmos*, 2d ed. (Leipzig: T. D. Weigel, 1850), p. 263.

German. His long essay on Darwinism appeared in a collection of papers and speeches in 1876. A comparison of the two sides reveals that the balance of translations was weighted clearly in favor of the Darwinists. Carus particularly contributed to this margin. Furthermore, the call for new editions shows that the translators' work was rewarded and that foreign authorities had good success in spreading the evolutionary idea in Germany.[2]

The Darwinist party among German biologists was by no means a homogeneous group. Yet they did differ from their opponents in certain notable ways. As a group, they were relatively young. When the *Origin* appeared in 1860, the ages of a number of leading Darwinists were as follows: Matthias Schleiden, 56; Bernhard von Cotta, 52; Moritz Wagner, 47; Carl Vogt, 43; Fritz Müller, 38; Wilhelm Hofmeister and Ludwig Büchner, 36; Carl Gegenbaur, 34; Julius Sachs and Carl Semper, 28; Ernst Haeckel and August Weismann, 26; Carl Claus, 25; and Anton Dohrn, 20. This youthfulness of the Darwinists was reflected in the academic positions they held. It was customary in this period for a man to take his first job at a small, remote university. If he won a name for himself, he might expect a call to a higher-paying position at a larger, more prestigious institution. Because many Darwinists were still in the early stages of their careers, they were still located in small, rural settings. Ernst Haeckel and Carl Gegenbaur were in Jena. Gegenbaur eventually accepted a call to Heidelberg in 1878. August Weismann spent his career at Freiburg in Breisgau, likewise a small university. Hermann Schaaffhausen taught at Bonn. Carl Claus, first situated in Marburg, won a position at Vienna after Ernst Haeckel turned it down.[3] Carl Semper was at Würzburg throughout his career. Julius Sachs, first located at an agricultural academy in Poppelsdorf, later moved to Würzburg also. Bernhard von Cotta taught at the mining academy at Freiberg in Saxony. Wilhelm Hofmeister was situated at Heidelberg. And Carl Vogt, exiled from Germany after the 1848 revolution, taught in Geneva. A significant number of Darwinists held no university chairs at all. Anton Dohrn, a successful scientific entrepreneur, built the Naples Zoological Station, where he spent his career as director. Fritz Müller lived as a

[2] See appendix for a list of German editions of important works on evolution.
[3] Carl Claus, *Hofrath Dr. Carl Claus* (Marburg: N. G. Elwart, 1899), p. 19.

school teacher in Brazil, and his brother Hermann was likewise employed in Westphalia. Moritz Wagner, the least conventional of them all, lived as a freelance journalist.

The Darwinists were distinctive in other ways too. A large number of them had abandoned the Christian religion. Carl Vogt and Ludwig Büchner were philosophical materialists, both of them authors of forthright publications against religion. Ernst Haeckel was, if anything, even more outspokenly opposed to Christianity. He was not, however, a thoroughgoing materialist. Haeckel believed in a materialistically flavored monism, derived from his somewhat superficial reading of German idealists. Anton Dohrn was a materialist as a young man but was converted to a neo-Kantian position by the philosopher Friedrich Albert Lange. Fritz Müller was a freethinker who declined a post in the Prussian school service because of its officially Christian policies. Bernhard von Cotta had employed evolutionary arguments against religion as early as 1848. Moritz Wagner believed also in the incompatibility of science and religion. August Weismann's views on the subject are difficult to ascertain, though he seems to have been indifferent to religion. These viewpoints are not surprising; nevertheless, not all Darwinists were unbelievers. Gustav Jaeger devoted an entire book to reconciling Christianity and Darwinism. Matthias Schleiden was a religious man who criticized Haeckel's use of evolution for antireligious purposes. Hermann Schaaffhausen had no hesitation about expressing religious views in his papers on Darwinism. Carl Gegenbaur was more reticent. A man of pronounced anticlerical views, he followed the Old Catholic sect, which deserted the church in reaction to the proclamation of papal infallibility.[4]

---

[4] Carl Vogt, *Köhlerglaube und Wissenschaft: Eine Streitschrift gegen Hofrath Rudolf Wagner in Göttingen* (Giessen: Ricker, 1855); Ludwig Büchner, *Kraft und Stoff* (Frankfurt am Main: Meidinger Sohn & Cie, 1855); Guenther Altner, *Charles Darwin und Ernst Haeckel: Ein Vergleich nach theologischen Aspekten*, Theologische Studien, ed. Karl Barth and Max Geiger, no. 85 (Zurich: EVZ-Verlag, 1966); Theodor Heuss, *Anton Dohrn*, 2d rev. ed. (Stuttgart and Tübingen: Reiner Wunderlich Verlag Hermann Leins, 1948), pp. 73–75, 407–413; Ernst Haeckel, "Fritz Müller—Desterro: Ein Nachruf," *Jenaische Zeitschrift für Naturwissenschaft* 31 (1897): 159; Cotta, *Briefe*, pp. 268–269; Hanno Beck, "Moritz Wagner in der Geschichte der Geographie" (Dissertation, University of Marburg, 1951), pp. 262–264; Karin Gebhardt, "Zum philosophischen Gehalt des biotheoretischen Werkes August Weismanns," *Deutsche Zeitschrift für Philosophie* 13 (1965): 1280–1292; Gustav Jaeger, *Die Darwin'sche Theorie und ihre Stellung zu Moral und Religion* (Stuttgart: Julius Hoffmann, 1869);

The political views of the Darwinists reflected the liberality of their religious opinions. Carl Vogt lost his job in Giessen after participating as a radical member of the National Assembly of 1848. He lived thereafter in Switzerland. Fritz Müller, depressed by the failure of the 1848 revolution, departed for Brazil, where he spent the rest of his life. Ludwig Büchner recorded his radical social views in numerous books. Ernst Haeckel, a patriotic Saxon, had great contempt for neighboring Prussia and its aristocratic, reactionary political system. However, like many liberal Germans, he was awed and thrilled by Bismarck's success in unifying the Reich and became extremely nationalistic as he got older. Weismann's nationalism was evident earlier; he approved of Prussia's German policy even before the unification. In general, one can describe the Darwinists as young, often unconventional in religious and social opinion, and located in smaller, less distinguished universities. Of course, they did not remain that way. The success of their scientific work and the acceptance of their theories brought them promotions and acclaim. With age, they too ascended into the scientific establishment. However, in spite of their professional success, some of them bore to their graves an air of radical unorthodoxy. The mood, once acquired, was never really lost.[5]

In contrast to the Darwinists, their opponents presented the image of older men, better established and more orthodox, at least in their religious views. Among the opponents of evolution, age seems to have been a significant factor. In 1860 a group of them were the following ages:

Matthias Schleiden, *Ueber den Materialismus in der neueren deutschen Naturwissenschaft* (Leipzig: Wilhelm Engelmann, 1863); idem, *Das Meer*, 2d ed. (Berlin: A. Sacco, Nachfolger, 1874), p. 142 n probably refers to Haeckel; Fr. Maurer, *Carl Gegenbaur* (Jena: Gustav Fischer, 1926), p. 14.

[5] Haeckel, "Fritz Müller," pp. 160–161; Ludwig Büchner, *Die Stellung des Menschen in der Natur in Vergangenheit, Gegenwart und Zukunft*, 3 vols. (Leipzig: Theodor Thomas, 1870), III; Ludwig Büchner, *Darwinismus und Sozialismus, oder der Kampf um das Dasein und die moderne Gesellschaft* (Leipzig: Ernst Günther, 1894); Ernst Haeckel, *Natürliche Schöpfungsgedanken* (Berlin: Georg Reimer, 1868), pp. 139–140; Georg Uschmann, ed., *Ernst Haeckel, Forscher, Künstler, Mensch*, 3d ed. (Leipzig: Urania, 1961), pp. 79–80; Georg Uschmann and Bernhard Hassenstein, eds., "Der Briefwechsel zwischen Ernst Haeckel und August Weismann," in *Kleine Festgabe aus Anlass der hundertjährigen Widerkehr der Grundung des Zoologischen Institutes der Friedrich-Schiller-Universität Jena im Jahre 1865 durch Ernst Haeckel*, ed. Manfred Gersh, Jenaer Reden und Schriften (Jena: Friedrich-Schiller-Universität, 1965), p. 18.

Dietrich von Schlechtendal, 66; Andreas Wagner, 63; Karl Schultz-Schultzenstein, 62; Heinrich Göppert and Heinrich Bronn, 60; Rudolf Wagner, 55; Hermann Burmeister, 53; Johann Lucae, 46; Hermann Hoffmann, 41; Rudolf Virchow and Albert Wigand, 39; Otto Volger, 38; Oscar Fraas, 36; and Adolf Bastian, 34. Individuals who accepted evolution without becoming Darwinists were somewhat younger: Alexander Braun, 55; Oswald Heer, 51; Albert von Kölliker and Carl Nägeli, 43; Ludwig Rütimeyer, 35; Alexander Goette, 20. However, the small size of this group does not allow any dependable generalizations. In accordance with their ages, the opponents of Darwinism held somewhat more distinguished teaching positions. Alexander Braun, Rudolf Virchow, Karl Schultz-Schultzenstein, and Adolf Bastian held chairs in Berlin. Heinrich Bronn taught in Heidelberg. Carl Nägeli and Andreas Wagner were in Munich. Rudolf Wagner held a chair in Göttingen. Albert von Kölliker taught at Würzburg. Ludwig Rütimeyer was located in Basel and Oswald Heer in Zürich. Christoph Giebel and Dietrich von Schlechtendal taught at Halle. Heinrich Göppert occupied a chair in Breslau. Albert Wigand was located in Marburg. Hermann Hoffmann was in Giessen. And Johann Lucae taught at the medical school in Frankfurt. Like the Darwinists, some of these scientists were also active outside academic life. Otto Volger taught at the Senkenberg Museum in Frankfurt. Oscar Fraas and Hermann Burmeister were museum curators, the former in Stuttgart and the latter in Buenos Aires.

The religious views of Darwin's opponents also differed from those of his supporters. Rudolf Virchow was not a believer, though he objected to scientists' debating religion in public. Albert Kölliker was apparently indifferent to religion too. However, they were probably a minority. Alexander Braun thought that investigating God's purpose was one of the goals of science. Ludwig Rütimeyer, who studied theology before he became a scientist, grew reticent about religion in his later years; but he does not seem to have abandoned his earlier faith. Oswald Heer thought that species transmutation required the direct intervention of the Creator; and Oscar Fraas, even more conservative, believed literally in the biblical flood. Rudolf Wagner and Andreas Wagner had once engaged Carl Vogt in a noisy public debate over the issue of materialism. Karl Schultz-Schultzenstein was also a man of religious feeling, though he seems to have been inclined as much toward *Naturphilosophie* as toward orthodox Christianity. Adolf Bastian expressed commitment to a religious outlook. And Albert Wigand was firmly orthodox. The religious opinions of Carl Nägeli are difficult to ascertain, but he seems to have had respect for religious opinion even if he did not fully share it himself.[6]

In political matters the difference between the scientific camps was not clearly marked. Albert von Kölliker held opinions noticeably more conservative than those of most scientists,[7] and none of the opponents of Darwinism held views as radical as those of Louis Büchner. However, as one might expect in a group of university professors, the opponents of Darwin reflected a generally liberal outlook. Rudolf Virchow was an outstanding liberal leader in the German Reichstag. Hermann Burmeister, like Carl Vogt, had participated in the 1848 National Assembly and had been obliged to emigrate. For the most part, it is difficult to establish the scientists' political views at all. The theories of evolution and natural selection did not affect political questions directly. Many scientists commented on the religious implications of Darwinism, but few of them mentioned the political ones. In drawing a balance between the Darwinists and their critics, one can say that the Darwinists included more young men, freethinkers, and materialists. The evidence on political beliefs is less decisive. There are reasons for suspecting that religious and political radicalism are associated to some degree; nevertheless, men

[6] Rudolf Virchow, "Ueber die mechanische Auffassung der Lebensvorgänge," in *Amtlicher Bericht der Versammlung deutscher Naturforscher und Aertze zu Carlsruhe* (1858), pp. 43–44; Remigius Stölzle, *A. von Köllikers Stellung zur Descendenzlehre: Ein Beitrag zur Geschichte moderner Naturphilosophie* (Münster in Westfalen: Aschendorff, 1901), pp. 6–9; Cecilie Mettenius, *Alexander Braun's Leben nach seinem handschriftlichen Nachlass* (Berlin: G. Reimer, 1882), pp. 653–657; Adolf Portmann, "Die Frühzeit des Darwinismus im Werk Ludwig Rütimeyers," *Basler Stadtbuch* 1965: 170, 178; Oswald Heer, *Die Urwelt der Schweiz* (Zurich: Friedrich Schulthess, 1865), p. 604; Oscar Fraas, *Vor der Sündfluth: Eine Geschichte der Urwelt* (Stuttgart: Hoffmann, 1866); Rudolf Wagner, *Der Kampf um die Seele vom Standpunkt der Wissenschaft* (Göttingen: Dietrich, 1857); Karl Heinrich Schultz-Schultzenstein, "Ueber die Stellung Blumenbachs zur Darwin'sche Schöpfungstheorie," *Amtlicher Bericht der Versammlung deutscher Naturforscher und Aertze zu Hannover* (1865), p. 47; Adolf Bastian, *Schöpfung oder Entstehung* (Jena: Hermann Costenoble, 1875), pp. 238–243; Albert Wigand, *Der Darwinismus und die Naturforschung Newtons und Cuviers*, 3 vols. (Braunschweig: F. Viewig und Sohn, 1874–1877), II.

[7] Ernst Ehlers, "Albert Kölliker zum Gedächtnis," *Zeitschrift für wissenschaftliche Zoologie* 84 (1906): iv.

felt the relevance of religion more keenly than that of politics. It is not likely that political beliefs had a significant independent influence on scientists' judgments of Darwinism.

Of course, the influence of age, religion, and political opinion is relatively limited in importance. Scientific opinions must be justified by scientific reasoning. And the best evidence for determining how purely scientific factors affected decisions about Darwinism may be found by analyzing the distribution of Darwinists in the different scientific specialties. A surprisingly large proportion of Darwinists, particularly among those engaged in active research, were specialists in invertebrate zoology. Ernst Haeckel, Fritz Müller, Carl Claus, August Weismann, Carl Semper, and Anton Dohrn all fall into this category. Age had something to do with this. Since the vertebrate animals bore a closer resemblance to man, they had been closely studied for a much longer time. By the mid-nineteenth century, they were already well known to zoologists. Young men of talent and ambition were more likely to turn to a field in which they could make their names by fresh discovery. Entomology, long an activity for hobbyists and collectors, was being transformed by serious new investigation into insect anatomy and embryology. And for some time, young zoologists had been making expeditions to Helgoland in the North Sea and to Italy in order to explore the unfamiliar realm of marine animals. But the invertebrate animals were important not only because they attracted bright young men but also because they were ideal subjects for evolutionary investigations. Existing in vast numbers of closely related species and having relatively short life cycles, they afforded the best possible material for the study of variation and for the comparison of larval forms, which were so vital to the study of species transformation. Darwin himself was an expert on barnacles; and it was no accident that Anton Dohrn, who founded the Naples Zoological Station, was a deeply engaged Darwinist. Primary among his concerns was the desire to provide library and laboratory facilities to further Darwinist research on marine invertebrates.[8]

---

[8] Both opponents and supporters agreed on the importance of the invertebrate animals for Darwinism. See Bastian, *Schöpfung*, pp. 211–212; Carl Claus, *Die Copepodenfauna von Nizza: Ein Beitrag zur Characteristik der Arten und deren Abänderung im Sinne Darwins*, Schriften der Gesellschaft zur Befoerderung der Gesammten Naturwissenschaften zu Marburg, vol. 9, supplementary issue 1 (Marburg and Leipzig: N. G.

By contrast, men in other scientific specialties tended to be less enthusiastic. Among paleontologists, the older men were imbued with catastrophist ideas and found it difficult to adjust to the idea of gradual species change. At best they might be willing to view evolution as a process of saltatory jumps. A second unenchanted group consisted of medical-school anatomists. Accustomed to the type concept of comparative anatomy and having little knowledge of field biology, they were reluctant to adopt evolution—or they preferred to seek the causes of it in some internal principle of growth. The third significant group of opponents were the botanists. Superficially, the plants seem to offer the same advantages for evolutionary study as the invertebrate animals. However, experimental studies on variation and hybridization often yielded results that looked unfavorable for Darwinism. There were exceptions, of course. The anatomist Carl Gegenbaur was a Darwinist, as were the botanists Matthias Schleiden, Julius Sachs, and Wilhelm Hofmeister. The geologists Friedrich Rolle and Bernhard von Cotta spoke out for Darwin, and, eventually, younger men like Leopold Würtemberger, Carl Zittel, and Melchior Neumeyer began unearthing form series showing the gradual transformations of the type Darwin had predicted. On the other side, Albert von Kölliker had done distinguished work in invertebrate zoology. Nevertheless, it is still true that most of the initial Darwinist arguments were framed by men who drew their inspiration from studying invertebrate animals. And the best arguments against these men were invented by botanists, anatomists, and paleontologists. The question of Darwinism was not decided merely on the basis of age or ideological predisposition. The most immediate influences bearing on the case were scientific, and only by reviewing the scientific arguments can one fully comprehend the conflict and its outcome.

The German scientists who had to consider Darwin's new theory faced the same tasks as scientists everywhere. They were obliged to overcome old modes of thinking, old prejudices and predilections, in order to understand Darwin on his own terms. They had to reinterpret or abandon a tradition of biological thought based on older principles.

Elwart, 1866), pp. 7–8; Anton Dohrn, "Der gegenwärtige Stand der Zoologie und die Gründung zoologischer Stationen," *Preussische Jahrbücher* 30 (1872): 147–149.

Fundamental among the axioms of traditional biology was the idea of type. It was basic to the natural system of taxonomy and to the discipline of comparative anatomy. It was also central to embryology, that branch of biology in which German researchers had most distinguished themselves. Here, however, it was supplemented by a second principle, that of development or *Entwicklung*. German scientists were the first to abandon the notion of preformation in favor of the dynamic assumptions of epigenesis. Form was no longer regarded as an unexamined given, transmitted mechanically from one generation to the next, but was considered the product of a teleological process, an inner *Bildungstrieb* or formative drive through which all other life processes gradually unfolded. Some anatomists interpreted this to mean that an animal's embryological development repeated in succession the forms of the scale of nature from the lowest up to its own position. However, biologists had come to doubt this. Most of them accepted Cuvier's view that the animal kingdom was separated into several major types or branches. The concept of type was supplemented by a physiological notion, that of correlation, according to which the functional adaptation of an animal is due to a harmonious inner correlation of its organs. The arrangement of organs is fundamentally different for each type, and no alteration can be made in one part without upsetting the functioning of all the others. In line with this opinion, Karl Ernst von Baer taught that the basic type of the embryo is laid down at the very first, and it grows by gradually differentiating itself from the path taken by all other related forms. The younger the embryo, the more it resembles all others of its type.

These ideas were all basic to the idea of design and to the various idealist theories of evolution that developed prior to Darwin. Of course, they were not accepted without question. Discoveries in cell theory, nerve physiology, and organic chemistry seemed to undermine the idealist assumptions on which this older biology was based. Young critics insisted that teleology be rooted out of biology, saying that the only true science is mechanistic. They wanted to reduce all explanation to some form of causal determinism. But, outside the realm of cell theory and physiology, the new mechanist philosophy was more preached than practiced. Taxonomy and comparative anatomy could not really get along without the type concept; nor was embryology able to dispose of the *Bildungstrieb*. And these were the very fields in which the question of evolution was most pertinent. In the minds of most biologists, both idealist and mechanist principles operated in uneasy coexistence. The fate of Darwin's theory depended on the way in which men interpreted it in terms of these two biological traditions.[9]

Among the three rival factions who contended over Darwin's revolutionary ideas, the first to state their case were the opponents of evolution. In his afterword to the *Origin of the Species*, Heinrich Bronn took issue with the book's principal thesis, and the first German edition was burdened from the start by an unfavorable review sewn within its own binding. Articles critical of evolution appeared in a number of German scientific journals during the early sixties. Furthermore, in speeches before the annual Convention of Natural Scientists and Physicians, the skeptics seem to have been the most vocal. Although the battle was essentially lost by the end of the decade, some opponents were still active in the 1870's. Books by Adolf Bastian and Albert Wigand and a notable address by Rudolf Virchow represented the last important resistance.

It should be noted that few of these men seriously defended the Mosaic creation as an alternative to Darwinism. Some of them may have been attached to the Biblical story for personal reasons, but they realized that it could not serve as a scientific account. Although Heinrich Bronn found no persuasive evidence for the origin of species either by evolution or by spontaneous generation, he was not attracted to theistic creation. Bronn considered the doctrine inconsistent with the practice of science. He preferred the idea of some plan or force, which might be attributed to the Creator, but which functioned according to purely natural means. Hermann Burmeister agreed closely with Bronn. According to Burmeister, if science did not wish to depend on miracles, it would be obliged to postulate the origin of life from the productive power of matter itself. He admitted that no such power seems to operate in the present world but insisted that it may have done so at an earlier period in the earth's history. The Berlin botanist Karl Schultz-Schultzenstein considered it a mistake to identify the Creator of the ethical commandments

[9] For the mechanist-idealist disagreement as the background to the German debate over Darwinism, see Pierce C. Mullen, "The Preconditions and Reception of Darwinian Biology in Germany, 1800–1870" (Ph.D. dissertation, University of California at Berkeley, 1964).

with the Creator of Nature. He thought that the secular aspects of the Old Testament could be separated from its theological message, making it possible to discuss creation without reference to religion. Among the later opponents of evolution, Adolf Bastian also insisted that science knows nothing of the origin of life and that it cannot deal with an anthropomorphic *deus ex machina*. Albert Wigand, less hesitant about describing the origin of things, suggested that the higher species arose separately through larval stages from original *Urcells*. These men resented the antireligious commitments of many Darwinists. For the most part they favored some teleological explanation of nature's development. However, they knew that science could not rely on religious texts, and they preferred to defend their position with traditional biological arguments.[10]

One of the most characteristic arguments against Darwin involved the problem of variation. According to the type concept, each species possesses certain essential characteristics. Species representatives may vary in nonessential characteristics, but the extent of possible variation is ultimately limited. Indeed, hybrid infertility indicates a physiological basis for these limits. Under the proper conditions, both variants and hybrids will eventually revert to type. For men accustomed to this view, Darwin's notion of species was hopelessly ambiguous. It seemed to Heinrich Bronn that the small, random changes postulated by Darwin should produce a chaos of forms, completely dissolving the observable order of species. This complaint was taken up by several individuals, but it should be noted that not all opponents of evolution stood inflexibly on the species question. The botanist Schlechtendal accepted the common descent of "some groups of very similar species, which also inhabit a limited area." But he denied that a few original forms could have brought forth the hundreds of thousands of living species. In other words, Schlechtendal was willing to expand the concept of type to include possibly several species. Nevertheless, he too insisted that a firm definition be established

---

10 H. G. Bronn, *Untersuchungen über die Entwicklungsgesetze der organischen Welt während der Bildungszeit unserer Erdoberfläche* (Stuttgart: E. Schweitzerbart, 1858), pp. 77–82; Hermann Burmeister, *Geschichte der Schöpfung* (Leipzig: Otto Wigand, 1854), pp. 325–326; Schultz-Schultzenstein, "Ueber die Stellung Blumenbachs," p. 47; Bastian, *Schöpfung*, pp. 5–10, 296; Albert Wigand, *Die Geneologie der Urzellen als Lösung des Descendenzproblems* (Braunschweig: Friedrich Vieweg & Sohn, 1872), pp. 37–38.

at some level. In the eyes of these men, types either existed or they did not. Hybrids either were infertile or they were not. If types existed, variation could not be unlimited. If they did not exist, all forms should flow together. There could be no other alternative.[11]

Closely related to the issue of type was the problem of the coexistence of closely related variants. Pointing to the paleontological evidence, Bronn wondered how a species and its successor could exist side by side in the same beds when the intermediate forms had been wiped out. Darwin's explanation that some variants were isolated could not be expanded into a general rule, Bronn thought. Nevertheless, the geographic isolation of the different human races seemed to support this idea. Furthermore, a few centuries of general contact had brought some of them to the brink of extinction. Bronn considered this fact one of the best arguments that could be advanced in Darwin's favor. The botanist Fritz Ludwig was not so optimistic. Variations arise easily, he maintained, but they are impermanent. Crosses with original types quickly erase them. Ludwig did not see how a new variant form could become established, no matter how useful it might be. As soon as it was exposed to crosses with members of the same species, it would inevitably revert to type.[12]

Species change is the most fundamental process in evolution. But Darwin's opponents did not limit their critique to his idea of species formation. They also objected to the notion that members of the higher systematic groups share a common evolutionary ancestry. For these groups, too, type assumed great importance. And, just as the type of a species was presumably based on the physiological principle of hybrid

---

[11] H. G. Bronn, "Schlusswort des Uebersetzers," in his translation of *Ueber die Entstehung der Arten im Thier- und Pflanzenreich durch natürliche Zuchtung, oder Erhaltung der vervollkommneten Rassen im Kampfe um's Dasein* by Charles Darwin (Stuttgart: Schweitzerbart, 1860), pp. 503–504; D. F. L. von Schlechtendal, "Ueber die Entstehung der Arten" (review), *Botanische Zeitung* 1863: 348. See also Rudolf Wagner, "Bericht über die Arbeiten in der allgemeinen Zoologie im Jahre 1862," *Archiv für Naturgeschichte* 29, no. 2 (1863): 14–15, 17–19; Bastian, *Schöpfung*, pp. 40–41, 44; Wigand, *Darwinismus*, I, 13–38, 208–215, 221–222.

[12] Bronn, "Schlusswort," pp. 506–508; Fritz Ludwig, *Die Befruchtung der Pflanzen durch Hülfe der Insekten und die Theorie Darwin's von der Entstehung der Arten* (Bielefeld: Velhagen & Klasing, 1867), pp. 31–32. See also Wigand, *Darwinismus*, I, 67–84.

infertility, the types of higher groups were supposed to be determined by the adaptive principle of correlation. The botanist August Griesebach likened the structure of an organism to a machine or a work of art. Borrowing a metaphor from Asa Gray, he compared the generic type to a blueprint of an implement from which various different species could be stamped out as needed or desired. Adolf Bastian assumed that correlation would make it impossible for the external environment to reorganize an organism. No accidental effects such as Darwin postulated could produce the necessary internal harmony. Furthermore, partly developed organs, such as lungs or wings, would be totally ineffectual before their completion. This excluded the possibility that natural selection had created new types. Bastian regarded correlation as a causal principle, opposing it to Darwin's accidental one. Albert Wigand employed virtually the same analysis. He thought it inconsistent of Darwin to employ both concepts. As in so many other matters, Darwin wanted to have it both ways—or so it looked to traditionalists.[13]

Behind the Darwinist belief that the members of a systematic group share a common ancestor lay the assumption that form relationships entail blood relationships. The opponents of evolution fought this assumption at every step. Particularly concerned were those anatomists who reacted against Huxley's thesis of a common ancestry for men and apes. Christian Aeby, Christoph Giebel, Heinrich Bischoff, and Johann Lucae did their best to emphasize the differences between human anatomy and that of the higher apes. They concentrated especially on the anatomy of the skull but also investigated the brain and the foot. Bischoff was finally obliged to concede that the fundamental structure of the human brain was the same as that of the orangutan and the chimpanzee and that the resemblance between them was greater than that between the higher apes and lower ones. Nevertheless, he pointed to other differences between man's brain and that of the anthropoid apes, especially in size, and insisted that their similarities did not constitute evidence for human descent from an ape. Albert Wigand expanded this viewpoint into a general principle. The notion that form resemblances

13 August Griesebach, "Die geographische Verbreitung der Pflanzen Westindiens," *Abhandlungen der königlichen Gesellschaft der Wissenschaften zu Göttingen* 22 (1864): 67; Bastian, *Schöpfung*, pp. 63–64, 103–109; Wigand, *Darwinismus*, I, 194–199.

are due to common descent is not, he said, a discovery of inductive research. Rather it is a purely arbitrary assumption.[14]

This attitude also carried over into anti-Darwinist considerations of embryology. Traditionalists found no virtue in the idea that the embryonic development of an organism repeats in brief the evolutionary stages by which it appeared. In opposition to this viewpoint, Bastian and Wigand cited the laws of von Baer, according to which development begins from a more or less identical basis for each major type and proceeds through gradual differentiation to the stage of unique individual identity. Incapable of an independent existence, an embryo cannot resemble any real living ancestor. Bastian and Wigand maintained that the development of animal form is due to an inner law of organization, not to the accident of a creature's evolutionary past. As Bastian expressed it, palaces are not simply extensions of peasant huts, though, in building either of them, one may start with a single room.[15]

This left one other significant area in which the functioning of evolution could be disputed, the geological record. Uniformitarian geology had been accepted only slowly in Germany, and there were many who still held to the old catastrophist presuppositions. Men like Heinrich Bronn and the paleobotanist Heinrich Göppert held moderate views, believing that the catastrophes were not absolute and that some species had survived from one period to the next. However, since they interpreted the geological record fairly literally, they felt that Darwin's theory lacked sufficient inductive demonstration. In opponents' eyes, this lack was not overcome by the discovery of a few isolated missing links. Andreas Wagner denied that the *Archaeopterix* possessed true feathers and classed it as a reptile. Rudolf Virchow persuaded many of his colleagues that the

---

[14] Christian Aeby, *Die Schädelformen des Menschen und der Affen* (Leipzig: F. L. W. Vogel, 1867); Th. L. W. Bischoff, "Die Grosshirnwindungen des Menschen," *Abhandlungen der mathematisch-physikalischen Klasse der Königlich bayrischen Akademie der Wissenschaften, München* 50, no. 2 (1868): 486, 491–492; idem, *Ueber die Verschiedenheiten in der Schädelbildung des Gorilla, Schimpanse und Orang-Outang* (Munich: Königliche Akademie, 1867); C. G. Giebel, "Eine antidarwinistische Vergleichung des Menschen- und der Orangschädel," *Zeitschrift für die gesammten Naturwissenschaften* 28 (1866): 401–419; idem, *Der Mensch, sein Körperbau, seine Lebenstätigkeit und seine Entwicklung* (Leipzig: O. Wigand, 1868); Johann Christian Gustav Lucae, "Affen- und Menschenschädel im Bau und Wachsthum verglichen," *Archiv für Anthropologie* 6 (1873): 13–38; Wigand, *Darwinismus*, I, 268–269.

[15] Bastian, *Schöpfung*, pp. 59–62, 67, 207–208, 220; Wigand, *Darwinismus*, I, 303–304.

Neanderthal skull segment was merely pathological. However, it was another matter to dispose of the gradually modified series of fossil shells that Darwinist geologists began to discover. As the evidence accumulated, ad hoc explanations became less persuasive. The wiser opponents of evolution relied instead on the same general principle that applied to comparative anatomy: form relationships do not entail blood relationships. The zoologist Rudolf Wagner pointed out that the *Archaeopterix* was really no more significant than a duck-billed platypus. Both had a certain intermediate character, but this proved nothing about evolution. Adolf Bastian declared that the difficulty of imagining the evolution of the eye was enough to outweigh millions of intermediate forms. For Wigand, of course, this argument was a matter of basic principle.[16]

In defining the intermediate forms out of existence, the opponents of evolution were not simply being perverse. In their eyes, Darwin's concept of evolution failed to meet the standards for scientific practice that traditional biology had established. To every essential question, Darwin's answer was mystifyingly ambiguous. Did Darwin believe in species or did he not? Did fertile hybrids exist or did they not? Were species formed by accident or was there a cause? Did correlation determine animal types or did it not? Was natural selection a plan or was it a cause? None of Darwin's explanations assumed a form that was recognizable to a scientist who had gotten his start in the first half of the nineteenth century. As a result, their estimation of him vacillated between extremes. Albert Wigand typified this mentality best. He could not reconcile Darwin's theory with either the causality of Newton or the typology of Georges Cuvier. In the end he concluded that Darwin believed in evolution because of the comparability of natural forms. But Wigand knew

[16] Bronn, "Schlusswort," pp. 500, 506; Heinrich Robert Göppert, "Ueber die Darwinsche Transmutationslehre, mit Beziehung auf die fossilen Pflanzen," *Jahresbericht der schlesischen Gesellschaft für vaterländische Cultur* 42 (1864): 39–42; idem, "Ueber Aphyllostachys, eine neue fossile Pflanzengattung aus der Gruppe der Calamarien, so wie über das Verhältniss der fossilen Flora zu Darwin's Transmutation-Theorie," *Nova acta Academiae Caesareae Leopoldino Carolinae germanicae naturae Curiosorum* 32, no. 1 (1865); Andreas Wagner, "Ein neues, angeblich mit Vogelfedern versehenes Reptil," *Sitzungsberichte der königlichen bayrischen Akademie der Wissenschaften, München* 1861, pt. 2: 146–154; Rudolf Virchow, "Untersuchung des Neanderthal-Schädels," *Verhandlungen der Berliner Gesellschaft für Anthropologie, Ethnologie und Urgeschichte* 1872: 157–165; R. Wagner, "Bericht," p. 15; Bastian, *Schöpfung*, p. 57; Wigand, *Darwinismus*, I, 268–269.

that this position was untenable. The laws of organic form could explain nature's regularities quite as well as evolution. Therefore, Wigand concluded that evolution was incapable of inductive proof or disproof. It was not a scientific theory at all, but a philosophical interpretation of nature—a revival of *Naturphilosophie*. Aside from all their misunderstandings and misinterpretations, the men who shared Wigand's opinion grasped one thing well enough: evolution was incompatible with biology as they knew it.[17]

This pessimism about the possibility of fitting evolution into the traditional structure of biological thought affected only a minority of scientists. Before long, most men accepted evolution, albeit in different ways. One group followed Darwin in holding natural selection, or at least external factors of some kind, to be the mechanism of species change. A second party rejected natural selection and sought to substitute some internal developmental principle as the primary motive force of evolution. They were adherents of the traditional concept of type, who drew on the older idealist tradition in German biology. However, in designating these men "idealists," one must introduce certain qualifications. They were not idealists in the sense that they intended to dispose of the usual notion of causality in science or abandon the discussion of nature in terms of physical laws. Oswald Heer thought that evolution could only be explained as a result of divine intervention, and Alexander Braun believed the laws of biological form transcended those of physics and chemistry. On the other hand, Albert von Kölliker regarded organic forms as subject to the same laws as inorganic ones. And Carl Nägeli cited approvingly Carl Vogt's thesis that thought is just as much a secretion of the brain as bile is of the liver. Wilhelm His and Alexander Goette likewise understood the laws of form in a strictly mechanical fashion. What united the idealists was not a program for smuggling metaphysics into science. Rather, it was their opinion that biological form could be explained independently of its historical adaptation to the external environment. In separating the laws of form from those of

---

[17] Wigand, *Darwinismus*, II, 79–80. For the view that Darwinism simply revived *Naturphilosophie* or the theories of Etienne Geoffroy St. Hilaire, see also Schultz-Schultzenstein, "Ueber die Stellung Blumenbachs," p. 52; R. Wagner, "Bericht," p. 20; Bastian, *Schöpfung*, p. xviii.

physical science, Braun and Heer were looking backward to the old romanticism. In connecting the laws of form to those of the physical sciences, Kölliker, Nägeli, His, and Goette were looking forward to the new reductionism. But all of them agreed on the inadequacies of the present Darwinism.[18]

Although the idealists did not deny evolution, they had their own reasons for believing in it. The chief reason was basically a negative one. They were unable to think of an alternative explanation for the facts of organic nature. The notion that spontaneous generation had abruptly populated the world with completed adult forms smacked too much of miracles. Of course, the lower animals and plants could have arisen by the spontaneous generation of germ cells, but the higher ones could not have. Mammals, for example, require a placenta for development; no mammal could be expected to grow from germ cells in a free living state. There were, naturally, additional reasons for believing in descent. Idealists were impressed by the increasing tendency of later fossil forms to resemble modern ones, by the recapitulation of forms in embryonic development, and by rudimentary organs. Alexander Braun considered the sequential ordering of plant groups noteworthy, as well as the geographical evidence for the common origin of members of these groups. The contrast with opponents of evolution is quite apparent. The idealists saw the need to explain the origin of species by transmutation. Consequently, they attributed evolutionary significance to the common morphological features in organisms—at least as long as the truth of evolution itself was under discussion.[19]

In their devotion to stability in organic forms, some idealists were attracted to rather extreme hypotheses. Oswald Heer and Albert von Köl-

[18] Heer, *Urwelt*, p. 604; Mettenius, *Alexander Braun's Leben*, pp. 583–585; Albert von Kölliker, "Anatomisch-systematische Beschreibung der Alcyonarien," *Abhandlungen, herausgegeben von der Senkanbergischen Naturforschenden Gesellschaft* 8 (1872): 208; Carl Nägeli, "Die Schranken der naturwissenschaftlichen Erkenntnis," *Amtlicher Bericht der Versammlung deutscher Naturforscher und Aerzte zu München* (1877), p. 37. For the opinions of His and Goette, see below.

[19] Alexander Braun, *Ueber die Bedeutung der Entwicklung in der Naturgeschichte* (Berlin: Gustav Lange, 1872), pp. 27–34, 40–46; Heer, *Urwelt*, pp. 591–592; Albert von Kölliker, "Ueber die Darwin'sche Schöpfungstheorie," *Zeitschrift für wissenschaftliche Zoologie* 14 (1864): 179–181; Carl Nägeli, *Entstehung und Begriff der naturhistorischen Art* (Munich: Franz, 1865), pp. 8–14.

liker held that species could only have changed by jumps or saltations. This theory was designed to avoid several problems that gave Darwin trouble. The paleobotanist Heer was led to this viewpoint by the breaks he found in the geological record. Unlike most evolutionists, Heer was still convinced of catastrophism, and he preferred to regard fossil preservation as fairly complete. The only way he could overcome the significant gaps that existed between fossils of major systematic groups was to postulate an abrupt restructuring of species. He thought of this process as similar to the changes that occur in metamorphosis or the alternation of generations.[20]

The anatomist Kölliker was less disturbed by breaks in the geological record than was Heer. Kölliker's primary motive for rejecting gradual transmutation was biological. He was bothered by the lack of intermediate forms among living species. He also doubted that the struggle for existence really functioned in a selective way. To him it seemed that the adaptation of every animal is already perfect and that the loss of life resulting from the struggle for existence is due to purely accidental factors. And he too pointed to the infertility of hybrids as evidence against gradual species transmutation. As an alternative Kölliker chose saltatory evolution, or, as he referred to it, heterogenesis. This process would not be dependent on the external environment, but would function according to a general developmental law. Kölliker explained the functioning of this law by analogy to the alternation of generations and embryonic recapitulation. It seemed to him that a mammal embryo required only a slight modification in order to produce a different form. This idea was somewhat vague, but Kölliker imagined that it might eventually be explained in terms of natural laws. He did not share Heer's disposition to assign the problem to the Deity.[21]

Of course a shortage of intermediate forms was not the only possible defect in natural selection. The botanist Carl Nägeli attributed much less importance to saltations, but he agreed with Kölliker on the need for an inner developmental principle. Nägeli was not entirely opposed to natural selection. To him it seemed necessary to explain environmental adaptation. However, he did not regard it as the primary cause of evolu-

[20] Heer, *Urwelt*, pp. 593–604.
[21] Kölliker, "Darwin'sche Schöpfungstheorie."

tion. Nägeli reasoned that, if variation occurred randomly in all directions and were controlled only by the effect of the external environment, every species transferred to a new environment would be transformed into a new race. On its return, it would then presumably revert to the original form. Indeed, two related species could not exist under the same environmental conditions without merging, since, under a given set of circumstances, there could be only one optimally adapted form. This led Nägeli to suppose that variation is in fact not random but predetermined in a given direction. Furthermore, he thought that this directed variation must be toward greater differentiation and higher organization. Otherwise mosses could just as easily evolve backward into algae as forward into ferns. Nägeli saw an additional need for directional change in the existence of plant characteristics that seemed to have no particular functional role. According to Darwin's theory, such purely morphological characters should be quite variable, since they are relieved of the pressure of natural selection. But, in fact, such apparently nonfunctional characters as leaf and petal arrangements are the most constant of all. To Nägeli, this was further evidence for a general law of form.[22]

The idealists' preference for an inner law of development was closely related to their attitude toward form. They were determined to combat the idea that form was simply a product of common descent. This led Kölliker in 1872 to reject the concept of a simple genealogical tree as the model for systematic relationships. If form results from the inner workings of physiochemical laws, he reasoned, why should it be uniquely determined by evolutionary history? Not only the higher systematic categories but even genera and species might have had multiple origins. A species might appear in not one, but several, separate genealogical sequences. This being the case, it would be pointless to erect hypothetical genealogical trees, because the specific course of evolution is intricately intertwined. The members of a single systematic group do not necessarily share a common bond of ancestry. Alexander Braun concurred. He pointed out that the concept of homology would thereby be separated from the idea of common evolutionary origin. Thus, by attributing form to an inner developmental law, the idealists attempted to save the idea of type and give it an autonomous position in biology, in-

22 Nägeli, *Entstehung und Begriff*, pp. 15–28.

dependent of the idea of descent. They also eliminated the embarrassing association between man and the ape by hiding the human past in a tangle of uncertainties.[23]

The autonomous type theory also found adherents among embryologists. Wilhelm His and Alexander Goette both believed in evolution, but neither of them thought that evolution should be used to explain developments in embryonic form. Both of them attributed embryological development to a purely mechanical law of form. Furthermore, His was able to give some indications about how such a mechanical law might function. He showed that the folding that occurs in a developing embryo is due to irregular rates of growth over its surface. By slicing and bending a rubber tube, he was able to approximate the form assumed by the developing brain. From this sort of demonstration, he concluded that similarities in embryonic form were due, not to common evolutionary ancestry, but to a common mechanical process of growth. This reductionist approach to embryology thus provided additional support for an autonomous concept of type, liberated from any connection with an organism's evolutionary past.[24]

Because the idealist theories of form were interesting and significant, historians have dwelt upon them to the exclusion of idealist theories of heredity. As a result, sufficient attention has not been paid to the fact that Darwin's idealist opponents were early critics of the notion of the inheritance of acquired characteristics. The belief that animal form was primarily due to inner mechanical laws worked not only against natural selection but also against any kind of environmental explanation of form, including inherited adaptations. In 1865 Carl Nägeli published the first of a series of papers on plant distribution and its relevance to the matter of variation and species formation. He found that two dif-

---

[23] Kölliker, "Allgemeine Betrachtungen," pp. 207–211, 217–223; Alexander Braun, "Die Frage nach der Gymnospermie der Cycadeen erläutert durch die Stellung dieser Familie im Stufengang des Gewächsreichs," *Monatsberichte der Preussische Akademie der Wissenschaften, Berlin* 1875: 245–249.

[24] Wilhelm His, *Ueber die Bedeutung der Entwicklungsgeschichte für die Auffassung der organischen Natur* (Leipzig: F. C. W. Vogel, 1870), pp. 30–37; idem, *Unsere Körperform und das physiologische Problem ihrer Entstehung* (Leipzig: F. C. W. Vogel, 1874), pp. 1–20, 96–99, 161–168; Alexander Goette, *Die Entwicklungsgeschichte der Unke (Bombinator igneus) als Grundlage einer vergleichenden Morphologie der Wirbeltiere* (Leipzig: Leopold Voss, 1875), pp. 855–902.

ferent varieties might often exist in the same locale and thereby share the same external conditions. He found also that a variety native to one locale might be artificially bred under very different conditions elsewhere. Neither of these facts accorded with the idea of transmutation by direct effects. Nägeli did not explicitly deny that acquired characters could be inherited. However, he insisted that no such characters could ever become fixed or constant. Nägeli shared the conventional belief of his day that a given variation becomes constant by being successfully inherited through a series of generations. For him the critical question was not the heritability of acquired characteristics but their heritability over an extended period. Since this was not possible, the direct effects of the environment could not have played a role in evolution.[25]

The most decisive opposition to direct effects came from the embryologists Goette and His. Both of them flatly rejected the inheritance of acquired characteristics in all its forms. In particular, they reacted against Darwin's theory of pangenesis. His saw no need for a mechanism that transmitted separate individual characters. He viewed heredity as a distinctive mechanical impulse transmitted to the germ material of the new generation. Assuming that the process of growth is initiated in the same way each time, it will proceed to the same inevitable goal. This would exclude any need for individual organs of the parent creature to exercise an influence on its germ material, as required by pangenesis. It also eliminated the teleology implicit in older theories of embryology. Most important, without pangenesis there was no basis for the inheritance of acquired characters. As His pointed out, children must still be taught language, and thousands of years of circumcision have not altered bodily form.[26] Not only did His dismiss the inheritance of acquired characters, but he also endorsed natural selection as "a far reaching key. to the understanding of the development and stabilization of particular forms."[27] If he had any other mechanisms of evolution in mind, he did not mention them; however, the endorsement of selection was quite perfunctory, offering no specific indication about how His conceived its

[25] Carl Nägeli, "Ueber den Einfluss der äusseren Verhältnisse auf die Varietätenbildung im Pflanzenreiche," *Sitzungsberichte der koniglichen bayerischen Akademie der Wissenschaften zu München* 1865, pt. 2: 228–232, 258–260, 277–284.

[26] His, *Unsere Körperform*, pp. 153–161.

[27] Ibid., p. 160. My translations.

workings and leaving the impression that he was not much interested in the subject.

Goette also identified the inheritance of acquired characters with Darwin's pangenesis. And he too thought it impossible for each organ of the body to contribute to the germ cells. In Goette's view, an animal is not a continuation of its parents in a material sense. It merely repeats in its development the laws of form that governed the parents. Variation has two causes. One lies in the parent, "which constructs the reproductive product under the changing influence of its physiological relations." The other lies in the varying conditions under which separated eggs exist. Goette thought it was a mistake to identify the cause of variation with the cause of natural selection, as he understood Darwin to be doing. Because he regarded the cause of variation as the primary cause of evolution, he discounted the importance of selection. It seemed to him that natural selection was restricted to choosing among the different available forms in the struggle for existence. It could not produce them. Thus, although Goette did not introduce any teleological basis for variation, he diminished the role of the external environment in favor of the mechanical laws of form.[28]

Nägeli, His, and Goette all rejected the inheritance of acquired characteristics; and all of them likewise acknowledged a role for natural selection. Yet none of them attached much importance to selection in his own theoretical scheme of things. Nägeli and Goette emphasized "inner" forces of biological change over external ones. Having disposed of one external force, the inheritance of acquired characteristics, they were not disposed to lay great weight on another, natural selection. His and Goette shared a common dislike for the Darwinist concept of recapitulation. It is understandable that they were reluctant to extol any other Darwinist positions. The primary concern of all three men lay in establishing an autonomous theory of type, not in expanding the applicability of natural selection. They all agreed that organic form went by its own rules, which were generally applicable to organic forms in all circumstances and at all times. This theory of type had grown up in close association with doctrines of metaphysical idealism; nevertheless, it had always had its own unique biological component and did not ab-

[28] Goette, *Entwicklungsgeschichte der Unke*, pp. 890–901.

solutely depend on specific philosophical assumptions. In accord with the prevalent mistrust of metaphysics and the consequent emphasis on mechanism in biology, Nägeli, His, and Goette all moved to reinterpret type in terms of a mechanistic reductionism. But this reductionism was no more appreciative of natural selection than was the most blatant metaphysics. Neither one accounted satisfactorily for the element of contingency, which was so critical to the Darwinists.

The difference between Darwinists and idealists was immediately apparent in the reasons advanced by the two parties for accepting evolution. Idealists emphasized the impossibility of spontaneous generation of higher creatures, the familiar difficulty of defining species, the comparability of living types, and their resemblance to fossils. Darwinists also utilized such evidence, but their assumptions cast it in a completely altered light. The Austrian entomologist Carl Brunner von Wattenwyl produced new research findings in favor of Darwinism as early as 1861. His paper dealt with problems in the taxonomy of Orthoptera, a group of creatures including grasshoppers, mantises, and crickets. Brunner concluded that species with rudimentary wings, which previous systemists had grouped together in common genera, were in fact scattered among many genera. Indeed, he found that rudimentary-winged types often bore a striking resemblance to other "twin" species with normal wings. However, the rudimentary-winged types lacked certain structures, common to the winged species, which would have interfered with their running and jumping. From this observation, Brunner concluded that the rudimentary-winged types were descended from the normal-winged species and had lost the impeding structures through natural selection. The idea of special creation according to plan seemed to provide no explanation of why some insects scattered through many genera should possess analogous characters, but this fact seemed perfectly reasonable, assuming evolutionary adaptations.[29] Brunner's neat analysis offers an unusually good example of an important Darwinist assumption. The evidence for evolution is inseparable from the evidence for natural selection. The strongest proof of evolution lies in the anomalies of organic nature, anomalies that may be traced to the adaptive effects of natural selection.

[29] Carl Brunner von Wattenwyl, "Orthopterologische Studien," *Verhandlungen der zoologisch-botanischen Gesellschaft, Wien* 11 (1861): 221–228.

Where the "plan" of nature is violated by natural selection, the case for evolution is presented in its most compelling form.

This argument against the "plan" of nature is fundamentally a biological one, but it could be applied in geology as well. Bernhard von Cotta showed this in his *Geology of the Present* (1866). His book was primarily a defense of uniformitarianism against the catastrophist beliefs still influential in German geology. However, Cotta also included a chapter in defense of Darwinian evolution, which he took to be an extension of uniformitarian principles to the organic world. In particular, Cotta took aim at the saltatory theory of Oswald Heer. Cotta insisted that the conventional geological formations were in no sense absolute boundaries. Species were known to appear and disappear throughout formations and to overlap them. He expressed his conviction that the apparent breaks in the record merely indicated the imperfect knowledge of the geologist. Cotta illustrated this point elegantly by citing fifty-three species of deep-sea fish found in the Eocene Flysch formation in Switzerland. Not a single one of these species had been found anywhere else in the world, although the formation was supposedly well known. Since this ancient sea could not have been isolated, the existence of the fish revealed a remarkable gap in geological knowledge of what was happening elsewhere in the Eocene world. Cotta went on to argue that many breaks in the record simply indicated the migration of forms in response to gradual climatic changes. For example, many tropical European plants of the Miocene have close relatives living in present-day America, Asia, Africa, and Australia. Thus, Cotta perceived how nature's irregularities testified to a gradually unfolding historical process, which stood in conflict with the neat idealized schemes of catastrophism.[30]

Perhaps the most spectacular application of this insight may be found in Fritz Müller's short but important book, *Facts and Arguments for Darwin* (1864). In investigating Darwin's theory, Müller selected for study a single animal family, the Crustacea. He considered a variety of evidence for evolution, including the existence of intermediate forms, the regressive adaptations of parasitic crustaceans, and the distribution of specialized adaptations among various systematic groups.[31] However,

[30] Bernhard von Cotta, *Die Geologie der Gegenwart* (Leipzig: J. J. Weber, 1866), pp. viii–xi, 196–215.
[31] Fritz Müller, *Für Darwin* (Leipzig: Wilhelm Engelmann, 1864), pp. 1–30.

the greatest part of his book was devoted to the stages of growth among crustaceans. Comparing the larval development of different crustaceans, Müller showed that its path was quite variegated, even within systematic groups.[32] According to Müller, animal eggs are not all identical. The organs are not laid down in the order of their importance. Animals in the early stages of growth do not necessarily resemble one another more than do their adult forms. Indeed, many characteristics of isopods and amphipods are most similar among adult creatures. In some cases, the initial and adult stages are the most similar, with dissimilar intervening forms, and in others it is the middle stages that show the greatest resemblance.[33]

Müller summarized his discoveries by saying that evolutionary adaptations can occur at any stage of the developmental process from the earliest larval stage to sexual maturity. The newly adapted forms may establish their form by undergoing alterations sooner or later in the process of growth or by adding an additional change at the end of the process. In the latter case, they will repeat the entire growth cycle of their ancestors; and, in so far as the species has evolved in this manner, its entire historical development will be recorded in the course of maturation. However, this historical record maintained in the growth process will be gradually effaced as the organism adopts a more direct course from the egg to the adult; it will also be falsified by the adaptations of free living larvae to changed conditions of existence. The longer the larval period lasts in the life cycle of the organism and the more its way of life has changed, the more the evolutionary history of the animal will be falsified in the growth process.[34]

The force of Müller's argument was directed against Baer's laws and their assumption that growth follows a planlike course of gradually increasing differentiation. Müller asserted that, on the contrary, change can intervene at any point, wherever the external conditions of existence require it. Furthermore, these changes have not occurred uniformly for all members of each systematic group. They are marked by anomalies and irregularities that can be explained only by reference to the accidents of evolutionary history. It is sometimes maintained that the Darwinists

[32] Ibid., pp. 31–65.
[33] Ibid., pp. 65–74.
[34] Ibid., pp. 75–80.

simply revived the old idealist theory of recapitulation and reapplied it to evolution. This is not at all correct. The idealist concept of recapitulation, as formulated by individuals like Johann Meckel and Louis Agassiz, made no provision for exceptions and anomalies. It assumed a planlike development, which, taken by itself, provided no genuine argument for evolution. In applying the concept of natural selection to the data of crustacean life cycles, Fritz Müller produced a thoroughly original idea and showed for the first time how the facts of animal growth yielded decisive information about the process of species change.

Studies like those of Brunner and Müller were vital for the development of evolutionary thought among German scientists. Carl Gegenbaur, Hermann Schaaffhausen, Carl Claus, Hermann Müller, Anton Dohrn, and others contributed articles and monographs designed to support the Darwinist cause. At the same time, Darwinism was encouraged by the popular books of Carl Vogt, Friedrich Rolle, Gustav Jaeger, Matthias Schleiden, and Ludwig Büchner. However, the man who attained the most stature among German Darwinists in the sixties was Ernst Haeckel. Like no other scientist, he was able to write for both popular and professional audiences, and his great ability to organize and present ideas enabled him to dominate the discussion of evolutionary theory. His *General Morphology* (1866) and the more popular *Natural History of Creation* (1868) provided the most comprehensive surveys of the Darwinist position authored by a German.[35] In addition he wrote numerous books and articles on more specialized aspects of the subject.

Haeckel's case for evolution benefited considerably from his interest in scientific methodology. He challenged those scientists who tried to hide behind the doctrine of inductivism in order to dismiss evolution as an unproved hypothesis. Darwinism was not a hypothesis at all, Haeckel insisted; it postulated no unknown forces and restricted its evidence to the factual information of biology. He called on John Stuart Mill's analysis of the importance of deductive arguments in science. In Haeckel's opinion, science rests on a balance between empiricism and theory. Darwin's theory was more than vindicated in Haeckel's eyes by the extraordinary variety of biological phenomena it was able to explain. Haeckel also laid great weight on what he understood to be the mechanistic

[35] Ernst Haeckel, *Generelle Morphologie der Organismen* (2 vols., Berlin: G. Reimer, 1866); idem, *Natürliche Schöpfungsgeschichte* (Berlin: G. Reimer, 1868).

character of Darwin's theory. The elimination of teleology from biology was of great concern to Haeckel; and the compromise with teleology that he saw in idealist theories of evolution impressed him as a betrayal of the scientific spirit. Only Darwinism could be reconciled with that quintessence of scientific thinking, the law of causality.[36] On the other hand, Haeckel set himself against the notion, popularized by Kant, that true science is based solely on mathematics. Biology, he insisted, is a historical science. Particularly, the studies of embryology, paleontology, and phylogeny are historical; and, as such, they form a link between the natural and the humanistic sciences.[37] In this regard, Haeckel deliberately opposed the kind of reductionism that lay behind the theories of Nägeli and His. Of course, with his urgent belief in the importance of causality, Haeckel sometimes offered hostages to reductionism; but he usually managed to interpret this causality in a historical sense, thereby keeping faith with his more basic Darwinist instincts.

One area in which Haeckel's mechanist faith led him astray was genetics. He could not admit that the cause of organic variability might remain unknown. A confession of ignorance at this point would have offered idealists a perfect opportunity to insert a scheme of directed variations, such as Nägeli's principle of perfection. Variation had to be explained in a completely mechanical way, and the solution Haeckel embraced was the inheritance of acquired characteristics. Haeckel defined reproduction as growth beyond the mass of the individual. From this it followed that the new generation was merely an extension of the old, possessing all the characteristics of the parents, both inherited and acquired. In adopting this view, Haeckel was not abandoning one teleology in order to embrace another. For the most part he still believed that the functional adaptation of organisms resulted from natural selection. However, his philosophy of science made no provision for any concept of chance or accident. There were only two possible alternatives; variations either were teleologically directed or they were caused. Of these two alternatives, Haeckel chose the latter.[38]

[36] Ernst Haeckel, *Generelle Morphologie der Organismen*, I, 64–88, 94–104.

[37] Ernst Haeckel, "Die heutige Entwicklungslehre im Verhältnis zur Gesammtwissenschaft," *Amtlicher Bericht der Versammlung deutscher Naturforscher und Aerzte zu München* (1877), pp. 15–16.

[38] Haeckel, *Generelle Morphologie*, II, 165–180, 186–217.

Haeckel fitted the idea of the inheritance of acquired characteristics into his general view of evolutionary history. If a creature evolved by inheriting its parents' acquired characteristics, would not the entire growth process of the animal consist of adding up the accumulated characters that its species had acquired over the course of evolution? This theory accorded nicely with Fritz Müller's findings, which Haeckel reformulated in the famous phrase "ontogeny recapitulates phylogeny." Haeckel knew that this was not an absolute rule, for he recognized that free living larvae could falsify the record by special adaptations to the struggle for existence. Nevertheless, Haeckel took much less notice of these falsifications than Müller had done. In fact, by emphasizing the relative accuracy of the record, Haeckel very nearly undermined Müller's original purpose. Müller regarded these exceptions as empirical disproof of Baer's laws. Haeckel, though, did not dispute the factual correctness of Baer's laws. He merely denied that they should be interpreted in Baer's teleological manner and asserted that recapitulation had at last provided a legitimate causal explanation for the phenomena Baer described.[39] By dwelling on the regularity of embryonic development, Haeckel gave away a potent Darwinian weapon. This was especially evident in his reply to the theories of Goette and His. Haeckel correctly observed that their interpretation of embryology undermined the theory of evolution by dispensing with the relevant embryological evidence. He ridiculed their belief in a purely mechanical explanation of growth, rather than one based on an animal's phylogenetic past. But, insofar as they relied on Baer's theories, Haeckel did not question them; he too was tempted by the prospect of a general law of growth.[40]

Haeckel did have a reason for applying the law of recapitulation as he did. Of all evolutionary problems, the one that interested him most was the effort to reconstruct a phylogenetic history of the living species. Impatient with the limitations of available fossil evidence, Haeckel determined that the course of life cycles could provide the data he wanted. His principle that form relationships entail blood relationships excluded at the onset a polyphyletic view, such as Kölliker's. In the first part of the *General Morphology*, Haeckel did consider the possibility

---

[39] Ibid., pp. 7–12, 180–186.

[40] Ernst Haeckel, "Ziele und Wege der heutigen Entwicklungsgeschichte," *Jenaische Zeitschrift für Naturwissenschaft*, Supplement 10 (1875).

that each of the phyla had arisen from a separate form of life. However, by the end of the work he was inclined to a completely monophyletic ancestral tree; and in subsequent works this conviction grew stronger.[41]

Haeckel's belief was strengthened by two important discoveries. The Russian embryologist Alexander Kovalevsky found that the larval stage of the ascidian, a creature previously regarded as a mollusk, possesses a rudimentary chorda dorsalis. In vertebrate embryos this structure presages the formation of the vertebrae; therefore, the ascidian looks like a perfect transitional form between vertebrates and lower animals.[42] The second discovery was Haeckel's own. As a result of his work with sponges, he was taken by the resemblance between the saclike body form of the simplest sponges and the gastrula that results from the invaginated blastula in the earliest steps of growth in higher animals. Haeckel concluded that the gastrula was a homologous form in all multicelled animals, pointing to their common descent from a "gastraea" that must have resembled these primitive sponges. There were some objections to this gastraea theory. Critics protested that some sponges acquired mouth openings in ways other than invagination. Haeckel countered that these creatures had probably made special adaptations to the struggle for existence. This was not an arbitrary rationalization, for it was perfectly consistent with his thesis that evolutionary history, not a law of development, had determined the patterns of animal growth. However, Haeckel's preoccupation with phylogeny led him to focus on the regularities rather than the anomalies in life cycles. Consequently, he lost sight of Müller's original argument, and his ideas assumed an exaggerated, one-sided character, with which subsequent generations of biologists have not felt entirely comfortable.[43]

It would be misleading to suggest that the debate between the Darwinist and the idealist-reductionist camps completely occupied the field of controversy in Germany. Particularly interesting, from the point of

---

[41] Haeckel, *Generelle Morphologie*, I, 197–198; II, 391–404, 411.

[42] Haeckel, *Natürliche Schöpfungsgedanken*, pp. 435–439.

[43] Ernst Haeckel, "Die Gastraea-Theorie, die phylogenetische Classification des Thierreichs und die Homologie der Keimblätter," *Jenaische Zeitschrift für Naturwissenschaft* 8 (1874): 1–55; idem, "Die Gastrula und Eifurchung der Thiere," *Jenaische Zeitschrift für Naturwissenschaft* 9 (1875): 454–456.

view of current evolutionary theory, was the contest over speciation initiated by the geographer Moritz Wagner. Wagner, writing in 1868, took up the classic weakness of Darwin's theory, the question of how advantageous variations could become established if their bearers were constantly exposed to crosses with unaltered types. Wagner thought this problem would be overcome by the migration of small numbers of individuals, thus allowing natural selection to work among them unimpeded beyond the bounds of the previous species range. He cited an impressive number of instances where natural geographic boundaries also set boundaries between closely related species or varieties. Two years later, though, Wagner decided that even this expedient was not enough to save natural selection. Only if all the newly isolated individuals were different from the basic population would change be preserved among them. Thus, the role of external influence was made negligible; and evolution, in Wagner's eyes, had to proceed almost solely by what amounted to genetic drift.[44]

This idea had a mixed reception. Fortified with extensive knowledge of alpine plant distribution, the Austrian botanist Anton Kerner agreed very closely with the view expressed in Wagner's first paper. To Kerner, it seemed to provide the solution to the vexing species question. He proceeded to define a species as the collective of identical forms having a given range and reproducing themselves unchanged over an observable period of time. The Baltic-Russian entomologist Georg Seidlitz was also sympathetic, but other observers were not so sure. Carl Nägeli and August Weismann subjected Wagner to extended criticism. Both men cited evidence indicating that species were indeed capable of dividing into separate varieties and ultimately new species on the same ground. They conceded that geographic isolation might facilitate species division but denied that it was essential. The issue was clouded by the failure of any of the disputants except Seidlitz to distinguish between simple evolutionary change in the character of a given population and the division

44 Moritz Wagner, "Ueber die Darwin'sche Theorie in Bezug auf die geographische Verbreitung der Organismen," *Sitzungsberichte der königlichen bayrischen Akademie der Wissenschaft zu München* 1868, pt. 1: 359–395; idem, "Ueber den Einfluss der geographischen Isolierung und Colonienbildung auf die morphologischen Veränderungen der Organismen," *Sitzungsberichte der königlichen bayrischen Akademie der Wissenschaft zu München* 1870, pt. 2: 154–174.

of a population into two separate groups. Furthermore, there was considerable disagreement about the significance of the available evidence. Weismann, particularly, was led astray by Franz Hilgendorf's study of the snail *Planorbis multiformis*, which seemed to have evolved through several parallel forms in the same lake bed. Geologists concluded, however, that the snail was simply an unusually plastic species, whose protean shapes did not necessarily reflect evolutionary development. In any case, the complex problem of speciation could hardly have been unraveled by the men of that day, though they did make a significant beginning.[45]

Because of the common confusion over the questions involved, the disagreement with Wagner did not lead August Weismann to any notable success. Fortunately, Weismann's critique of the idealists had more positive results. In *Studies on the Theory of Evolution* (1876) he undertook to refute Nägeli's claim that many plant characteristics are purely morphological and hence can be explained only by positing an internal developmental principle. Weismann suggested that many apparently nonfunctional characteristics actually had important functions, a fact he demonstrated for the camouflage patterns of many caterpillars. However, he admitted that many other characteristics were indeed nonfunctional. Weismann met this problem by resorting to the inheritance of acquired characteristics. He carried out experiments on butterflies that indicated that their presumably nonfunctional wing colors had been produced by the direct effects of the environment. Thus, he hoped to show that even those characteristics not affected by natural selection could still be explained without recourse to an internal teleological principle.[46]

Another point defended by Weismann against his adversaries was the thesis that form relations entail blood relations. Kölliker had disputed

[45] Anton Kerner, *Die Abhängigkeit der Pflanzen von Klima und Boden* (Innsbruck: Wagner, 1869), pp. 18–26, 46; Georg von Seidlitz, *Die Darwin'sche Theorie: Elf Vorlesungen über die Entstehung der Thiere und Pflanzen durch Naturzüchtung* (Leipzig: Wilhelm Engelmann, 1871), pp. 26, 149–150; August Weismann, *Ueber die Berechtigung der Darwin'schen Theorie* (Leipzig: Wilhelm Engelmann, 1868), pp. 32–39; idem, *Ueber den Einfluss der Isolierung auf die Artbildung* (Leipzig: Wilhelm Engelmann, 1872); Carl Nägeli, "Die gesellschaftliche Entstehung neuer Species," *Sitzungsberichte der methematisch-physikalischen Classe der königlichen bayrischen Akademie der Wissenschaften zu München* 2 (1872): 305–344.

[46] August Weismann, *Studien zur Descendenztheorie*, 2 vols. (Leipzig: Wilhelm Engelmann, 1875–1876), II, 85–110.

this idea, while proposing a polyphyletic theory of descent in opposition to Haeckel's monophyletic one. Weismann's reply exhibited a greater flexibility and a more sensitive grasp of priorities than Haeckel usually showed. In a review article of 1873, Weismann agreed that, if a single species might have originated in more than one individual, higher systematic groups might have originated in more than one species. He drew the line only at reducing the phylogenetic system to an indecipherable tangle or surrendering the principle of the identity of form and blood relationships. Nevertheless, in his *Studies* Weismann softened this principle too. Following the lead of Fritz Müller, he decided to test the possibility of a teleological law of growth by comparing the form relationships among butterflies to those among caterpillars. The results were decidedly incongruent. Like Müller before him, Weismann discovered that evolutionary change proceeded entirely separately in the adult and larval stages. This finding told heavily against the possibility of a general plan of growth, and Weismann saw that it also told against the principle of the absolute identity of form and blood relationships. Weismann was willing to recognize exceptions to this popular Darwinist notion in the interests of a more important principle.[47]

He showed similar flexibility in confronting the old idealist view that variation is not in fact random. Weismann met the idealists halfway. He admitted that variation is not random, but he denied that the direction involved is in any way predisposed to yield increasingly higher forms. Variation can just as easily lead to degenerate results like parasites or rudimentary organs. It can also lead to an evolutionary dead end and extinction. Weismann insisted that the influences determining variation are purely mechanical. He was thinking primarily of the law of correlation. Weismann cited an example first suggested by the biologist Rudolf Leuckart. Leuckart had pointed out that animals of a spherical form can respire from the surface only if they are microscopic. Any increase in size occurs as the cube of the volume, but only as the square of the surface area. At this rate, their respiration would quickly become inadequate. Correlation understood in this way was perfectly acceptable to Weismann. He recognized it as an important determining factor in the

[47] August Weismann, "Bericht über die Weiterentwicklung der Descendenztheorie im Jahre 1872," *Archiv für Anthropologie* 6 (1873): 124–126; idem, *Studien*, II, 142–144, 185–192.

course of evolution. Thus, he endorsed the idea of a law of form, although not a teleological one.[48]

This willingness to listen closely to the idealists and even to adopt some of their ideas was an important feature of August Weismann's scientific practice. Nägeli had been the first scientist to oppose the inheritance of acquired characteristics as an evolutionary mechanism, and this viewpoint was widely associated with the idealist position. The Darwinist Georg von Seidlitz, who agreed with Nägeli, found himself stranded between two camps, unable to get a sympathetic hearing from either.[49] But, when Weismann came to agree with him in 1873, he was following a precedent that he had already established some time earlier. He had already acknowledged virtues in the idealists' theory of form; it was consistent of him to acknowledge virtues in their theory of heredity as well. He did not simply appropriate these ideas or capitulate to the idealist way of thinking; as always Weismann remained a Darwinist. However, he knew better than most Darwinists how to distinguish between essential and nonessential aspects of Darwin's position, where he could compromise and where he could not. Haeckel remained imprisoned in a tight ideological system; but Weismann, picking his way carefully, tested the intricacies and paradoxes of evolutionary theory and finally emerged as the leading spokesman for the Darwinist position.

In conclusion I would like to review the doctrine on which Weismann did not compromise, the canon that separated the German Darwinists from their adversaries most decisively. The German Darwinists often asserted that natural selection provided them with a mechanist key to evolution, as opposed to the teleological one of their opponents. However, this statement tells more about Darwinist propaganda than about Darwinist practice. After all, most idealists granted a certain limited role to natural selection; and some of them were second to none of the Darwinists in their insistence on mechanistic causality. The conflict between causality and teleology was in some ways a superficial one. As Weismann himself was obliged to admit, causality pursued far enough is indistinguishable from teleology.[50] The real weight of the Darwinist

---

[48] Weismann, *Studien*, II, 285–290.
[49] Seidlitz, *Darwin'sche Theorie*, p. 109.
[50] Weismann, *Studien*, II, 323–324, 329.

innovation fell elsewhere. Its greatest impact on German biological practice lay in the introduction of historical modes of explanation for the observable phenomena of living nature. This historical approach to nature was rejected, not only by the opponents of evolution, but also by the idealist evolutionists. Whether they favored a teleological or a reductionist biology, the idealists could not see the point of a theory that emphasized the irregularities and exceptions in the organic world. In the timeless realm of idealist thinking, unchanging laws worked out an inevitable destiny. But Darwin taught his followers to look at living beings one by one. Thus prompted, they recognized, as if for the first time, the surprising fact of anomaly and the wisdom of an open-ended theory. Into their science of life they received the unique effects of time.[51]

APPENDIX

German Editions of Important Works on Evolution

German editions of Darwin's works are listed in R. B. Freeman, *The Works of Charles Darwin: An Annotated Bibliographical Handlist* (London: Dawsons of Pall Mall, 1965). Other German translations include *Spuren der Gottheit in der Entwicklungs und Bildungsgeschichte der Schöpfung*, trans. A. Seubert (Stuttgart: Becher, 1846), which includes translations of William Whewell's *Indication of the Creator* and Robert Chambers's *Vestiges*; Chambers, *Natürliche Geschichte der Schöpfung des Weltalls, der Erde und der auf ihr befindlichen Organismen, begründet auf die durch die Wissenschaft errungenen Thatsachen*, trans. Carl Vogt (Braunschweig: Vieweg & Sohn, 1851; rev. ed., 1858); T. H. Huxley, *Zeugnisse für die Stellung des Menchen in der Natur*, trans. J. Viktor Carus (Braunschweig: Vieweg & Sohn, 1863); Huxley, *Handbuch der Anatomie der Wirbelthiere*, trans. F. Ratzel (Breslau: Kern, 1873); Huxley, *Grundzüge der Anatomie der Wirbellosen Thiere*,

---

[51] A fine account of the methodological issues at stake in the debate between evolutionists and reductionists is Richard C. Lewontin's "The Basis of Conflict in Biological Explanation," *Journal of the History of Biology* 2 (1969): 35–45. The role of historical thinking in Darwin's own work is discussed in Michael T. Ghiselin's *The Triumph of the Darwinian Method* (Berkeley: University of California Press, 1969), pp. 28–29, 68.

trans. J. W. Sprengel (Leipzig: Engelmann, 1878) ; Alfred Russel Wallace, *Der Maley Archipel: Die Heimath des Orang-Utan und des Paradiesvogels; Reiseerlebnisse und Studien über Land und Leute*, trans. Adolf Bernhard Meyer (Braunschweig: Westermann, 1869); Darwin and Wallace, *Charles Darwin und Alfred Russel Wallace: Ihre ersten Publicationen über die "Entstehung der Arten" nebst einer Skizze ihres Lebens und einem Verzeichniss ihrer Schriften*, ed. and trans. Meyer (Erlangen: Besold, 1870); Wallace, *Beiträge zur Theorie der natürlichen Zuchtwahl: Eine Reihe von Essais*, trans. Meyer (Erlangen: Besold, 1870); Wallace, *Die geographische Verbreitung der Thiere: Nebst eine Studie über die Verwandschaft der lebenden und ausgestorbenen Faunen in ihrer Beziehung zu den früheren Veränderungen der Erdoberfläche*, trans. Meyer, 2 vols. (Dresden: von Zahn, 1876); Charles Lyell, *Das Alter des Menschengeschlechts auf der Erde und der Ursprung der Arten durch Abänderung, nebst eine Beschreibung der Eis-Zeit in Europe und Amerika*, trans. Ludwig Büchner (Leipzig: Thomas, 1864) (this and another 1864 edition follow the third English edition; Büchner produced a new edition in 1874, following the fourth revised English edition); Louis Agassiz, *Die Classification des Thierreichs*, trans. Chr. Hempfing (Marburg: Erhardt, 1866); Agassiz, *Der Schöpfungsplan: Vorlesungen über die natürliche Grundlage der Verwandschaften unter der Thiere*, trans. Christoph G. Giebel (Leipzig: Quandt & Händel, 1875); Armand de Quatrefages, *Das Menschengeschlecht*, Internationale wissenschaftliche Bibliothek, nos. 30, 31 (Leipzig: Brockhaus, n.d.); Karl Ernst von Baer, "Ueber Darwins Lehre," in *Reden gehalten in wissenschaftliche Versammlungen und kleinere Aufsätze vermischten Inhalts*, 3 vols. (St. Petersburg: H. Schmitzdorff, 1864–1876), II, 235–480.

FRANCE

ROBERT E. STEBBINS

To a Frenchman, even in the 1880's, "Darwinism" and "evolution" were still basically foreign terms. The preferred French word was *transformisme*. A French biologist was no more likely to consider himself a "Darwinist" than a physicist would have been to call himself a "Newtonian" or an astronomer a "Copernican." There was discussion of transformism, and there were many transformists in France from 1859 to 1882, but little Darwinism and fewer Darwinists.

Darwinism and the Darwinian Revolution have dramatic qualities that attract careful study and attention in the Anglo-Saxon world. However much they may be interested in the immediate change in biological science, intellectual historians are attracted to the Darwinian Revolution by the excitement of the entire drama of human concern, confrontation, and argument. The Huxley-Wilberforce encounter at the British Association for the Advancement of Science meeting in 1860 may, after the fact, symbolize the feeling of drama and excitement that makes the Darwinian impact much more an object of historical investigation than the effects of the law of thermodynamics, for instance, have ever been.

The French, too, had a dramatic encounter over transformism. The excitement generated by the clash of the titans in a single meeting of the Academy of Sciences electrified the French and sent powerful shock waves far beyond the French borders. Nor did the stormy seas immediately subside. For decades afterward the French scientific community remembered and was deeply influenced by the impact of the initial shock.

The French drama was thirty years ahead of its English counterpart. The single most dramatic event in French evolutionary discussion took place on July 19, 1830. Georges Cuvier, the famous paleontologist, squelched the relatively mild evolutionary position of Etienne Geoffroy Saint-Hilaire. Geoffroy, following Lamarck, had attempted to defend the mutability of species and the continuity of development in life forms. Cuvier ridiculed Lamarck's arguments and insisted that Geoffroy's own investigations of Egyptian remains over six thousand years old had proved that no transformations of species had taken place in that long period and, therefore, that no extensive changes could have taken place in the years then assumed to have passed before that time. In addition, Cuvier's eulogy on Lamarck before the Academy of Sciences was so bitter and so full of violent language that it was not printed until after Cuvier's own death and then only after the more offensive portions had been deleted.[1] Cuvier's prestige and the bitterness of his attack have often been credited with virtually silencing discussion of transformism in France, where the question had long been debated.

Although Cuvier's attack on Geoffroy and transformism had a definite chilling effect, it would be wrong to say that transformism was totally demolished and only rose from its ruins after 1859. Geoffroy, even in defeat, still received some honors, serving as president of the Academy of Sciences in 1833. His son, Isidore, likewise highly respected, was elected to the Academy of Sciences in 1833 and served as its president in 1857. Many other Frenchmen, including two others cited by Darwin in his "Historical Sketch" as French precursors, continued to think and write along evolutionary lines between 1830 and 1859. Nevertheless, there is almost universal acknowledgment that evolutionary thought was advanced only under a heavy cloud, and those who wrote in favor of evolution condemned Cuvier for his wilful obstruction. For instance, Henri Favre's enigmatic introduction to his protransformist *Développement de la série naturelle* (1856) seemed clearly to indicate that it was difficult to get a hearing. He magnanimously said he would make no recriminations, but he nevertheless downgraded Cuvier and praised Geoffroy as a much greater man.

The drama of 1830 cast a long and pervasive shadow over the ques-

---

[1] R. A. Muttkowski, "A Centenary of Lamarck," *Thought* 4 (1929): 396.

tion of the mutability of species. Cuvier remained a most prestigious scientific figure for decades after 1859 and from his tomb may be considered a leading opponent of Darwin.

The echoing thunders of Cuvier's condemnation of Geoffroy and Lamarck tended to put every scientist contemplating evolutionary thought on his guard. The definitely ascertained facts did not justify positing too great an age for the earth, and the constancy of life forms for six thousand years did not encourage belief that the necessary changes could have taken place in the time extending beyond the earliest Egyptian records. The oppressive conclusions could be surmounted only with boldness, and the French scientific climate of opinion was not conducive to boldness.

Between 1830 and 1859 and beyond, French philosophy of science actively opposed speculative ventures and bold hypotheses not immediately verified by empirical means. Francis C. Haber has traced the religious side of this opposition back to Abbé Noel Antoine Pluche, who denied that scientific generalizations could correct or go beyond the truth of revelation. In his *Histoire du ciel*, Pluche proclaimed that "the natural conclusion of the comparison we have made of the thoughts, either of the ancients or the moderns, on the origin and end of all things, with what Moses teaches us is that NOT ONLY IN RELIGION, BUT ALSO IN PHYSICS, WE MUST RESTRICT OURSELVES TO THE CERTAINTY OF EXPERIENCE AND THE MODERATION OF REVELATION."[2] Pluche's *Spectacle de la nature*, which first appeared in 1732 and was "perhaps the most widely read book of its kind in France during the eighteenth century,"[3] reached a twelfth edition in 1770 and was published again in 1885. More significantly, an abridgement of the work appeared in 1844, went to a ninth edition in 1863, and had ten more impressions from 1866 to 1888. Whether cause or reflection, Pluche's continuing popularity provides a good illustration of a deep-seated and pervasive attitude among those not inclined to oppose religious opinion.

The positivists offered a complementary stricture on speculation and hasty generalization. While the word "positivist" has been much used

[2] Quoted by Francis C. Haber, in "Fossils and the Idea of a Process of Time in Natural History," in *Forerunners of Darwin*, ed. Bentley Glass (Baltimore: Johns Hopkins Press, 1959), p. 227.

[3] Haber, "Fossils and the Idea of a Process of Time," p. 227.

and abused in relation to the French scene during the Second Empire, virtually all who have been called positivists were united on the need for facts and caution in generalization. Auguste Comte had condemned Lamarck, and his followers took the same position either out of filial piety or because they felt that Lamarck had made a mockery of positive science. They could not stand to be accused of being soft on metaphysics and speculation. If the broader meaning of positivism as a general philosophy of science in France during the second half of the nineteenth century is accepted, the stricture still applies; the positivists were all critical of speculation and of going beyond the immediate data of experience. This caution was fully applied to transformism.[4]

If the general idea of evolution was under a Cuvierian cloud, neither the influence of Pluche and his religious friends nor that of the positivists of whatever stripe was likely to produce any additional blue sky. Research and scientific thought were more likely to be channeled into more circumscribed fields than the broad reaches of evolution.

Two such areas for closer scrutiny deeply occupied French scientists and the French scientific community in the 1850's and 1860's: human prehistory and spontaneous generation. At the same time that the French eschewed the larger issue of evolution as not open to demonstration, these two studies were intimately involved in the whole evolutionary discussion. They were the two termini of transformism: the development of man and the origin of life.

Maurice Caullery has claimed that "pre-history is a science of French creation."[5] However this may be disputed, many of the most exciting finds of prehistoric man were made in France in the 1840's and 1850's, and the stimulus to French scientific endeavor cannot be denied. The finds were the more engaging because of their geographical proximity and national identity. Here was an area for concentration in which the researchers could dwell upon the facts. Speculations and conclusions might be advanced at times, but always on the basis of discovered evi-

---

[4] *La Philosophie Positive*'s editorial policy, enunciated in early 1868, said there was no reason for this positivist organ to lean toward or away from the transformist ideas (2: 27).

[5] Maurice Caullery, *French Science and Its Principal Discoveries since the Seventeenth Century*, trans. Henri Dupont (New York: privately printed for the French Institute, 1934), p. 150.

dence. The origin of man was a prime topic for consideration, but the inspiration behind it was French and antedated 1859.

In 1858 Félix-Archimède Pouchet presented to the Academy of Sciences his researches on spontaneous generation. Although such ideas had been condemned in the eighteenth century and had generally passed out of favor, the development of the achromatic compound microscope in the nineteenth century opened up a new world of life. Pouchet was convinced that life was currently being generated, and he believed he had developed demonstrations to prove it. This proof of spontaneous generation could accord well with Cuvier's picture of successive destructions and renewed creations of life on the globe. Pouchet and many of his followers at first put the ideas in this framework.[6] But, most important, Pouchet was concentrating on a directly empirical question. His main view may be called naïve spontaneous generation, as opposed to a philosophical assertion that life must have been generated at one time or another from not-living matter by some natural means. Later Pouchet and his followers espoused a philosophical as well as a naïve view of spontaneous generation, but in 1858 almost the entire focus was on naïve spontaneous generation and the attempt empirically to prove it. The Academy of Sciences was involved in numerous ways and occasions with the question of spontaneous generation between 1858 and 1870. Louis Pasteur became the hero, while Pouchet was ridiculed by many. Nevertheless, Pouchet had surprising support and retained the loyalty of many of his followers even after his arguments had supposedly been demolished by Pasteur.

Here, then, were two scientific questions coming into notable prominence in France before 1859. Each had a special relationship to France. Each met the criterion of being immediately demonstrable or refutable on the basis of evidence and investigation. Each dealt with an important issue in the development of life forms. Above all, each had a dramatic and exciting quality about it. Both provided the French scientific community with enough material to work on without paying too much attention to the writings of an Englishman.

[6] Some also associated vitalistic doctrines with spontaneous generation. In general, beliefs in spontaneous generation, or heterogeny, were quite diverse in scientific and philosophic associations. No one doubted that they needed to be discussed in relation or in opposition to evolution.

When Darwin published his *Origin of Species* in 1859, the French stage was not empty and awaiting his production; the time was not "ripe" in France. The evolutionary plot had already been explored, exploited, and driven off the stage in the dramatic encounter of 1830 and subsequent developments. The climate of opinion was set against broad generalizations, but the termini of evolutionary thought were considered to be more amenable to exact study, and Frenchmen had already captured the spotlight for their valuable issues. They did not need Darwin's works to stimulate their research. Darwin was to be relegated to the wings by the greater attraction of the very two issues he sought to avoid in his *Origin of Species* as being too dangerous to treat at length.

Jean Rostand, in his history of evolutionary ideas, said that, "in France, the Darwinian work ran into a systematic hostility."[7] T. H. Huxley was perhaps more accurate when he said that Darwin's ideas were met in the Academy of Sciences by a "conspiracy of silence."[8] The same phrase could be applied to much of the rest of the French scientific community and publications. Having been shunned by a national scientific community for which he had great respect and initially some hopes, Darwin could only rail against the "horrid unbelieving Frenchmen."[9]

The conspiracy of silence was not quite perfectly achieved at any time. The English publication of the *Origin of Species* received at least five reviews in the French periodical press, and only one of these was totally condemnatory. Alfred Sudre, in the August 1, 1860, *Revue Européenne*, roundly attacked Darwin's book as "purely and simply a reproduction of Lamarck's system."[10] Nothing showed that transformations between species had taken place. Darwin's ideas were materialistic; they naturally fit in with other materialistic views such as Félix-Archimède Pouchet's theory of spontaneous generation and Georges Pouchet's anthropology, which accepted an animal origin of man. Of the five reviews, Sudre's was the most flatly opinionated and negative toward Darwin and also

---

[7] Jean Rostand, *L'Evolution des espèces: Histoire des idées transformistes* (Paris: Hachette, 1932), p. 118.

[8] T. H. Huxley, "On the Reception of the 'Origin of Species,'" in *Life and Letters of Charles Darwin*, ed. Francis Darwin (New York: D. Appleton and Company, 1925), I, 539.

[9] Darwin, *Life and Letters*, II, 255.

[10] Alfred Sudre, "Des origines de la vie et de la distinction des espèces dans l'ordre animé," *Revue Européenne* 10 (1860): 599.

showed the least acquaintance with and related the fewest facts about the ideas.

In the May–June, 1860, issue of *Revue Contemporaine*, Henri Montucci reviewed the *Origin of Species* together with Pouchet's book on spontaneous generation, again illustrating the relationship of the two ideas in France. He was firmly convinced "that to the degree that our knowledge is extended, the truth of Darwin's system will become more and more manifest."[11] Montucci showed that Darwin's ideas were compatible with the Biblical account of creation, and he would accept any interpretation that would fit with the facts and the Biblical accounts. In essence, he thought, Darwin's work was the same as what Charles Bonnet had advanced in 1779, but he assured his readers that he was not tracing the genealogy of the ideas in order to belittle them; rather, it was to give them stature and authority they might otherwise lack. Darwin's genius and the brilliance of the natural-selection theory in overcoming previous difficulties thus did not greatly impress Montucci.

A short review in the *Revue de Géologie* in 1861 offered some positive support for Darwin. The Swiss author, Edouard Claparède, wrote a still more positive and encouraging review in the *Revue Germanique*, a publication with Protestant religious interests. Claparède was concerned about possible theological objections and carefully indicated that Darwin adequately recognized the ultimate mystery of all life. Arguments on the animal origins of man were arguments from sentiment; Claparède said his sentiments led him to prefer being a perfected ape rather than a degenerated Adam. Further, natural selection had a remote Biblical parallel in that many were called but few chosen.[12]

Probably the most important review of the *Origin of Species* was by Auguste Laugel in the April 1, 1860, *Revue des Deux Mondes*. Laugel had frequently traveled in England, knew many of the leading English Darwinists, and was significantly described by Huxley as "an accomplished writer, out of the range of academical influences."[13] Laugel per-

---

[11] Henri Montucci, "Notes critiques sur la marche et le développement des sciences," *Revue Contemporaine* 15 (1860): 165.

[12] Edouard Claparède, "M. Darwin et sa théorie de la formation des espèces," *Revue Germanique* 16 (July–August 1861): 535, 546 ff.; ibid. (September–October 1861): 259.

[13] Thomas Henry Huxley, "On the Reception of the 'Origin of Species,'" p. 539.

ceptively understood his readers and approached his subject cautiously. He argued that scientific explanation of the facts of natural history was necessary. Though he was not inclined toward religious orthodoxy, he was careful to obviate religious objections in advance. He thought Darwin would do better to admit some causes not yet known or currently operating; Charles Lyell's and Darwin's strict uniformitarianism was unfortunate. In an article in *Revue Germanique* in July–August, 1859, and without reference to Darwin, Laugel had already shown himself quite sympathetic to the notion of species transformation, a question that he considered to be quite well known at the time. He reviewed the *Origin of Species*, but it had not presented him with or convinced him of anything drastically new. Nevertheless, Laugel's whole approach was exceedingly positive, and Darwin himself was well pleased with this key review in an important publication.

Two of the men Darwin considered as his French precursors were still living in 1859. The aged Isidore Geoffroy Saint-Hilaire had published the first two volumes of his three-volume *Histoire naturelle générale* before the *Origin* appeared. The third volume, partly written after 1859 and posthumously published in 1862, referred to Darwin only once. Geoffroy asked:

Is natural selection, as Darwin believes and as he has ingeniously undertaken to demonstrate, the means generally employed by nature of creating new types? There is *at least* reason to doubt it; but even granting this, one could still, despite generally incontestable analogy, refuse to accept the parallel between *natural selection* and the selection practiced by our agriculturalists, which Darwin utilizes to explain the multiplication of species. The work of the agriculturalists is so little natural, that not only has nature never brought new races to birth in this way, but it also tends ceaselessly to cause such variations to disappear.[14]

Geoffroy accepted only those things that could be proven; Darwin's system was too conjectural and oversimplified; for him the *Origin of Species* offered no significant break-through. Charles Naudin, who had utilized the concept of natural selection in an article in the *Revue Horticole* in 1852, apparently did not write on evolution again until 1867

---

[14] Isidore Geoffroy Saint-Hilaire, *Histoire naturelle générale* (Paris: V. Masson, 1862), III, 522–523.

and then did not praise or condemn Darwin as such. Finally in 1874 he did discuss Darwin, only to distinguish his own views from Darwin's false notions, primarily by reference to a prior cause that Darwin had wrongly omitted.[15]

Apparently no French books were written on Darwinism prior to the translation of the *Origin*. Of the thirty-four articles on evolution or related topics appearing from 1859 to 1862 in eleven of the periodicals of which this study took note, thirteen made no mention of Darwin, three were insignificant, eight made only indirect reference to Darwin, and ten dealt directly but not exclusively with Darwin. Darwin received passing mention or was noted as one who had stimulated a renewed look at the questions of transformism, but before 1862 there was not a single nonreview article directly on "Darwinism."

Nevertheless, it is important to note that virtually all of the important reactions to Darwin's writings had at least a forerunner in arguments presented before Darwin appeared in French translation. The French had been considering transformism for some time and had developed most of the lines of argument before Darwin appeared in English and especially before he appeared in French.

It irritated Darwin that he was unable immediately to find a French translator and publisher for his *Origin of Species*. The general hostility to transformism doubtless discouraged any scientist's interest in making a translation. After two negotiations had fallen through, arrangements were finally made in September, 1861, for publication by Guillaumin and Masson of a translation of the third English edition by Mlle Clémence Royer. Darwin had never heard of her before, nor, apparently, had she heard of him. Darwin said that she must be a very clever woman, while Ernest Renan is reported to have said that she was "almost a man of genius."

Born in Nantes to a family both religious and legitimist, she left both the home town and the familial religious and political positions fairly early. She spent some time in England and in Switzerland. Her writings and interests seem to have mainly been in economics and social-science

---

[15] Charles Naudin, "Cas de monstruosités devenus le point de départ de nouvelles races dans les végétaux," *Académie des Sciences: Comptes Rendus* 64 (May 13, 1867): 929–933; idem, "Les Espèces affinés et la théorie de l'évolution," *Bulletin de la Société Botanique* 21 (November 13, 1874): 240–272.

fields. She did not know anything special about biology, nor was she well known in scientific circles. Her general attitude seems to have amused and confused many who met her. Two men who aided her with the translation testified that she needed much help but was reluctant to receive any. She had been convinced of transformism before Darwin's book appeared.

When he finally found a translator, Darwin apparently got both less and more than he had hoped for. He did not get the support of a noted scientist, but Clémence Royer offered a nearly fifty-page introduction. Her numerous translator's footnotes disputed with the author and freely offered her own opinions. The introduction was more suitable for an antireligious polemic than for a serious scientific work, as the first and last two paragraphs make clear:

> Yes, I believe in revelation, but in a permanent revelation of man to himself and by himself, in a revelation which is only the result of the progress of science and of contemporary knowledge, of a revelation which is always partial and relative and affected by the acquisition of new truths and even more by the elimination of ancient errors. It is even necessary to assert that the progess of truth gives us even more to forget than to learn, and teaches us to deny and to doubt as often as to affirm.
>
> The doctrine of M. Darwin is the rational revelation of progress, pitting itself in its logical antagonism with the irrational revelation of the fall. These are two principles, two religions in struggle, a thesis and an antithesis of which I defy the German who is most proficient in logical developments to find a synthesis. It is a quite categorical yes and no between which it is necessary to choose, and whoever declares himself for the one is against the other.
>
> For myself, the choice is made: I believe in progress.[16]

The idea of progress was more important to Royer than the idea of natural selection. She even changed Darwin's title to read *De l'origine des espèces, ou des lois du progrès chez les êtres organisés*.[17] She said Darwin's book was especially fertile in its moral and humanitarian consequences. He offered a vast and universal synthesis of economic laws and social and natural science. She was sure that all the remaining doubts

[16] Clémence Royer, "Préface de la première édition," in her translation, *De l'origine des espèces* (Paris: Guillaumin and V. Masson, 1866), pp. xv, lix.

[17] "On the origin of species, or the laws of progress among organisms."

about evolution would be cleared up. Where Darwin was uncertain, she had no doubts at all.

If Darwin himself was hesitant about applying his knowledge and his questionings to the actual origin of life, Royer shared no such hesitancy. She pronounced from the beginning, in conformity with Lamarck's thoughts, that the original form of life must surely have come from non-living substance and must be ultimately reducible to its component elements in some less-than-mysterious way. Thus she maintained and confirmed the kind of identification that several Frenchmen had already made between Darwin's ideas and spontaneous generation, then being debated in French scientific circles.

Royer willingly admitted that Darwin had not solved all the problems, but, building on the work of others, Darwin had finally advanced to the place that the general transformation of species was virtually an impregnable doctrine. The details had yet to be filled in, and in some instances Darwin's own specific conclusions would have to be revised, but on the whole the victory was won.

Later, when opponents of the Darwinian views wanted to object to the social and moral implications of his ideas, they could refer to Royer's introduction to see just what those implications were in the minds of some. To be sure, pernicious motives had already been attributed to Darwin before he appeared à la Royer, and disclaimers could be insisted upon, but the formal introduction was probably one of the best-noted parts of the book for many a French reader, whether for or against Darwin.

In the second French edition (1866) Royer repeated the first introduction and added a buoyantly optimistic second preface. She exaggerated various scientists' support of Darwin but said that she refrained from naming them for fear of causing embarrassment. In a letter to Lyell, Darwin said that "the introduction was a complete surprise to me, and I dare say has injured the book in France."[18]

Friction between author and translator was already apparent, and in the third edition (1870) it became even more obvious. Royer repeated her first preface. She expressed her deep disappointment that Darwin

---

[18] Darwin, *Life and Letters*, II, 255. Francis Darwin's dating of this letter applies the statement to the second edition, but it seems more appropriate to the first.

had disregarded proper empirical standards and had entered into speculation for which there was neither evidence nor need. She thought his idea of pangenesis was worthless and unsubstantiated. Although a publisher's notice indicated that the third edition was published with the author's permission, Darwin wrote to Joseph Hooker that Royer had brought it out without telling him. Apparently he was offended by her strong reactions against pangenesis, and he authorized a new French translation from the fifth English edition by a Swiss, Jean-Jacques Moulinié, which was published by C. Reinwald in 1873. After Moulinié died in 1873, Reinwald published a third translation (1876) by Edmund Barbier. Neither Moulinié nor Barbier was an eminent biological scientist, but they did provide French editions with much shorter, more positive, and more scientifically oriented introductions. Although Barbier's translation was based on the sixth English edition and therefore might have had some claim to finality, Royer published her translation again in 1883.

The *Origin of Species* had to wait over two years before appearing in French, but its appearing in eight editions between 1862 and 1883 indicates that it was not a total failure in sales.

Moulinié translated *The Variation of Animals and Plants under Domestication* in 1868, and a translation of the second English edition was made by Barbier in 1879. The preface to both editions was written by Carl Vogt, a German who lived in Switzerland and who was mentioned more often by Frenchmen than perhaps any other person outside England and Germany as a supporter of Darwin. Vogt's stature was due to his scientific work, but he was also noted for his antireligious stance. His preface flattered Darwin, recognized the attacks being made upon Darwinian theories, but indicated that it was necessary to clear up many of the false assumptions still commonly held.

The Descent of Man and Selection in Relation to Sex was translated by Moulinié and published by Reinwald in 1872. Apparently the translation was not very effective, for a new edition appeared in 1873 with Edmund Barbier serving as "editor." Barbier translated the third edition in 1881. All three editions of *Descent of Man* carried the same preface, mainly consisting of part of a speech delivered by Carl Vogt in 1869. Somewhat like Royer's first preface, it showed that the heavy hand of tradition was preventing many people from looking at life scientifically.

Progress would be slow, but Darwin's work would help free the human mind from past errors.

The translators of *The Expression of the Emotions in Man and Animals* (1877) cited the fact that they were not offering a voluminous and argumentative preface as one of the virtues of their work. To those mainly interested in Darwin's work as science, it may well have seemed so.

While the French professed to be more eager to accept Darwin's empirical observations than his theories, the book Lyell considered to be Darwin's second most valuable work, *On the Various Contrivances by Which British and Foreign Orchids Are Fertilized by Insects*, was not translated until 1870—eight years after its English publication. While one review was offered on the English edition, none has been discovered for the French.

Edouard Heckel, who published his translation of Darwin's *The Power of Movement in Plants* in 1882, commented that this was a purely factual work. Those who wished to see minute empirical investigations without theories and speculation should pay special attention to this publication of Darwin, who was better known for another kind of work.

Significantly, none of the French translations or prefaces to the major Darwinian works was by a noted French man of science. Furthermore, the prefaces to both the *Origin* and the *Descent* tended to exacerbate rather than to allay theological tensions. Theoretically, the explicit and head-on disagreement with theological traditions might seem to damage the possibilities of a good reception. In practice it seems as though, even without the introductions, the reactions might have been virtually the same, as was shown in the two years before any translation appeared.

The lack of a leading scientist's backing was serious in itself. Darwin had said that if Huxley and two or three others thought him right he would "not care what the mob of naturalists" thought.[19] In France he faced not only the mob, but the leaders as well. Pierre-Marie-Jean Flourens, permanent secretary of the Academy of Sciences, in some ways typified the opposition to Darwin, as his predecessor and inspiration, Cuvier, had led the opposition to both Lamarck and Geoffroy.

[19] Ibid., I, 527.

Flourens continued Cuvier's efforts and spoke out against both transformism and spontaneous generation in his *Cours de physiologie comparée* (1856). That relatively more of his time before 1864 was spent on the spontaneous-generation issue than on evolution reflected the interest of the times in France. Flourens discussed the general issues of transformism but did not mention Darwin until late 1863, when he presented a series of three articles in the *Journal des Savants*. The same ideas were repeated in *Examen du livre de M. Darwin* (1864), apparently the first book published in France on Darwin. Flourens attacked Darwin on three main lines: (*a*) Darwin failed to define species adequately; (*b*) while noting what everyone else knew—that species vary—Darwin failed to note the limits to variability; and (*c*) Darwin fooled himself and many of his readers with figurative and metaphorical language. This last was the "radical vice" of the book. Darwin personified nature and let nature do all his work for him. "Nature chooses, nature scrutinizes, nature works and works without cease, and works to what end? . . . to change, to perfect, to transform species." Either natural selection was nothing, or it was nature, and the nineteenth century had rightly abandoned metaphors and personifications. Flourens went on to say that Darwin had ostensibly written a book on the origin of species; yet nowhere did he discuss the *real* origin. There were really only two possible alternatives for the origin: spontaneous generation, or the hand of God. Since spontaneous generation was "thoroughly discredited," the second alternative must be accepted. "Here," said Flourens, "we are no longer left at the low level of the personifications of nature . . . we *know*."[20] Although both Darwin and Huxley were as amused as they were irritated by such arguments, Flourens's book received some good French reviews and was republished in 1881.

Flourens insisted that he and Darwin reached different conclusions on the species question because Darwin followed a system, while Flourens followed the facts. There was no experimental evidence for species transformation; so no one should conclude that they had ever been transformed. This claim for pure experimentalism, right after acceptance of "the hand of God" as an explanation for the origin of species, is certain-

---

[20] Pierre-Marie-Jean Flourens, *Examen du livre de M. Darwin sur l'origine des espèces* (Paris: Garnier frères, 1864), pp. 66–68.

ly not without parallel in the French scientific community of the time. That a man in a leading position, ready and able to use his influence, should present such an argument without fear of challenge may well illustrate the obstacles to free discussion of the species question.

In contrast to Flourens's critical, naïve, and bitter attacks were the fair and ubiquitous writings of Jean-Louis-Armand de Quatrefages de Bréau (1810–1892), professor at the Sorbonne in 1847, professor of anthropology at the Museum of Natural History in 1855, elected to the Academy of Sciences in 1852, president of the academy in 1873. De Quatrefages was almost universally respected as a gentleman and as a scientist. His scientific interests and knowledge ranged widely, and his numerous publications included many articles and two most important books on Darwin: *Darwin et ses précurseurs français: Etude sur le transformisme* (1870 and 1892) and *Les Emules de Darwin* (1894).

Darwin had very high respect for de Quatrefages. He had sent him a prepublication copy of the *Origin of Species*, realizing that de Quatrefages would be a very important man to have on his side. Considering de Quatrefages's many writings and his stature among French scientists, it is tempting to suggest that the conversion of no other single man could have been more important for the French reception of Darwin. There is a striking parallel between Huxley's statement that Darwinism's "logical foundation was insecure so long as experiments in selective breeding had not produced varieties which were more or less infertile"[21] and de Quatrefages's assertions of the current lack of proof. Huxley made his "act of philosophical faith," but de Quatrefages did not.

De Quatrefages's interest in the species question and his intellectual honesty can perhaps be indicated best by a paragraph appearing in the fourth of five articles he wrote for the *Revue des Deux Mondes* in 1869:

Such is the last word of this long study. It is not without regret that I write it. I would not be of my own time, if I did not understand and share the anxious curiosity with which so many educated or common intelligences today question the creation on the secrets of its origin and of its end in the name of science. To avow that human knowledge cannot even yet approach these problems is as painful to me as to any other person. However, one thought sweetens this otherwise galling feeling of impotence. I like to believe that we

---

21 Huxley, "On the Reception of the 'Origin of Species,'" pp. 550–551.

are opening out the happier route, and we are perhaps preparing the distant solution of these questions which are unfathomable for us. As humble as it may appear to certain minds, this task abundantly has its grandeur and its charms. It is that which our fathers have accomplished for us; let us accomplish it for our sons; but if we want to leave them a genuine inheritance, let us not dream about that which can be, but let us look for that which is.[22]

De Quatrefages's consistency of argument over more than thirty years is remarkable. He saw the arguments for evolution as clearly as any person could. It was even claimed that he presented Darwin's arguments better than Darwin himself did. He deeply respected Darwin's empirical work and even saw a remarkable correspondence between Darwin's theories and reality. He certainly saw much value in the ideas of natural selection, struggle for existence, many types of adaptation to the environment, etc. He was pleased that Darwin's later works allowed for factors in addition to natural selection. Nevertheless, Darwin had not demonstrated a single change from one species to another, and de Quatrefages adamantly withheld assent to transformism pending such empirical demonstration. In his last book, published posthumously, he said in a mood of at least mild disappointment: "Each day, in the crowd of publications of many sorts and on the most diverse subjects, it is affirmed that transformism presently reigns as master in science, and that it has the assent of all somewhat well informed minds and those of all savants truly worthy of the name."[23] He seemed to suggest that the idea of transformation of species would probably be accepted in the future on the basis of more knowledge than was yet available, but such commitment would be premature at the time he wrote.

He explicitly defended the right and necessity for propounding theories, but said that, in the very important question of evolution, the matter should be kept much more open than either the pro-Darwinists or the anti-Darwinists wanted. While the evolutionists insisted on more variation than could be substantiated by the facts, the antievolutionists were too dogmatic and went beyond the facts in denying any change. In con-

---

[22] Armand de Quatrefages, "Histoire naturelle générale: Origines des espèces animales et végétales; Discussion des théories transformistes; L'Espèce et la race," *Revue des Deux Mondes*, 8th ser. 80 (1869): 672.

[23] Armand de Quatrefages, *Les Emules de Darwin* (Paris: Alcan, 1894), pp. 1–2.

trast, de Quatrefages underlined the fact of human ignorance. Toward transformism he remained agnostic and was critical of the true believers on either side of him.

A friend and colleague of de Quatrefages, Henri Milne-Edwards (1800–1885), had grown up in a household and in a scientific atmosphere in which transformist questions had been thoroughly discussed. He was well acquainted with the Geoffroy Saint-Hilaires and continued to speak favorably of them. He was coeditor of the second edition of Lamarck's eleven-volume *Histoire naturelle des animaux sans vertèbres* (1835–1845), the only republication of Lamarck appearing before 1859. Likewise, in the few comments he made about Darwin, he was relatively positive. Most notably, he was an outspoken proponent of electing Darwin as a corresponding member of the Academy of Sciences. He noted that even Darwin's critics admitted his solid contributions to empirical science, for instance, in the field of geology. In addition, while he admitted that Darwin had made some serious mistakes, had been too bold in his conceptions, and had exaggerated some points, he supported Darwin, even before the Academy of Sciences, for his hypotheses as well as for his empirical observations. He was confident that time would take care of the errors in the bold hypotheses offered; in the meantime, some useful fruits might grow. Milne-Edwards insisted that it was wrong to equate the ideas of Lamarck and those of Darwin; there were significant differences between the two, and it was neither fair nor accurate to castigate Darwin for Lamarck's sins.

Perhaps Milne-Edwards's most favorable comment on transformism can be seen in his saying that natural science could present only three possible hypotheses for the origins of animal forms discovered in various geological epochs: (*a*) catastrophism, which claims that each period was interrupted and life renewed by a new creation by God; (*b*) spontaneous generation of the various forms; or (*c*) a hypothesis that says that "the animals with new forms are the descendants of other animals which have preceded them on the surface of the earth, but which were organized in a more or less different form." In evaluation, Milne-Edwards said: "The first of these hypotheses seems to me to be incompatible with the ideas of the grandeur and stability that the majestic spectacle of the universe inspires in us; the second is in disaccord with all that we know about the birth of living beings; the third, on the contrary, seems to me not to raise any grave objection."[24] Milne-Edwards said he did differ from Darwin, inasmuch as Darwin had incorrectly tried to explain all species changes by natural selection acting under biological conditions like those of the present day. But he defended Darwin's right to offer hypotheses and even suggested that some of these were basically fruitful.[25]

Nevertheless, Milne-Edwards said relatively little on evolution in public. No article was discovered from his pen dealing directly with evolution and Darwin. He was coeditor for the *Annales des Sciences Géologiques*, which produced no article directly on transformism from its first publication in 1869 to 1882. Furthermore, he edited the *Annales des Sciences Naturelles: Zoologie*, which had only one article directly on evolution, a short review of *On the Origin of Species*—by a Dublin professor of medicine, H. Freke. The review stated that Freke's ideas had "an analogy with" Darwin's views but came to quite different conclusions.

Even more silent on the question of evolution was the great Louis Pasteur. Although it was frequently maintained that Pasteur opposed transformism, this study has uncovered no book or article by Pasteur containing any antitransformist argument and no secondary article anywhere producing a single direct quotation from Pasteur specifically and directly denying the possibility of evolution. Pasteur's position may best be seen by a statement made in lectures delivered in 1864: "There are many great questions being discussed these days: unity or multiplicity of human races; creation of man many thousands of years or centuries ago; fixity of species or slow transformation of species from one to the other; matter reputed to be eternal rather than created; the idea of God being useless, etc. These are all questions that cannot be solved. I take a much more

[24] Henri Milne-Edwards, "Sur les travaux de Ch. Darwin," *La Revue des Cours Scientifiques* 7 (1870): 591.

[25] Edmond Perrier reported after Milne-Edwards's death that he had once been sent to Milne-Edwards as the only man who could help him get one of his works presented before the Academy of Sciences. Milne-Edwards promised his support only if Perrier would point out that his work was very like one Milne-Edwards himself had presented in 1826. Perrier said he was accused of all sorts of exaggerations and unworthy ideas but was nevertheless expected to advance his ideas as a kind of plagiarism of a respected man (Perrier, "Préface," to *Les Emules de Darwin*, by de Quatrefages, pp. lxix–lxxi). Milne-Edwards may not have been as antitransformist as he usually appeared.

humble role in tackling a problem that can be solved experimentally."[26] Many French scientists and antitransformists were quick to utilize this sort of statement to prove that Pasteur believed the problem of transformism could not be solved experimentally and therefore should not be considered.

There was an important sense, however, in which Pasteur's work in discrediting naïve concepts of spontaneous generation was important in also denying some concepts of evolution.[27] The most outspoken advocates of transformism in France tended to be materialists, eager to dispense with any supernatural agency. The opponents of the materialists could therefore grasp at Pasteur's discrediting spontaneous generation and say that, lacking the initial origin of species via spontaneous generation, the materialists were powerless to explain life forms without a creator. Even in silence, then, Pasteur was taken as a leading opponent of evolution.

Claude Bernard (1813–1878), the leading physiologist, was cited in the Darwinian controversies even more than Pasteur. Bernard's leading position was a strong insistence upon experimental evidence. Those who sought to discredit Darwin or evolution took pleasure in asserting that Darwin did not measure up to Bernard's high standards of experimental proof and therefore had not made any important scientific advances. Nothing but facts could be accepted as science, and Darwin's hypotheses were totally out of place. Bernard avoided scientific controversies by a philosophical agnosticism often associated with positivism. He said: "I ignore final causes because I ignore the initial cause. . . . I abstain from both; for I am reduced, in both cases, to invent, to imagine without being able to prove. . . . I support ignorance. There is my philosophy. I have the tranquility of ignorance and faith in science. . . . Others cannot live without faith, without belief, without theology; I do without all of these. I do not know, and I shall never know; I accept this fact without tor-

---

26 Louis Pasteur, "Chimie appliquée à la physiologie: Des générations spontanées," *Revue des Cours Scientifiques* 1 (1864): 257.

27 But Alphonse de Candolle indicated that the way some interpreted spontaneous generation as the main support for catastrophism should lead to the discrediting of catastrophism and open the way to evolutionary thought ("Etude sur l'espèce à l'occasion d'une révision de la famille de cupulifères," *Annales des Sciences Naturelles*, 4th ser. 18 [1862]: 102).

menting myself about it."[28] In specifically applying this to Darwinism, Bernard said: "Whether one is a Cuvierist or a Darwinist matters little; these are two different ways of understanding the history of the past and the establishment of the present regime; they cannot furnish any means of ruling the future."[29] He also stated that "In place of making unrealizable hypotheses on the origin of things, on which one can discuss or experiment only in a sterile and blind manner, the experimenter proceeds otherwise."[30]

Since evolution was not important to his work in physiology, Bernard was not inclined to spend much effort on the question. He, like Darwin, wanted to see every aspect of life explained by natural causes, but he was temperamentally indisposed to enter into needless or fruitless argument. He therefore found it much more convenient to speak little on the transformist questions. Nevertheless, Bernard kept an open mind and had high hopes for future discoveries of science. For instance, he may have asserted that man would some day be able to create new life.[31] Bernard, like de Quatrefages, believed that evolutionary thought might be more firmly established in the future, but this study has uncovered only one direct indication of any single person claiming Bernard in the protransformist camp. Given the widespread tendency to claim supporters, this tends to substantiate the other evidence that Bernard was certainly not an active proponent of transformism.

Jean-Albert Gaudry (1827–1908) wrote his magnum opus on *Les Enchaînements du monde animal dans les temps géologiques.* The work was definitely transformist, but, when he presented the volumes to the Society of Geology (1877 and 1883), he made little comment on transformism and no mention whatsoever of Darwin. In other comments made on Darwin's contributions to his own field of paleontology, Gaudry made it clear that he considered Darwin an outstanding scientist who had opened the way for scientific explanations of life developments.

[28] Quoted in Henri Cotard, *Pour connaître la pensée de Claude Bernard* (Grenoble: Editions Françaises Nouvelles [1944]), p. 53.

[29] Claude Bernard, *Claude Bernard: Morceaux choisis,* ed. Jean Rostand (Paris: Gallimard, 1938), p. 213.

[30] Ibid., pp. 149–150.

[31] Henri de Lacaze-Duthiers, "Direction des études zoologiques," *Archives de Zoologie Expérimentale et Générale* 1 (1872): 64.

Gaudry accepted much of organic evolution as a useful and necessary hypothesis and based most of his life work on transformist assumptions, but he said he would leave the explanation of transforming mechanisms to the evolutionists. It was sufficient for him that he could trace different strata by means of their life forms. In an article on *Les Enchaînements* in the *Revue des Deux Mondes* he asserted:

If it is the proper function of paleontologists to supply some proofs to the doctrine of evolution, it does not fall to them to explain the processes by which the author of the world has produced the modifications. That study of processes is what is called Darwinism from the name of the illustrious savant who has been its principal promoter. Assuredly it is a subject quite worthy of the attention of those naturalists that study the causes of the modification of beings; but it is up to the physiologists, who experiment on living creatures, to teach us how the changes are produced today and must have been produced formerly; using the expression of M. Claude Bernard, I will say that it is up to them to tell us the determination of species, genuses, and classes. That is to say, it is up to them to show us the secondary causes which have determined their formation. On this subject, a paleontologist can avow his ignorance. All that we can say is that the discovery of vestiges buried in the bowels of the earth teach us that a constant harmony has presided at the transformations of the organic world.[32]

Evolution was a fact, but no one had yet offered a compelling explanation. Gaudry seldom argued for transformism, he did not believe that Darwin had provided even a majority of the needed answers, and most of his presentations to the Academy of Sciences and elsewhere made little or no mention of Darwin.

Gaudry, like many other figures in French science of the day, knew the meaning of discretion. Although some already considered him a "declared partisan of the mutability of species," he was named to the chair of paleontology at the Sorbonne late in 1868. Victor Duruy, the minister of public instruction, appointed him despite the reported objections of scientists and others.[33] Having previously gotten in trouble over his ap-

[32] *Revue des Deux Mondes*, 9th ser. 23 (1877): 183.
[33] Victor Meunier, "La Mutabilité de l'espèce professé à la Sorbonne," *Cosmos*, 3d ser. 3 (1868): 645–646.

pointment of Ernest Renan to the Collège de France, Duruy was doubt-less eager to see that Gaudry's lectures at the Sorbonne should not cause a public uproar. They did not, and Gaudry assumed the chair of paleon-tology at the Museum of Natural History in 1872 with no difficulty.

Gaudry might believe in transformism, but he was careful not to pro-fess beliefs contrary to church doctrines. It was claimed that Gaudry's support of evolution was less harmful in the eyes of the Academy of Sci-ences because he was Roman Catholic.[34] He was elected to the Academy of Sciences in 1882 and served as its president in 1903. Gaudry could be a transformist without being condemned for it at least partly because he said relatively little about the subject and never saddled himself with being a partisan of Darwin.

Perhaps the best examples of the growing impact of Darwinism on some French men of science are provided by Henri de Lacaze-Duthiers and two of his students, Edmond Perrier and Alfred Mathieu Giard. Lacaze-Duthiers, a leading zoologist, was professor at the Museum of Natural History (1865), professor at the Sorbonne (1868), and was elected to the Academy of Sciences in 1871. Though he corresponded with T. H. Huxley from 1857[35] and considered himself a close friend of Carl Vogt,[36] perhaps the leading exponent of Darwin writing in French, his relatively minor comments on Darwin's "so famous" book in 1864[37] suggest that he was mainly interested in examining varying points of view. If any interpretation of his own appeared, it was a tendency to re-ject transformism. At the Sorbonne on April 10, 1869, Lacaze-Duthiers reputedly said: "I am not an antagonist to Darwinism; only I find the facts entirely as difficult to explain while admitting the theories as I find them inexplicable while not admitting them."[38] In 1872 he followed the frequent French line that "science ought to remain strictly in the sphere

[34] Emile Jourdy, "Le Darwinisme, par Emile Ferrière—Darwin devant l'Académie des Sciences: Joachim Barrande," *La Philosophie Positive* 9 (1872): 295.

[35] See Leonard Huxley, ed., *Life and Letters of Thomas Henry Huxley*, 2 vols. (New York: D. Appleton and Company, 1916), I, 162.

[36] *Archives de Zoologie Expérimentale et Générale* 2 (1873): ixl.

[37] Henri de Lacaze-Duthiers, "Zoologie: De la série animale," *La Revue des Cours Scientifiques* 2 (1865): 86.

[38] Quoted by Georges Petit and Jean Théodorides, "Quatre lettres inédites," *Janus* 48 (1959): 209.

of observed facts and to remain completely independent of all theories, which would tend to involve it in the sphere of imagination."[39]

In introducing the first review of a book relative to transformism in the *Archives de Zoologie Expérimentale et Générale,* which he founded in 1872, Lacaze-Duthiers set out a neutral editorial policy:

The theories on the transformation of beings have for several years once again had the privilege of attracting naturalists' attention.—Presented in their entirety in a form which people attempt to relate to the present state of natural sciences, utilizing new expressions if not absolutely new ideas, they are supported by some facts which their advocates constantly make great efforts to multiply. They create powerful interest in a renewal of debate on the fixity or mutability of species, which has been taken up many times and abandoned many times.

When occasion presents itself, we will recall notable works either for or against transformism without following step by step a discussion which too often degenerates into a polemic.[40]

At least a willingness to publish both sides of the issue showed some belief that transformism was a worthwhile question. In 1873 Lacaze-Duthiers praised Reinwald for bringing out Ernst Haeckel's *Histoire de la création* and offered perhaps the highest praise he ever gave to Darwin: "Darwinism has incontestably determined a considerable movement in zoological and paleontological studies; and some certain progress has been the consequence of this movement."[41] In the French scientific climate of opinion, this was relatively high praise; yet nearly everything in the record seems to point to the fact that Lacaze-Duthiers was attentive and appreciative, but also noncommittal, on Darwinism. He, like so many others, apparently felt no great enlightenment descending upon France by means of Darwin's works and felt no compulsion to advocate the Darwinian gospel.

Jean Octave Edmond Perrier (1844–1921), at first noncommittal like Lacaze-Duthiers, gradually increased his agreement with the transformist

[39] Henri de Lacaze-Duthiers, Review of Joachim Barrande's *Epreuves des théories paléontologiques par la réalité, Archives de Zoologie Expérimentale et Générale* 1 (1872): xxxiv.

[40] Ibid., p. xxvi.

[41] *Archives de Zoologie Expérimentale et Générale* 2 (1873): L.

positions. His report to *La Revue Scientifique* on the meeting of the British Association for the Advancement of Science at Edinburgh in 1871 said that the main concern and discussion could be summarized in one word: "le darwinisme." But he completely withheld his own judgment.[42] In a review of *The Descent of Man* in 1873 he tried to analyze rather than to criticize. He said: "It is not true that we share Darwin's doctrines, nor that we wish to become the mouthpiece for his school. We do not feel that we have yet the right to speak for or against. We see in Darwinism a gigantic effort of the human mind to explain phenomena which have for a long time been considered as closed to us."[43] But he concluded: "No doubt Darwinism will perish; but Darwin will remain as one of the great names of modern philosophy and science."[44] He certainly did not accept Darwin's full argument, but he did take Darwin seriously.

In 1878, when Ernst Haeckel attended the Congress of the French Association for the Advancement of Science, his admirers gave an banquet in his honor. Perrier reportedly offered a toast praising Haeckel for hastening the day when reason alone would govern men.[45] This and lectures he gave in the spring of 1879 at the Museum of Natural History on the topic of "Transformism and the Physical Sciences" make him appear already to be a transformist. Though he was still confronted by the received opinion of the immutability of species, he argued that the whole direction of science was to connect everything in determinism and chemicophysical reality. If this was correct, fixity must be an illusion. He went on to describe more of Darwin's ideas without committing himself to them. They were important, suggestive, and possibly correct, but they still could not be directly affirmed as his own.

Finally, in the fall of 1879, Perrier announced his conversion to transformism. That he had appeared so close to the transformist ideas only a few months earlier and yet felt it necessary to announce a formal conversion may further illustrate how Perrier or any scientist could be ex-

[42] Edmond Perrier, *La Revue Scientifique*, 2d ser. 1 (September 23, 1871): 301.

[43] Edmond Perrier, "Le Transformisme en Angleterre: L'Origine de l'homme d'après Darwin," *La Revue Scientifique*, 2d ser. 4 (February 1, 1873): 717.

[44] Ibid. (March 15, 1873): 875.

[45] "Variétés: Un Banquet transformiste—M. Haeckel," *La Revue Scientifique*, 2d ser. 9 (1875): 24.

ceedingly close to evolutionary views and yet fail to espouse them on his own. The words of Perrier before the Museum of Natural History in December, 1879, are, in the materials for this study, unique. Perrier began his new series of lectures by commenting that one of the great privileges of teaching at the museum was the considerable influence one exercised over his students. There was likewise a reciprocal influence on the professor both from the students and from the fact that the lectures themselves forced the professors to stretch their minds to the limit:

I avow without embarrassment, gentlemen, that I have undergone that influence. I began a series of studies on transformism last year; I had no bias relative to that powerful doctrine. If the general ideas that I had the honor of exposing to you in one of my first lectures attracted you to it, I also presented to your minds the objections, repeated without cease, which the most illustrious French naturalists have made to it, and among these men are those I like and venerate the most. It seemed to me, however, in the course of my lectures, that these objections were not insurmountable, that they attacked some fashions of conceiving the evolution of beings which had nothing necessary about them, and left the foundation of the doctrine perfectly intact.[46]

However slow he had been in accepting transformist ideas as fully his own, Edmond Perrier went on to become one of the leading advocates of transformism in France.[47]

Another student of Lacaze-Duthiers, Alfred Mathieu Giard (1846–1908), was perhaps the leading Darwinist in France in the nineteenth century.[48] A laboratory assistant for Lacaze-Duthiers at the Faculty of Sciences of Paris in 1871, he held various positions at Lille from 1872 to at least 1880. From 1887 to 1892 he was Maître des Conférences de Zoologie at the Ecole Normale Supérieure and then went on to the Faculty of Sciences at Paris.[49] While only two of his numerous publications between 1869 and 1882 were solely on transformism,[50] many others were directly related to that subject.

---

[46] Edmond Perrier, "Rôle de l'association dans le règne animal," *La Revue Scientifique,* 2d ser. 17 (1879): 553.

[47] Caullery, *French Science,* p. 147.

[48] Rostand, *L'Evolution des espèces,* p. 118 and passim.

[49] *Exposé des titres et travaux scientifiques (1869–1896) de Alfred Giard* (Paris: Imprimerie générale Lahure, 1896), p. 3.

[50] Alfred Mathieu Giard, "Les Controverses transformistes," *La Revue Scientifique,*

In 1874 Giard said that he stood with Huxley in preferring to take a bold stand for the future even if he might not yet have all the experimental evidence available.[51] Even without final proof, he announced to the French Association for the Advancement of Science in September, 1874, that he expected to see the doctrine of evolution established.[52] For France, a milestone on that road came in 1888 when Giard himself was chosen as the first occupant of the chair of "évolution des êtres organisés" at the Sorbonne.

The much less than enthusiastic response to Darwin by the leading scientists was reflected and perhaps even amplified in the leading scientific societies. The conspiracy of silence may not have been total or consciously motivated, but lack of comment and apparent indifference were in many ways more impressive than open opposition might have been.

The Société Zoologique and the Société Botanique might both be expected to have been interested in Darwinian questions. But the *Bulletin de la Société Zoologique*, founded in 1876, revealed no discussion of Darwinism from 1876 to 1882. The Société Botanique paid more attention to Darwin, but neither Darwin nor transformism was prominent in its deliberations, despite several French and English writers' indications that French botanists tended to be more favorable to Darwin than other French scientists were.[53]

The *Bulletin de la Société Botanique* devoted about three hundred pages annually to a listing and review of new publications, both French and foreign, but it did not mention any edition of the *Origin of Species* or the *Descent of Man*. It did take notice of the more directly empirical and descriptive works of Darwin, such as *On the Various Contrivances by Which British and Foreign Orchids Are Fertilized by Insects.* The at-

---

2d ser. 7 (July 11, 1874): 25–35; idem, "Les Mathématiques et le transformisme," *La Revue Scientifique*, 2d ser. 12 (February 10, 1877): 771–774.

[51] Alfred Mathieu Giard, "Les Controverses transformistes: L'Embryogénie des ascidées et l'origine des vertébrés: Kowalevsky et Baer," *La Revue Scientifique*, 2d ser. 7 (1874): 33–35.

[52] Alfred Mathieu Giard, "Laboratoire de zoologie maritime à Wimereux," *La Revue Scientifique*, 2d ser. 7 (1874): 221.

[53] E.g., Emile Jourdy, "Le Transformisme devant le positivisme," *La Philosophie Positive* 14 (1875): 37; Gustave-Frédéric Dollfus, *Principes de géologie transformiste* (Paris: F. Savy, 1874), p. 26.

titude of the society toward entertaining a broad synthesis, such as Darwin was advancing, may be suggested by the reactions to a published letter from Joseph Decaisne to the Abbé Chaboisseau in April, 1860, indicating the need for some synthesis of biological views. The existing scientific outlook presented too many separate species; there had to be some better way of handling, categorizing, and relating information.[54] Decaisne did not mention Darwin or advocate Darwin's particular hypotheses, but even his suggestion of hypothesis or generalization was sufficient to induce strong negative criticism. His critics pointed out that, while he might wish to simplify things, nature happened to be complex, and nothing could be done about it.

The first direct mention of Darwin or his ideas in the Société Botanique was by Adolphe Gubler in a series of three articles beginning in April, 1862. Like Decaisne, Gubler argued that the classification systems needed revision, that the mass of confusing facts presented by the growing body of botanical knowledge required a new synthesis; yet scientists hesitated. Lamarck had seen the need for synthesis, and Decaisne had indicated somewhat the same problem.

Gubler presented three broad outlooks on the species problem: (*a*) limited variation, which Gubler accepted for himself along with the Geoffroy Saint-Hilaires and de Quatrefages; (*b*) absolute fixity, accepted by "the majority of naturalists today"; and (*c*) transformism, which was accepted by "a small number of distinguished men whose opinion . . . merits being taken into serious consideration" and which was most recently and faithfully presented by Darwin.[55]

Significantly, Gubler humbly admitted that discussion of Darwin's evolutionary views was out of place in the society and thereby justified not bothering to criticize all Darwin's theories accompanying his valuable empirical findings. Nevertheless, he had some basically high praise for Darwin:

The work of the eminent naturalist is not less than one of the most remarkable of our epoch and is most useful to consult for the excellent materials it con-

[54] *Bulletin de la Société Botanique* 7 (1860): 262–264.

[55] Adolphe Gubler, "Préface d'une réforme des espèces fondée sur le principe de la variabilité restreinte des types organiques, en rapport avec leur faculté d'adaptation aux milieux," *Bulletin de la Société Botanique* 9 (1862): 264.

tains. It teems with fine observations, with ingenious perceptions, and each page emits a sense of loyalty and of reflective conviction which gives the best impression of the author. It is to be understood that with such qualities this interesting book has captured for the doctrine of selection a good number of the most distinguished minds, principally of those which loathe supernatural things and that are wrongly frightened at the word of creation.[56]

If it did not express his own views, the last phrase was possibly calculated to soothe religious susceptibilities, and Gubler's assertion that indefinite immutability "escaped all direct demonstration" doubtless pleased other critics. Despite his noncommittal treatment, Gubler devoted half of his three articles to elucidating Darwinian notions. He did this, presumably, in pursuit of his stated goal of insisting that botany had to advance from a descriptive to an experimental science,[57] reforming the classification of species and better using hypothesis and generalization.

The first definitely favorable discussion of transformism appeared in the society's bulletin in 1869. True to a common French approach, the discussion was undertaken as part of an article on spontaneous generation. Ernest Germain de Saint-Pierre indicated that the idea of the variability of species was satisfying because it accounted for most of the facts without doing violence to other knowledge. It explained otherwise inexplicable mysteries. The idea of variation of species, he said, logically led him and others to admit that present types must have originated from spontaneous generation or what he preferred to call the "first organic production" of simple cells. He thus used transformism as an entering wedge for discussion of spontaneous generation or *protogénie*, which he later maintained had been demonstrated as a contemporary phenomenon by indisputably precise experimentation.[58] Interest in transformism was hardly profound when it was first advocated in the society by a man who treated it as a secondary concern; its credibility could hardly have been enhanced for those who considered Germain de Saint-Pierre's first concern, spontaneous generation, as pseudo-scientific trumpery.

Except for a short review article excerpted from the *Revue des Deux Mondes* without comment and a simple notice on the French translation

[56] Ibid., p. 274.
[57] Ibid., pp. 394–395.
[58] Germain de Saint-Pierre, "L'Evolution de l'espèce végétale," *Bulletin de la Société Botanique* 23 (1876): xxviii.

of *On the Various Contrivances by Which British and Foreign Orchids Are Fertilized by Insects,* the *Bulletin de la Société Botanique* was silent on transformism and Darwin from 1869 to 1874. In 1874 it offered a short unsigned review of a negative, unoriginal, and hardly important article on "Transformism: Its Origins, Its Principles, Its Impossibilities," by M. A. Malbranche.

Also in 1874, Charles Naudin finally published his views, carefully separating himself from Darwin. His presentation was important inasmuch as it introduced, on its own merits, the general discussion of evolution for the first time in the society. But, in 1876, Germain de Saint-Pierre led the discussion back to the question of spontaneous generation. Although de Saint-Pierre admitted that transformism could not be absolutely demonstrated and that evolutionary action was doubtless somewhat weaker than formerly, he spoke of selection as being "so expertly studied and demonstrated by M. Darwin."[59] Natural selection was an important mechanism, and its elucidation was a great scientific milestone. But the discussion following focused almost exclusively on spontaneous generation; neither Darwin nor his theory was discussed at length.

After 1876 the bulletin was again silent on transformism until 1882. Then it reviewed an article published by Thury in the *Archives des Sciences Physiques et Naturelles,* indicating that Thury had anticipated Darwin by eight years and had explained most things as well as Darwin and some better. Darwin, the review showed, was no giant among men. He was one transformist among many, and his observations and theories could readily be equalled and surpassed.

A silence and coolness like that witnessed in the Botanical Society, representing a science so intimately affected by the Darwinian Revolution, was perhaps even more profoundly obvious in the Academy of Sciences. The importance of this fact can hardly be overestimated. Despite some necessary and reasonable qualifications, it can still be said that the French scientific community in the nineteenth century was amazingly centralized in Paris and its apex well symbolized in the membership of the Academy of Sciences. Prestige and advancement had to be related to this center. No one could reasonably argue that the leading scientists

[59] Ibid., p. xxx.

favored transformism. The "leading" French scientists were in the academy, and it would have been ridiculous to maintain that a majority of the academy was for Darwin.

Huxley's assertion that the academy greeted Darwin with a "conspiracy of silence" is essentially correct. A search of the indexes of each volume of the Academy's *Comptes Rendus*, of the two indexes covering the whole period from 1859 to 1882, and of the *Tables générales des travaux contenus dans les memoires de l'Académie* reveals no item specifically on Darwin outside of facts relative to his nomination and election to the academy, his death, and simple notices of a very few of his publications. It likewise reveals few specific discussions of transformism. A detailed search will uncover a few articles on topics related to transformism and others that mention Darwin and his theories in passing or at least give some idea that there was some discussion on the species question. On the whole, however, the "conspiracy of silence" was maintained.

By the most generous standard, classifying anything distantly related to the subject as "evolutionary" discussion, it can be said that twelve items on transformism have been discovered from 1859 to 1870 and six more from 1871 to 1882. Jean Rostand, being less generous but more reasonable, finds only two.[60] The few presentations that mentioned Darwin at all tended to be opposed. Most articles that discussed transformism and the species question concentrated on minute points at least purporting to be substantiated by careful empirical observations, while general philosophical arguments were lacking.

Having said so little in their scientific sessions, the members of the Academy of Sciences revealed their viewpoints on Darwin somewhat better, or at least more clearly, while considering him for possible election as a corresponding member.[61] Since he was nominated at least eight times before finally being elected, there was ample opportunity for the members of the academy to display their opinions of Darwin. Unfortunately, the *Comptes Rendus* reported only the persons nominated and the

60 Rostand, *L'Evolution des espèces*, pp. 129–130.

61 The academy, itself one of the four divisions of the Institut de France, was composed of eleven sections. Each section had six members and generally chose six to ten corresponding members. Nominations were made by the sections, but election was by the entire academy.

number of votes received.[62] The nominations themselves and the discussions preceding the elections were made in secret. Generally speaking, the requirements of secrecy were kept, and comments made in the academy concerning elections were not known. The controversy over the proposed election of Darwin, however, was sufficient that secrecy was not complete, and various periodicals reported some of the discussion.[63]

Darwin was first nominated as a corresponding member in the Section of Anatomy and Zoology on June 7, 1870. First-place nomination was given to Jean Frédéric de Brandt. Darwin's name was listed, as was customary, in alphabetical order among six other names placed in second-line nomination. Two other well-known evolutionists, Carl Vogt and T. H. Huxley, were included. Brandt received 19 votes, Darwin 16, Huxley 3, and S. L. Loven 1. In the second ballot, required because no one had received an absolute majority of the 39 votes cast, Brandt got 22 and Darwin 16. Two of Huxley's three votes were given to Brandt rather than to Darwin.[64] Considering the amount of opposition to Darwin, the distaste for his theories in general, and the fact that he had only received the second-line nomination, Darwin had come very close to being elected.

The most extensive debate continued over several weeks, following the second nomination on July 11, 1870. The real question being debated, according to the *Revue Scientifique*, was whether or not "the transformists would be struck with ostracism" in the persons of Darwin and Huxley.[65] This claim does not seem to be far wrong, although de Quatrefages is reported to have said on August 1, 1870, that Darwin's views had gained the support of many of the great men of science, including Lyell, Hooker, Huxley, Haeckel, and even Brandt, who had been elected to the Academy of Sciences in preference to Darwin.[66] Darwin on July 11 again received the second-line nomination. The same four members of the Section of Anatomy and Zoology who led the discussion in June

[62] For several years after 1870 they reported only the results of the voting, thereby eliminating those who were nominated but received no votes.

[63] The best source of information was the *Revue Scientifique* (known as the *Revue des Cours Scientifiques* until 1871) which later admitted that its own editorial policy had been protransformist for several years.

[64] Reported by Emile Alglave, *Revue des Cours Scientifiques* 7 (1870): 513.

[65] Ibid.

[66] "Darwin before the French Academy," *Nature* 2 (1870): 298.

were prominent again in July and August. Milne-Edwards and de Quatrefages, the chief advocates of Darwin's election, enjoyed great prestige
in the Academy of Sciences, but neither accepted Darwin's theories.
Emile Blanchard and Charles Robin were the main opposition.[67]

Emile Blanchard, professor of zoology at the Museum of Natural History and at the Ecole Normale Supérieure, professor-administrator at the
museum from 1862 to 1895, and president of the academy in 1883, argued that Darwin was only an intelligent amateur, not really a savant,
and lacked the true scientific spirit. To honor such a man by election to
the academy would be a mistake. Darwin had claimed that man descended from the apes, and this sort of opinion should disqualify him.
Most important, Darwin had risen to fame mainly on the theory of transformation and not only was this a false theory, but Darwin was not even
its originator.[68] Charles Robin, as a positivist, argued that Darwin should
be evaluated on the basis of the demonstrable facts he had added to science; in this regard Darwin was not sufficiently outstanding.[69] Other
comments reported from the nominating committee were similar to
Robin's arguments.

The second 1870 election was first postponed and then, because of the
war, canceled. By 1872, Lacaze-Duthiers had replaced F.-Achille Longet
in the Section of Anatomy and Zoology, and J.-J. Coste was seriously
ill and unable to vote. This made Lacaze-Duthiers, Milne-Edwards, and
de Quatrefages a majority of three out of five in the section voting to
give Darwin a first-line nomination. Apparently Darwin's supporters
thought they had the necessary votes in the entire academy as well.[70]
Three long meetings in Comité Secret were devoted to the 1872 nomination, with Milne-Edwards, de Quatrefages, and Blanchard again being

[67] Also included among the opponents were François-Achile Longet (of the Section
of Anatomy and Zoology), Elie de Beaumont (permanent secretary), and Adolphe-
Théodore Brongniart (a prominent member of the Botanical Section).

[68] Reported by Emile Alglave, Revue des Cours Scientifiques 7 (1870): 563–564.

[69] Ibid., p. 577.

[70] Revue Scientifique, 2d ser. 2 (1872): 645. Lacaze-Duthiers confirmed this. Though
he declined to betray the secrecy of the academy, he certainly implied that some members had gone back on promises of support for Darwin. He said he would not divulge
"the considerations which have caused the failings and the defections that have had to
be recorded with no less pain than astonishment" (Lacaze-Duthiers, "Une Election à
l'Académie des Sciences," Archives de Zoologie Expérimentale et Générale 1 [1872]:
xlix–l).

the main speakers. But, to the surprise of most, Darwin received only fifteen votes in comparison to the thirty-two cast for Loven, while one ballot was left blank. Although Lacaze-Duthiers tried to gloss over the vote by indicating that the number of academicians was "far from being complete,"[71] the total of forty-eight votes was higher than the thirty-eight cast in Darwin's first defeat of 1870 or the thirty-nine cast in his eventual election in 1878. Nonetheless, Darwin received one less vote than he had previously. The rejection was even more serious than the votes indicated, for it was alleged that, even of those who voted for him, not one really "shared his philosophical doctrines."[72] An analysis of the positions defended by the known supporters of Darwin's election would seem basically to substantiate this claim.

In 1873 there were four vacancies among corresponding members of the Section of Anatomy and Zoology. Darwin received no votes in the first election, and, though he was nominated to the other three positions, in each case he received fewer votes than in the 1872 election.

The American Darwinist Asa Gray was elected correspondent of the Botanical Section on July 29, 1878. Although the *Comptes Rendus*, departing from standard practice, made no record of it, Darwin received six votes in this election.[73] On August 5, 1878, Darwin was elected to the Botanical Section, having received twenty-six out of thirty-nine votes. With four other candidates receiving between one and four votes each and five ballots left blank, continuing protest seems evident. Moreover, the preponderance of the vote for Darwin suggests prearrangement. *Le Moniteur Scientifique*, which appears to have made no comment on any of the previous elections in which Darwin's name figured, simply expressed what many must have felt: "M. Ch. Darwin, having received the absolute majority of votes, is proclaimed elected. Finally!"[74]

---

[71] Lacaze-Duthiers, "Une Election à l'Académie des Sciences," p. xlix.

[72] Hyacinthe Charles de Valroger, *La Genèse des espèces: Etudes philosophiques et religieuses sur l'histoire naturelle et les naturalistes contemporains* (Paris: Didier et Cie, 1873), pp. 380–381.

[73] "Bulletin des sociétés savantes," *Le Revue Scientifique*, 2d ser. 15 (1878): 141. One may suspect that something unusual surrounded this election, but no explanation has been discovered for it.

[74] *Le Moniteur Scientifique* 20 (1878): 1027. While Darwin had been nominated six times for the Section of Anatomy and Zoology, his two nominations and his election in 1878 were to the Botanical Section. Although this study found no contemporary

Neglect of Darwin in the Academy of Sciences is perhaps appropriately symbolized by the fact that more space was devoted in the *Comptes Rendus* to his death than had been devoted directly to his theories in all the previous years. His eulogies were longer than those of any but the greatest French men of science, possibly a tribute in itself. In announcing Darwin's death at the May 1, 1882, meeting of the academy, the president said that the academy should pay its regrets to "such a sagacious observer, and one who was an honor for it. . . . His works of natural history have been welcomed in the entire world with a marked and merited favor. His ideas of natural philosophy have raised impassioned controversies. They have not kept his compatriots from bestowing upon him, at Westminster Abbey, the honors of a tomb reserved for great citizens; they ought not to keep the Academy from rendering a magnificent justice to his rare merit."[75]

De Quatrefages, called upon to offer more extensive comments, said essentially the same thing he had said so many times: "There are two men in Charles Darwin: a naturalist, observer, experimenter as the case may be, and a theoretical thinker. The naturalist is exact, sagacious, patient; the thinker is original and penetrating, often just, often also too daring."[76] In addition, he claimed an unconscious wisdom of the academy. It had at first rejected Darwin's candidacy because he was supported by too many people on the basis of his theories, which were wrong and unproven. Then the academy, by electing Darwin, had shown that it knew how to recognize the valuable and the permanent in the illustrious English naturalist's works and "to render justice to his true merits. It has thus fulfilled at every point its duties as scientific tribunal

---

evidence to explain the change, two possibilities may be suggested. J.-J. Coste, whose absence in 1872 allowed a three-to-two majority for first-line nomination for Darwin within the Section of Anatomy and Zoology, died in 1873 and was replaced by François-Louis Paul Gervais in 1874. This may have meant that there was no majority in the appropriate section to give Darwin the nomination. Or, it may be suspected that the academy was eager to avoid giving Darwin any credit for his speculations; they were willing to honor him only for his empirical observations—his "real" contributions to science. They had examined his credentials for the Section of Anatomy and Zoology and had found him wanting. It could save face, then, if they noted his botanical observations and elected him on that new basis.

[75] *Académie des Sciences: Comptes Rendus* 94 (1882): 1215.
[76] Ibid., p. 1216.

with a high impartiality."[77] The atmosphere in which such claims could be made and heard without self-conscious feelings of hypocrisy was indeed unusual.

A revealing footnote to the academy's final reception of Darwin may be reflected in an appeal in December, 1882, for funds for the erection of a monument to Darwin in England. De Quatrefages, speaking for the fund-raising committee, felt it necessary to assure the academy just what contributions did and did not mean: "It is almost unnecessary to add that in answering the appeal of the English savants, the French committee means to remain absolutely outside of all appreciation of the general scientific or philosophical ideas of the departed illustrious one. These homages are addressed uniquely to the man who consecrated his entire life to scientific work, who approached with felicity several of the most arduous problems that the study of living beings presents and who, by the entirely special direction of his researches and the success which often crowned them, has rendered some magnificent services to positive science."[78] Such caution and delimitation of meanings suggest how far the Academy of Sciences still was from favoring Darwin's ideas. In the academy, at least, transformist theories seemed to attract as little support as Darwinism.

When the shock of the Franco-Prussian War of 1870–1871 convinced many that French science badly needed rejuvenation, the French Association for the Advancement of Science was founded in an attempt to popularize, broaden, and decentralize the French scientific establishment. To the degree that the French Association broke from tradition, it might have been more free and might have offered more positive discussion of Darwin and transformism. The potentialities for Darwinian discussion may have been present, but little took place. Only ten articles dealing

---

[77] Ibid., p. 1222.

[78] "Hommage à Darwin," *La Revue Scientifique*, 3d ser. 4 (1882): 829. Georges Pasteur of the University of Montpellier has helpfully suggested that the Academy of Sciences possibly elected Darwin because of popular pressures in the sense that, while scientists scorned him in France, the educated general public was more positively inclined toward transformism and considered Darwin's rejection a national disgrace. Several hints in the contemporary literature could lend support for such an interpretation.

directly with transformism appeared in the association's publications in
its first ten years, and none of these was explicitly and primarily concerned with Darwinism. Rather, they took up specific factors in the evolutionary question—especially the descent of man. It might have been
somewhat encouraging to pro-Darwinists that only two of these ten articles were directly opposed to transformism, while two firmly argued
for natural selection as an important cause of evolutionary development.

Yet relative silence on the broad issues of transformism was still the
rule, despite many protransformists among the membership and among
the speakers and commentators at association meetings. In 1873, for instance, Gaudry, Gaston de Saporta, Vogt, Charles Martins, Paul Broca,
Paul Topinard, and Royer—all transformists—presented papers, but not
one commented on Darwin. Furthermore, at least half the papers presented in favor of transformism or dealing favorably with the topic were
presented by members of the Society of Anthropology, the one French
scientific society in which transformism seemed to be a central issue and
an accepted idea.

The Society of Anthropology of Paris, founded in 1859 by Paul Broca
and others,[79] seems not to have discussed transformism until 1864, primarily following a report in its bulletin of the contents of the *Anthropological Review of London*, including Wallace's paper, "The Origin of
Human Races and the Antiquity of Man Deduced from the Theory of
'Natural Selection.' " Darwin was no more than mentioned. By late 1867,
leading members were commenting that it was becoming increasingly
difficult to deal with facts without having a transformist or antitransformist theory presented as "evidence."[80]

Unless accounts of the society's proceedings are quite misleading,
Clémence Royer was not strictly correct in her comment in the preface to
the second edition of *De l'origine des espèces* in 1866 that Darwin's
theory reigned "without contest" among the most influential and com-

---

[79] The circumstances of its founding were perhaps conducive to making it most open
to transformist arguments. A dispute over the species question and the freedom of
science from theological presuppositions really provided the impetus for the new society.
For details, see Robert E. Stebbins, "French Reactions to Darwin, 1859–1882" (Ph.D.
dissertation, University of Minnesota, 1965), pp. 181–186.

[80] Paul Broca, Discussion of "Mesures recueillies sur les individus des races diverses
par les Docteurs Scherzer et Schwartz discutées par le Docteur Weisbach," *Bulletins de
la Société d'Anthropologie*, 2d ser. 2 (1867): 639–640.

petent members of the society.[81] The records of the bulletin seem to suggest that the major discussion began after a paper read before the society at its July 2, 1868, meeting by J. P. Durand de Gros, who concluded: "Thus honor, honor to the creators of philosophical anatomy! Honor also to the great names of Lamarck and Darwin."[82] On November 19, 1868, Eugène Dally presented the first address whose title mentioned the word *transformisme*. It was mainly a reaction to Huxley's *Man's Place in Nature*, but it presented the whole development from monad to man. During the remainder of 1868, 1869, and 1870, the society's discussions concentrated on transformism and frequently mentioned Darwin.

But, while they mentioned Darwin, the members were not Darwinists. Their initial discussion in 1864 was based on an article in the *Anthropological Review of London*;[83] Durand de Gros carefully mentioned Lamarck's name before Darwin's; Eugène Dally was directly responding to Huxley and also said: "It is important to separate the cause of transformism from that of Darwinism. Without misunderstanding the considerable services rendered to science by the illustrious author of the *Origin of Species*, one can say that his views are only one of the explications that can support the undeniable facts of variation."[84] Dally put it more succinctly later when he said that Darwinism was "only an adventitious branch" of transformism.[85] When Paul Broca, the founder and leading member of the society, finally entered the discussion with his own fully developed statement in April, 1870, he declared that transformism was a fact that far transcended any particular presentation of it; Darwinian hypotheses were valuable but not all correct and certainly not all original or final. Natural selection was not as significant as many people had claimed. Clémence Royer, in contrast to all the other members who discussed transformism, drew many of her arguments and examples

[81] Clémence Royer, "Avant-Propos," in her translation, *De l'origine des espèces* (1866), p. ii.

[82] J. P. Durand de Gros, "La Torsion de l'humérus et les origines animales de l'homme," *Bulletins de la Société d'Anthropologie*, 2d ser. 3 (1868): 537.

[83] For an interesting argument that the London Anthropological Society itself had initiated its own evolutionary discussions before receiving any influence from Darwin, see J. W. Burrow, "Evolution and Anthropology in the 1860's: The Anthropological Society in London, 1863–71," *Victorian Studies* 7 (1963): 137–154.

[84] *Bulletins de la Société d'Anthropologie*, 2d ser. 3 (1868): 675.

[85] Eugène Dally, "Discussion sur le transformisme," *Bulletins de la Société d'Anthropologie*, 2d ser. 5 (1870): 149.

from Darwin's works. She also criticized him, but she cited Darwin on every page and gave him far more attention than she gave to any other source of biological data. Nevertheless, her argument was for transformism alone, using Darwin's information and ideas as tools.

Darwin was elected as a foreign associate of the Society of Anthropology on December 7, 1871, without difficulty or fanfare. A majority of the society was apparently protransformist. The issues that were important had been debated and decided, and from that time on the society settled back to do its work, utilizing transformist assumptions as general presuppositions. The mechanism of transformism—which was Darwin's main contribution but the very aspect of his work that the society rejected or accepted only in part—was only a secondary matter for them. They could leave this to the biologists. The essential factors of change for their discipline had already been solved.

Some of the society's leaders later organized an annual Conférence Transformiste, which was really an annual meeting of the society "consecrated exclusively to demonstrating that, without the mutability of species, knowledge of natural history is impossible."[86] The society was thus at least annually reminded of the importance of the transformist question.

The anthropologists actively promoted transformism in many ways. They were instrumental in encouraging the Municipal Council of Paris to establish in 1888 the first academic chair in France specifically devoted to the teaching of evolution.[87] They established the journal *Matériaux pour l'Histoire Positive et Philosophique de l'Homme*, often carrying a section specifically on *transformisme*. Paul Broca's *Revue d'Anthropologie*, founded in 1872, also advanced the cause of transformism. Furthermore, many of the articles in the general press and books favorable to transformism and Darwin were written by members of the society. The Society of Anthropology's exhibits at the expositions of 1868 and, especially, 1878 presented the story of man in evolutionary terms, much to the disgust of some.

At Darwin's death, the society's bulletin offered its profound respects

86 P. G. Mahoudeau, "Transformisme," in *Dictionnaire des sciences anthropologiques* (1889), p. 1070.
87 Ibid.

and gave great credit to Darwin, whose doctrines counted in the society
"numerous partisans, and, at the least, admirers." It is significant that,
with all the high praise, the death announcement said:

Darwin was the continuator of a grand tradition; he applied his genius to
demonstrating a theory that Diderot had presented and announced, and that
Lamarck had developed scientifically. Darwin knew how to make the marvelous discoveries of the savant [Cuvier], who with the most ferocity and
authority denied the transformism of Lamarck and of Geoffroy Saint-Hilaire,
serve his own demonstration. In creating paleontology, Cuvier, the defender
of the fixity of species, has furnished valuable arguments to the transformist,
Darwin.

The thought of this savant has stirred the civilized world, and the impul-
sion has been so powerful, that humanity still shakes from it.[88]

Humanity might still shake, but the Society of Anthropology's opinion
was revealed by the fact that Darwin was called a "continuator of a
grand tradition" that was essentially French and had provided various
explanations for the basic fact of transformation. Interpreting Darwin's
success where Lamarck and Geoffroy Saint-Hilaire had failed as being
based heavily upon the discoveries set in motion by Cuvier could again
only underline the conception that French science was behind what was
of value in Darwin.

Although Clémence Royer contended that the Society of Geology had
likewise been won over to transformism, it is difficult to ascertain this
from its bulletins. The Société Géologique de France had chosen Darwin
as a life member in 1837; yet its bulletin hardly paid attention to him
after 1859. Except for three books presented to the society with very minor transformist comment, evolution and Darwin were apparently
avoided.

The thought that perhaps some of the provincial societies might have
been more favorable to Darwin does not seem to be substantiated by their
publications. If anything, it would seem that the provincial societies and
academies discussed Darwin less, offered less information on the theories
they were combatting, and were more consistently and flatly opposed to
the views than were their Parisian counterparts. Silence was the general

[88] "Mort de Ch. Darwin," *Bulletins de la Société d'Anthropologie*, 3d ser. 5 (1882):
348.

rule, only infrequently broken and then with a heavy preponderance of opposition over favor.[89]

The dominant tone of French science insisting that transformism was not empirically proven provided a strong and comfortable position for the Roman Catholics. Their basic approach was simply to assert that the church would see no problems and had nothing to worry about as long as the leading scientists had all said that transformism was mere speculation, not at all proven. Virtually all the explicitly Catholic presentations during at least the first fifteen years after the appearance of the *Origin of Species* repeated this over and over. Granted, many unthinking people refused to be guided by the wisdom of the leading scientists and made rash and irresponsible assumptions and conclusions from unfounded theories. The resulting irresponsible writings were cause for concern and were attacked. But their refutation was on the philosophical and theological, rather than the scientific, plane. Catholicism had no need to attack true science as found in the Academy of Sciences.

The limited sampling of Protestant thought discovered in this study seems to have been more open to theories of progress and evolution, but of course did not accept the negative religious conclusions of the materialistic transformists. Protestants were no more inclined than Catholics to believe that Darwin had provided final answers.[90]

[89] Stebbins, "French Reactions to Darwin," pp. 240–246. A tempting refutation of this picture is seen in a letter from Hugh Falconer to Darwin on November 3, 1864. Falconer reported that Gaspard-August Brullé, professor of zoology and comparative anatomy at Dijon, had told him "in despair" that he could not get his students "to listen to anything from him except à la Darwin . . . all young Frenchmen would hear or believe nothing else" (*More Letters of Charles Darwin*, ed. Francis Darwin and A. C. Seward [New York: D. Appleton and Company, 1903], I, 256–257).

[90] See a leading Protestant writer, Edmond de Pressensé, *A Study of Origins; or, the Problems of Knowledge, of Being, and of Duty* (New York: James Pott and Co., 1882). While distinguishing between Darwinism as "a simple theory of natural history, and transformism as a materialistic explanation of the origin of things" and saying that "so far from being opposed to a purposive cause, Darwinian biology implies it," Pressensé added: "with the reservation that it still lacks scientific demonstration" (pp. 180–181). Nevertheless, Pressensé's overall-attitude made him strongly favor Wallace over Darwin, more than once calling Wallace the precursor of Darwin. He vehemently denied the original formation from the nonliving. The natural origin of instincts and the animal origins of man he likewise considered impossible. Pressensé seemed to believe that Charles Naudin offered the best biological picture of evolution, Paul Janet

Until around 1880, when transformist thought seemed to be gaining more ground and more public recognition as possibly being true, Roman Catholic thought in France was far more concerned about geological questions of the age of the earth as conceived in science, as opposed to the picture presented in the Biblical narrative, than they were about Darwin. Genesis and geology bothered the Biblical scholars; the dating of earliest man disturbed them; but the unproven idea of transformation of species seemed to be of little immediate concern. A calculated silence was sufficient defense.

A most startling breaking of the silence came in 1877 with the popular Paris physician Constantin James's *Du Darwinisme, ou l'homme-singe* (editions in 1882 and 1892 bore a slightly different title). James presented a highly prejudicial and popularized book attacking Darwin and the transformists for all the evils of the day. By far the most scurrilous attack discovered in any French book of the period, it was also the only one printing within its covers a letter of profuse approbation from Pope Pius IX.[91]

Religiously inclined scientists usually professed to make their scientific judgments empirically and without subjection to religious prejudgments. They were frequently accused of admitting religious biases, just as Pasteur was accused in his studies of spontaneous generation, but they denied the accusations. Were they not just being careful and empirical, as the positivists were? With the positivists taking the firm stand that transformism could not be accepted without definite proof that one species did change into another, the religionists could reasonably indicate that persons could hold such a view entirely independent of dogmatic considerations. Catholic scientists did reject transformism, but they did it as scientists accepting the general scientific climate of opinion and not—ostensibly at least—in the name of their religion. The nonscientist opponents of transformism merely followed the considered judgments of the French scientific community.

---

the best philosophical context, and above all he lauded Claude Bernard's methodology. Spencer and Haeckel were the real problems, while Darwin remained in a relatively insignificant intermediate position.

[91] Constantin James, *Moïse et Darwin: L'Homme de la Genèse comparé à l'homme-singe, ou l'enseignement religieux opposé à l'enseignement athée* (Lille: Desclée, de Brouver et Cie, 1892), pp. 340–341.

Whatever evaluations might be placed on the scientific value of Darwin's theories, however, Clémence Royer spoke for many in emphasizing their social implications. Many who, like Royer, did not venture to defend the whole Darwinian position scientifically at least found the new movement stemming from the discussions a fertile field for their own thought and argument. The Darwinian impetus was far more acceptable and effective in France in nonscientific areas than in science. This fact was asserted *ad nauseum* by those who sought to disparage Darwin and transformism; it was also admitted by those seeking to utilize the new intellectual stimulus potentially provided by Darwin's works.

For both the proreligious and the antireligious positions, the Darwinian works predominantly fit into a long-standing French encounter usually characterized as the materialist-spiritualist issue. The framework was not provided by Darwin, and the discussion proceeded along existing French lines rather than taking new directions from Darwin. Haeckel, Herbert Spencer, and Vogt were more prominently utilized and directly attacked in this regard than Darwin, for their philosophical conclusions were more exciting to the materialists and more disagreeable to the spiritualists.

The French shared the idea of progress with other Europeans in the nineteenth century. The concept was certainly conducive to transformist views, and transformism fit in well with the general idea of progress. For instance, Charles Dollfus, Protestant editor of the *Revue Germanique*, proclaimed his views in 1863:

As for progress itself, I do not see that it is really possible to deny it. The study of nature imposes it on us. It is incontestable in the history of humanity, and it is incontestable in the history of individuals; it governs physiological evolution from individual embryonic forms to the definite development of life and the living organism. . . . Everything attests in its turn to the ascending forms. But nature is one: progress cannot be here, if it is not there. It does not reign anywhere if it does not reign everywhere. It is one law; the law and the form of movement—which is universal.[92]

The idea of progress and the wider ramifications of evolution were

92 Charles Dollfus, "Essai sur le XIXᵉ siècle: La Crise religieuse," *Revue Germanique* 27 (1863): 406.

well illustrated in the works of three well-known writers: Edmond About, Ernest Renan, and Edgar Quinet.

Edmond About, whose writing was frequently pleasing to Louis Napoleon, wrote *Le Progrès* in 1864. Although About did not mention the English scientist, at least one writer thought the book identified About as a disciple of Darwin.[93] About introduced his general topic with biological references, especially the descent of man from lower forms. Considering the specific mechanism of evolution as an indifferent matter, About did not need Darwin or any other biologist to show him the general truth of progress. But he did feel that his introduction deserved a biological base and could be more comprehensive and justifiable if it took the whole realm of evolution into account.

Ernest Renan commented, in his famous letter to the organic chemist Marcelin Berthelot in 1863: "It may be that Darwin's hypotheses on the subject can be judged to be insufficient or inexact; but undoubtedly they are on the road to the great explication of the world and of true philosophy."[94] More than one writer commented that Renan had taken Darwin's views on biological evolution and applied them to the cosmos as a whole.[95] It would be more accurate to say that Renan already had the ideas and merely used Darwin where it was convenient. Renan's collected works certainly do not reveal any great attention to Darwinism per se. Renan was an evolutionist; he was trying to supplant the traditional world view with an alternative system; but Darwin was only a minor contributor to his whole argument.

Similarly, Edgar Quinet explicated his beliefs in a general evolution, of which transformism would be a very important part. His two-volume work, *La Création* (1870), sprang from a philosophic faith that everything was related. Transformation of species was a necessary piece in the whole picture puzzle of the universe. The world, life, and history could be viewed as a whole; geological developments, spontaneous generation of life, evolution of life forms up to and including man, and the general development of human history were all worked together into a unity in which human history might suggest certain analogies for previous devel-

---

93 F. Herincq, "Causerie scientifique," *Le Monde Illustré* 14 (1864): 318.

94 *Revue des Deux Mondes*, 8th ser. 47 (1863): 765.

95 E.g., Félix Ravaisson, *La Philosophie en France au XIXe siècle*, 4th ed. (Paris: Hachette, 1895), pp. 107–108 (first published, 1867).

opments, and vice versa. Admittedly, his whole system lacked proof, but this did not concern Quinet. He agreed with most of Darwin's ideas in the narrower biological sense and with the Darwinists in the larger implications of Darwinian theories; yet his two volumes had only four footnotes to Darwin's works. When he gave general credit, he lumped Darwin's name with those of the two Geoffroy Saint-Hilaires, Lyell, Pictet, Alphonse de Candolle, and Oswald Heer without giving Darwin any special prominence.[96] Quinet readily acknowledged that the origins and development of life were the greatest questions of the day and predicted that they would be explained minutely in due time, but he certainly revealed no sense of Darwin's having basically changed anything. Quinet did not attempt to prove transformism, nor did he seem to need Darwin.

Both Renan and Quinet would have denied that they were pure materialists, but they sought to obviate the usual spiritualist interpretation of the universe and life; so they were branded as materialists by their critics. They and the other transformists sought to explain the cosmos, all life, and humanity in general in natural, rather than in supernatural, terms. Without saying so, they were like the astronomer Pierre Laplace, saying, "Sire, I have no need of that hypothesis." They belonged to a long French tradition; they contributed to an old French quarrel.

Countless other literary, social-scientific, and philosophical works were related to evolutionary discussion, but in most cases their direct relationship to Darwinism was distant. The twenty-two volumes of the Goncourts' journals, for instance, contain two brief and insignificant references to Darwin. In his second edition of *Essai sur l'inégalité des races humaines* (1884), Joseph Arthur de Gobineau mentioned that Darwin's ideas on selection "were drawn from" the first de Gobineau edition. Léon Daudet's play *La Lutte pour la vie* (1889) could be viewed as a popularized argument against Social Darwinism, but it did not attack Darwin or any specific Darwinian influence in France. Théodule Ribot, the founder of experimental psychology in France, seemed quietly to assume transformism as the foundation of his work, mentioned Darwin fairly often, but still said that transformism was not finally proven. The Académie des Sciences Morales et Politiques said about as little directly on Darwinism as the Academy of Sciences itself. Paul Janet, perhaps

96 Edgar Quinet, *La Création* (Paris: Librairie Internationale, 1870), I, 53.

most frequently cited as a philosopher dealing with evolution before 1890, saw no strong objection to Charles Naudin's evolutionary views or to any other position as long as final causes were clearly admitted. Henri Bergson's *Creative Evolution* was hardly an antievolutionary document, but it was not Darwinist. Evolution, even when accepted, was usually not a Darwinian evolution.

The reception given to Darwin did not radically change over the twenty-five years following publication of the *Origin of Species*, nor were there any well-marked transitional points. A "reaction pattern" was basically established by 1862, and the variations from it were not especially notable. Lucien Cuenot, reflecting back from 1951 on developments in his younger days, said that when he began his studies at the Sorbonne in 1883 the atmosphere was still full of Cuvier, Henri de Blainville, and the two Geoffroy Saint-Hilaires. He was confident that during the two years he sat through courses and lectures, carefully noting every word, he had never heard either the name of Darwin or the slightest allusion to the phenomenon of evolution. Seldom was anything introduced into the curriculum that might even suggest change of species. It was known that somewhere in the provinces a heretic, Alfred Giard, taught daring and seductive opinions, but it was considered better to disregard him.[97]

Cuenot admitted that things were transformed in the 1880's and 1890's. The findings of this study contradict Cuenot's claims of silence on Darwin around 1883. Nevertheless, it must be granted that silence continued to reign in many areas. For instance, the *Dictionnaire de botanique*, appearing in four volumes from 1876 to 1892, had very little to say on evolution and Darwin. Volume 2, which appeared in 1886, devoted five lines to Erasmus Darwin but had no article on Charles Darwin. A four-line article on "evolution" merely said that that word was used to refer to *transformisme*, to which article the reader was referred. But Volume 4, in which *transformisme* should have been found, was published in 1892 without any article on that subject. The silence in the schools was often profound. Edmond Perrier called for a general up-

[97] Lucien Cuenot, *L'Evolution biologique: Les Faits, les incertitudes* (Paris: Masson et Cie, 1951), pp. v–vi.

dating of the educational system in 1882. He professed not to be able to understand why the schools still taught outmoded ideas on the absolute separation of various classes of species.[98] Georges Pasteur has commented that even in the mid-twentieth century Darwin was not prominent in French schoolbooks. In fact, he said that he could well recall how surprised he was at discovering the many French editions of Darwin's works; he "had been under the impression that Darwin was ignored" in nineteenth-century France.[99]

French transformism changed only slowly, but, nevertheless, the intellectual-scientific stage did not remain stationary. Some new figures entered, and some of the older ones passed from the scene. It was this process of substitution of players that made the greatest change, inasmuch as those who altered their own basic position were few and not easily identified.

The real development, however, was in the increasing numbers of scientists and laymen who accepted the general idea of transformism and the naturalistic explanation of all life processes. This development took place on a broad front, only indirectly influenced by "Darwinism," while drawing its main inspiration from Frenchmen. It was not fundamentally influenced by Lamarck, although many Frenchmen were flattered when Haeckel suggested that transformism should really be called "Lamarckism" after its real founder, and Félix Le Dantec, perhaps the leading French writer on transformism from 1895 to 1910, followed Lamarckian lines. Some say the same for Giard and Perrier. Lamarck was frequently mentioned, but only two editions of any of his works appeared between 1846 and 1906. Although the Darwinian centennial year in 1959 was not totally ignored in France, it was at least prominently noted that this was the sesquicentennial year of Lamarck's *Philosophie zoologique*. National honors were properly bestowed upon Lamarck.

By 1900, the transformist position was clearly predominant in France, as it was in most other countries. Some residual battles were fought, but it has been said that the last biological work to speak of transformism as an illusion was produced by Louis Vialleton in 1929.[100] Maurice

98 See "Les Programmes relatifs à l'enseignement de la zoologie dans les lycées," *La Revue Scientifique* 29 (1882): 722–725.

99 Georges Pasteur, University of Montpellier, letter to author, March 2, 1972.

100 However, religious polemics still insist that transformism is not yet proven.

Caullery seems to speak for a great deal of continuing French opinion when he emphasizes, in *Le Problème de l'évolution* (1931), that transformism is a fact; only its explanation remains uncertain.

The French still receive Darwin as one who offered one, but not the only, explanation for the idea of transformism, of which they more than the English were the originators. With a different sense of proportion than the Anglo-Saxons, they still perceive less sense of high drama in Darwin and Darwinism than do their foreign counterparts.

There was no "Darwinian Revolution" in France, just as there were no political revolutions in England in 1789, 1830, 1848, and 1871. Yet England experienced profound social, economic, and political changes from the eighteenth to the twentieth centuries, not always totally without French influence. And France accepted transformism, perhaps not totally without influence from across the Channel. Darwin was just one more factor in the evolution, rather than the revolution, of transformist ideas in France.

FRANCE
Bibliographical Essay

ROBERT E. STEBBINS

In England, Darwin was criticized for speaking of "my Theory" without adequate distinction between his own contribution of the concept of natural selection and the more general theory of evolution. Such criticism would have been unnecessary in France, for French readers and audiences made the distinction quite clearly for themselves; there was no confusion, whatever Darwin had said. Evolution-transformism was not new and certainly was not Darwin's theory. Natural selection was not a significant break-through in understanding transformism.

Most French literature on transformism, then, was not concerned with Darwinism. Darwin was only one of many foreigners to be considered in a question that was basically French in inspiration. Since Darwinism was not the focus of attention, research on the French reception of Darwin must often concentrate on incidental references in general discussions on transformism.

Less than a dozen books directly on Darwinism appear to have been published in France before 1885. Almost all of them were negative. The books published from 1885 to the present do not seem to be much more favorable toward Darwin or to treat natural selection much more positively.

Although inferior as a scientific or intellectual product, Pierre-Marie-Jean Flourens's *Examen du livre de M. Darwin* must be considered important because Flourens was permanent secretary of the Academy of Sciences and because the book was cited frequently. Armand de Quatrefages's books and articles are perhaps the most valuable for seeing the arguments of an insightful critic. His *Darwin et ses précurseurs français* and *Les Emules de Darwin* together provide perhaps the best contemporary French account of Darwin's theories. A series of lectures at the Ecole d'Anthropologie by Mathias Duval in 1883–1884 were published in a book, *Le Darwinisme*, in 1886. In over five hundred pages, Duval gave a good coverage of transformism in general, the history of the ideas in France, and Darwin's contributions and limitations. He summarized much of the information that can otherwise be found only in many scattered sources. *Le Darwinisme*, written from an anthropologist's viewpoint, thereby illustrated the views of the group of scholars most favorable to transformism, while at the same time showing Duval's appreciation of Darwin's contributions to anthropology.[1]

Félix Alexandre Le Dantec's *La Crise du transformisme* can very well illustrate the French scene of the early twentieth century: increasing acceptance but continuing opposition to transformism, criticism of Darwin, favor toward Lamarck. Jean Rostand's *L'Evolution des espèces* and *La Genèse de la vie* together offer good coverage on the transformist and spontaneous-generation questions, a combination essential for understanding the French reception of Darwin.[2]

The *Bulletins* and *Memoirs* of the scientific and learned societies provide an excellent resource for tentatively confirming the "conspiracy of silence" thesis offered by T. H. Huxley. An amazing number of these publications of a broad spectrum of French learned and scientific societies are available in American libraries. The climate of French scientific opinion was such that a reader, simply reviewing the learned journals, might underestimate the transformist discussion in France. When scientific and intellectual respectability requires that certain ideas not be seriously entertained or at least not be mentioned in a favorable light, there is danger that the discussion may never get to the floor of scientific meetings or that if it gets to the floor it may not be considered worthy of publication at a later time. Carl Jung's reflections on his own early career and

[1] Pierre-Marie-Jean Flourens, *Examen du livre de M. Darwin sur l'origine des espèces* (Paris: Garnier frères, 1864); Armand de Quatrefages, *Darwin et ses précurseurs français* (Paris, 1870, 1892); idem, *Les Emules de Darwin* (Paris: F. Alcan, 1894); Mathias Duval, *Le Darwinisme* (Paris: A. Delahaye et E. Lecrosnier, 1886).

[2] Félix Alexandre Le Dantec, *La Crise du transformisme* (Paris: F. Alcan, 1909); Jean Rostand, *L'Evolution des espèces: Histoire des idées transformistes* (Paris: Hachette, 1932); idem, *La Genèse de la vie: Histoire des idées sur la génération spontanée* (Paris, 1943).

the beginnings of Freudianism suggest the problem. Jung said that, while the papers read at scientific conventions were on more acceptable topics, the corridors were buzzing with excited discussion on Freud's views. If there was any analogous excitement over Darwinism in France, it must have been reserved for the recesses of the back stairs. The silence in the journals seems to reflect pretty accurately the scientific climate of opinion. The researcher in French journals and magazines is handicapped by not having any guidance from a general index. However, almost all periodicals had rather extensive topical indexes covering each volume; these can be quite helpful.

The general periodical press discussed Darwin more than the learned journals did. Periodicals of the day reflected the emphasis of the leading French scientists, insisting that there was no proof for transformism and that Darwin had not offered an adequate explanation, but they did discuss the questions. Often, if not usually, transformism was discussed mainly in its relation to philosophy and social thought.

Much of the upper- and middle-level press of France followed the lead of the French scientists in denying that any crucial change of interpretation and explanation had been offered. Those that saw reason for excitement were, like Clémence Royer, often more interested in philosophical speculations than in the scientific questions. The *Revue des Deux Mondes*, of the nonscientific periodicals, offered the greatest number of extensive arguments on the transformist issue and the most frequent mention of Darwin. The editorial policy shifted more and more toward favoring transformism, though never supporting Darwinism per se. The *Revue Scientifique* (*Revue des Cours Scientifiques* from its founding in 1863 to 1871) was probably the most widely read general scientific periodical and the best source of information on science and scientific activities in France. Though not Darwinist, the *Revue Scientifique* offered the most articles on transformism and by 1880 publicly admitted that it was transformist.

The French popular press is generally not available in American libraries and has not been canvassed in this study. Many French antitransformists, afraid of transformism for its religious and philosophical implications, asserted that the masses and the radicals were entirely too willing to accept Darwin's ideas and all ideas of transformation of species. A thorough study should be made of the popular and low-level periodical literature and the pamphlet broadsides published in the two decades following the publication of the *Origin of Species*. What is known of the circulation figures of most of the periodicals used in this study suggests that the "most important" periodicals reached relatively few readers in comparison with the more popular periodicals. What may have lurked beneath the surface may be suggested by two mass-circulation "picture" magazines, *Le Monde Illustré* and *L'Illustration*. These magazines had said almost nothing about Darwin previously, but, when Darwin died, *L'Illustration*'s cover page was devoted to a picture of Darwin, and *Le Monde Illustré*'s lengthy article on Darwin featured a large picture. Neither provided negative commentary in its obituary. They expected their readers to be familiar with transformism, though their magazines had hardly mentioned the subject.

Whatever sources may be utilized to illustrate the French reception of Darwin in a positive way, none can be as impressive as the countless books and articles where silence alone stands testimony to the French intellectual developments. If a single author must be selected to illustrate this silence, Jean Henri Casimir Fabre (1823–1915) should be chosen. Fabre published an incredible number of books, dealing with many different sciences. Seventeen columns in the *Bibliothèque Nationale Catalogue* are devoted to Fabre's works. Many of Fabre's books were textbooks which were widely used in French schools. Fabre corresponded with Darwin, but he never accepted Darwin's theories. He remained interested only in empirical work and minute descriptions, and he never mentioned evolutionary theories as even a possible explanation for the oddities of nature that he loved to note. Neither a defender nor an opponent of Darwin, he pursued his researches and writing apparently undisturbed and unenlightened by the "Darwinian Revolution." He likewise had no interest in disturbing the school children of France with the Darwinian controversies. He reveled in the wonders and mysteries of nature, and even an unequivocally successful solution to the question of the sphinx would have tarnished rather than brightened his whole scientific outlook and work. Silence may be harder to document than the trumpet fanfare, but in its own way it is equally impressive.

UNITED STATES

EDWARD J. PFEIFER

The most obvious fact about the American reception of Darwinism, which must still be mentioned, is that Americans could read Darwin's books in his own language and as they appeared. They thereby escaped the accidents of translation and were less subject to the misinterpretations of commentators than readers in non-English-speaking countries. None of this means that all Americans had a clear understanding of Darwin's work from a reading of his books at the moment of publication. Nor does it mean that they regarded Darwinism as a purely scientific question. Americans were certainly influenced by factors beyond Darwin's control. Of these, the struggle over slavery was very important, as were the maturing of an American scientific community and the development of American institutions after the Civil War.

In the response to Darwin, Americans were also influenced by the fact that they were playing the same role that they had played since colonial times and through the first half of the nineteenth century. It was only about twenty years before the *Origin of Species* appeared that Emerson declared that our cultural vassalage to Europe was ending. By 1859 the United States had indeed produced writers and scientists of international stature. Yet through colonial times American laborers in the field had supported the great European scientists. John Bartram won fame from the praise of Carolus Linnaeus, and Isaac Newton welcomed Thomas Brattle's observations. Furthermore, the ability of Americans to assist these scientists and American acceptance of new scientific thought —for example, Galileo's at seventeenth-century Harvard—have been taken by historians as the mark of a sophisticated, but still colonial, culture.

Americans were, however, always in a favorable position toward European science because of the insights that the American environment might yield for scientific theory. Sir Charles Lyell, after visiting the United States in 1841–1842, made the point by stating: "In the course of this short tour, I became convinced that we must turn to the *New World* if we wish to see in perfection the oldest monuments of the earth's history, so far at least as it relates to its earliest inhabitants. Certainly in no other country are these ancient strata developed on a grander scale, or more plentifully charged with fossils."[1] In this situation there was room for both national pride in scientific achievement and an inferiority complex over the lack of it.

While Americans would respond to European thought again in the Darwinian controversy from a position historically established, they had already been influenced against evolutionary theories by ideas transmitted through the same channels of communication. Lyell's condemnation of Lamarck in the *Principles of Geology* did much to discredit evolutionary theories in the United States. But even more important in conditioning Americans toward the *Origin of Species* was the furor over the *Vestiges of the Natural History of Creation*, published anonymously by Robert Chambers in 1844 (American edition, 1845).[2]

Chambers pictured the world and everything in it as the product of natural law. The first stage in his history was the existence in space of a heated nebulous matter, "the universal fire-mist." Next a nucleus formed in the fire-mist and began to rotate. Its spin sucked in other particles and eventually built up a huge incandescent mass, parts of which were thrown into orbit by centrifugal force. One of these satellites was the earth. When the world was ready for life, spontaneous generation occurred. These first living forms then developed into the plants and animals of the nineteenth century.

Chambers thus accounted for the universe without creative act of God. Yet he emphatically declared that physical laws are divine regulations that manifest God's wisdom and goodness. In spite of this profession, critics insisted that the *Vestiges* was atheistic. Invariably they

[1] Sir Charles Lyell, *Travels in North America in the Years 1841–42*, 2 vols. (New York: Wiley and Putnam, 1845), I, 15.

[2] Robert Chambers, *Vestiges of the Natural History of Creation* (New York: Wiley and Putnam, 1845).

identified Chambers's ideas with those of Democritus and the Epicurean philosophers, who made the world a "fortuitous concourse of atoms." Proceeding from this point, they usually made the related charge that Chambers was attempting to exclude God from his own universe by the processes of natural law.

In the United States, the *Vestiges* came under bitter attack. A leader in the warfare against Chambers was Francis Bowen, editor of the *North American Review*. Bowen could not see why Chambers brought God into the *Vestiges* at all. If the fire-mist could form worlds and generate life, its power might as easily be an essential property of matter as a gift of God. "If there is no need of a bricklayer," Bowen decreed, "we may discard also the brick-maker."[3] Underlying this conclusion was the conception of God generally accepted in 1845. The deism of the eighteenth century had faded in the nineteenth before William Paley's *Natural Theology*. By arguing that organic adaptations proved divine intervention, Paley popularized what was later irreverently called "the carpenter theory of creation." Bowen's insistence that God must be a "bricklayer" or no God at all was a clear espousal of Paley. One must understand that by 1859 Paley's theology was deeply rooted in American Protestantism to understand the controversy over Darwinism in the United States.[4]

Scientists were as ready as theological writers to condemn Chambers's views, and they showed no sympathy for his doctrines of evolution. They admitted as generally true his statement that fossils of successive geological periods form a series progressing toward contemporary plants and animals. But, like Georges Cuvier in his debate with Etienne Geoffroy Saint-Hilaire, they cited the absence of transitional forms between fossils of successive strata and on paleontological evidence would not forsake their belief in separate creations. Asa Gray, Darwin's later champion, conceded that fossil ferns appear in the rocks precisely where Chambers's theory would place them but showed that the same rocks contain fossils of leaf-bearing plants.[5] If everything had developed in

[3] Francis Bowen, "A Theory of Creation," *North American Review* 60 (1845): 426–478.

[4] Wendell Glick, "Bishop Paley in America," *New England Quarterly* 27 (1954): 347–354.

[5] Asa Gray, "Explanations of the Vestiges," *North American Review* 62 (1846): 465–506.

direct gradation, how could Chambers explain the higher forms that appeared too soon? Nor were scientists impressed by the fact that embryos in their development appeared to climb the ladder of being from simpler to more complex animals, though Chambers found this relationship the most convincing evidence of evolution. Neither were they swayed by homologies of animal structure, rudimentary organs, or families of six-fingered people, all of which Chambers cited.

Scientists also attacked the process by which Chambers thought that animal evolution proceeded. His hypothesis was that the simplest form of life gave birth to the form next above it until the highest forms appeared. This would mean that, among vertebrates, a fish gave birth to a reptile, a reptile to a bird, and a bird to a mammal, although the type changed so gradually that the more advanced form involved only slight modification. Chambers explained the metamorphosis on embryological grounds. He believed that the embryos of all vertebrates began life as germs and developed through common stages up to a point. If germs of the four classes of vertebrates began life simultaneously, the fish would be the first to depart from this common development, the reptile second, and the bird next. When each embryo left the general path, it then began the growth peculiar to its own class. For a fish to produce a reptile, the fish embryo had only to move past its normal point of divergence and turn off where reptile development commenced. A consequence of this doctrine was the animal origin of man, in which Chambers did believe.

The *Vestiges* contained much that had been offered previously under respectable scientific auspices. Occasionally, however, Chambers credited as science what others were dubious about, as in accepting as true the story that two English experimenters had generated insects by passing electric currents through chemical solutions. His book thus had a naïveté that kept specialists from taking it seriously. T. H. Huxley could not remember having read it when it came out, though he supposed he had, and Joseph Hooker was amused by it. Darwin, whose opinion of the *Vestiges* in 1844 contrasted with his later statement in the preface to the *Origin*, was less amused and thought Chambers's geology bad and "his zoology far worse."[6] Darwin worried too that his own ideas might be-

[6] Francis Darwin, ed., *Life and Letters of Charles Darwin*, 2 vols. (New York: D. Appleton, 1887), I, 302.

come known and incorporated into a book like the *Vestiges* and that he would "have to quote from a work perhaps despised by naturalists."[7] No doubt specialists would pity or despise an author ready to spin out an all-inclusive theory of the universe on inadequate evidence. Certainly, like Darwin, they might label such a book "incautious."

If the *Vestiges* is judged by its acceptance, it must be considered a failure. Unwelcome for theological reasons and attacked by scientists, the book had little chance of making converts. But its very failure had important consequences. The conclusion is unavoidable that the failure of the *Vestiges* prepared critics for the *Origin of Species*. Since only about fifteen years separated the books, some critics reviewed both. It was thus very easy for a reviewer to identify the *Origin* with the *Vestiges* and condemn it on the same grounds. Moreover, the temptation was strong to use in 1860 the weapons of ridicule and denunciation that had proved effective in 1845.

Another aspect of this reception that became meaningful in 1860 was the insistence that Chambers's book was atheistic. In hopes of refuting this charge he even wrote a second book, *Explanations: A Sequel to "Vestiges of the Natural History of Creation,"*[8] but critics would not retreat. They still felt that he had displaced God from the universe. From this, one can see how ready opponents would be to denounce the *Origin* as atheistic, especially since Darwin, unlike Chambers, did not protest a belief in God, but confined his attention to the scientific question.

In the *Explanations* Chambers also replied to the objections drawn from science, and some of his replies were admitted to be able and pertinent. But the *Explanations* was ineffectual. Those who attacked the *Vestiges* by theological argument continued to do so, and scientists were further alienated by Chambers's position in the *Explanations* that the *Vestiges* must be judged by a nonscientific tribunal. The man of science, Chambers said, was too much engrossed in his specialty and too timid to be a competent judge of such questions. When "the awakened and craving mind," Chambers wrote, asks what science can do to explain the purposes of the Creator, man's relation to him, good and evil, life and eternity, "the man of science turns to his collection of shells or butter-

[7] Ibid., p. 478.
[8] Robert Chambers, *Explanations: A Sequel to "Vestiges of the Natural History of Creation"* (New York: Wiley and Putnam, 1846).

flies, to his electrical machine, or his retort, and is mute as a child, who, sporting on the beach, is asked what lands lie beyond the great ocean which stretches before him."[9]

This was not the charge that men of science were accustomed to hearing. Francis Bowen had been in the usual vein when he declared that a fertile imagination and a bold face were among the more striking peculiarities of some geologists. "Grant to one of this character," he continued, "a few modest postulates,—give him certain millions of years, a sufficient number of earthquakes, a whole battery of volcanoes, a few dozen deluges, and the rise and fall of half a dozen continents,—and he will frame a theory off-hand, which will account for the most perplexing phenomena."[10] Bowen concluded the passage by placing Chambers "at the very head of this class of speculatists." Yet, whatever scientists might be charged with, they thought themselves the proper judges of scientific questions, and Chambers's attitude did not dispose them favorably toward his book.

Chambers and Bowen thus ironically agreed in their distrust of scientists. But ten years later the scientists themselves were no more inclined to accept the dictates of outsiders when a new controversy broke out in the United States that also influenced the American reception of the *Origin*. The principals in this episode were Tayler Lewis, professor of Greek at Union College, and James Dwight Dana of Yale, easily the most prominent geologist in the United States. Lewis, a Biblical literalist, offended Dana by his *Six Days of Creation*.[11] Lewis started with the premise that only Moses had left an account of creation, which Lewis received as inspired. A philological study of Genesis was thus the only reliable way to learn about the beginning of the world, a method that placed the Bible above science. Still, Lewis's conclusions were not as traditional as might have been expected. He denied that the days of creation could have been days of twenty-four hours. He interpreted them instead as long periods of time and concluded that an exercise of divine power began each day. So far, there was nothing very startling. But, when Lewis took up the creation of animated beings, he stated that the original

[9] Ibid., p. 126.
[10] Bowen, "A Theory of Creation," p. 448.
[11] Tayler Lewis, *The Six Days of Creation* (Schenectady: G. Y. van Debogert, 1855).

language of Scripture was compatible with a doctrine of biological development. He was even willing to admit the hateful teaching that man's body evolved from that of a lower animal.

Lewis knew that evolution presented theological difficulties that he must solve. The foremost was the question of atheism. He granted that versions of evolution that denied a divine origin were atheistic and that a version that acknowledged "only one divine origination, and this from the logical necessity of getting a starting-point for physical speculation, is as near to atheism as it can be." But Lewis believed that a development theory might include as many divine interventions as any other theory and might rationally be regarded as God's method of working. He thus retained the approved concept of God.

A second difficulty turned on the origin of men. On this point Lewis found the Bible specific "beyond all cavil" that the "present human race was from one single pair." He thought that the Bible "*seems* to imply" an immediate creation of men and that the story of Eve's coming supports this interpretation. But he still thought that the Biblical language also warranted the view that man's body was developed over a long period of time.

Though Lewis only offered evolution as a possibility, his argument for its compatibility with the language of Genesis might indicate a preference for this construction of the text. In any event, it was all too much for Dana, who attacked Lewis through four long articles in the Congregationalist *Bibliotheca Sacra*.[12] Dana identified Lewis's views with the *Vestiges*, which, he exulted, geology almost alone had put down. He offered the counter arguments that geology discovered no missing links but proved that separate creations followed catastrophes destructive of all life on earth. He cited Louis Agassiz against Lewis, who noted in reply that Dana eulogized an authority who taught the multiple origin of man. Dana quoted from Lewis's attack on the *Vestiges* in the *American Whig Review* to show Lewis's inconsistency. He scored Lewis's book as infidel because it slurred science and scientists and taught a "degraded and degrading development theory."

The violence of Dana's attack seems uncalled-for. What seemingly happened is that Dana was angered by Lewis's rejection of geology as

12 James Dwight Dana, "Science and the Bible," *Bibliotheca Sacra* 13 (1856): 80–129, 631–656; 14 (1857): 388–413, 461–524.

unreliable on this question and that he was further angered by some of Lewis's remarks about scientists. With this motivation Dana then used the unpopular evolution theory as a club. Dana was thus identified as an opponent of evolution shortly before the *Origin* appeared. His attitude could only strengthen the distaste for an already unpopular hypothesis.

Dana's performance won the approval of Louis Agassiz, the great Swiss scientist, who had visited the United States to lecture but had found a home instead.[13] By staying he enhanced American self-esteem and emphasized the growth of American science. Agassiz's fascination with American biology also won him the affection of Americans. He repaid their generosity by organizing the Museum of Comparative Zoology at Harvard, which, in his view, was to be one of the great scientific institutions of the world. His labors to produce a "natural history" of the United States similarly attracted Americans, who displayed their approval by contributing funds and collecting specimens for it. It is particularly significant that the "Essay on Classification" formed Part One of the first of ten projected volumes.[14] As Agassiz made clear, he intended to use American living forms to demonstrate the universal laws of biology stated in the "Essay on Classification." No wonder that Americans, conscious of the youth of their culture and institutions, loved Agassiz, who promised the world a true science based upon American specimens. The "Essay on Classification" was published in 1857. Thus, just before the *Origin* appeared, Agassiz had come forward with an opposing system, which he believed should guide the work of future scientists.

Darwin by this time had also looked to the United States for support of his views. In his quest for information from Asa Gray, he also gained a champion who would challenge even Agassiz. How Gray did so and on what issues are too well known to need elaboration here. A point about the debates between Agassiz and Gray, which A. Hunter Dupree emphasizes, should still be emphasized here.[15] Their main confrontation

13 Edward Lurie, *Louis Agassiz: A Life in Science* (Chicago: The University of Chicago Press, 1960).

14 Louis Agassiz, *Contributions to the Natural History of the United States*, 4 vols. (Boston: Little, Brown and Co., 1857–1862). Only four of the projected ten volumes were published.

15 A. Hunter Dupree, *Asa Gray, 1810–1888* (Cambridge: Harvard University Press, Belknap Press, 1959).

took place before Darwin published the *Origin*. Darwin's paper, read before the Linnean Society along with Wallace's, was the signal for Gray to come forward.

After the *Origin* appeared and provided friends and enemies alike with a statement of the theory, it was William Barton Rogers, to be the first president of the Massachusetts Institute of Technology, who debated Darwinism with Agassiz through four meetings of the Boston Society of Natural History. Gray did argue in Darwin's behalf at the American Academy of Arts and Sciences after publication of the *Origin* and did brush with Agassiz in doing so. Gray's main effort at the American Academy, however, was directed against Francis Bowen, Robert Chambers's old nemesis, who opposed Darwin at great length on philosophical grounds.[16]

The importance of the clash between Agassiz and Rogers can hardly be overestimated. In it Darwin's views were scrutinized by two scientists of international stature, and one of these was his most famous opponent. If anyone in the world could do to Darwin what Cuvier had done to Lamarck, Agassiz was the candidate most likely to succeed. The other papers written for this symposium show no equivalent event in any other country. The famous confrontation at Oxford between Huxley and Samuel Wilberforce was not so extended, not so carefully organized, and did not present opponents of comparable standing in science. The debates in Boston were unique.

Despite their importance, William Barton Rogers has not received justice in the history of the controversy. Edward Lurie, in his superb biography of Agassiz, discusses their clash and the issues that divided them.[17] But Lurie's concern seemingly is to clarify Agassiz's thought rather than to show the outcome of the debate. There is no doubt that Agassiz got much the worst of it. Jules Marcou, his protégé and admirer, conceded this.[18] Rogers himself was satisfied that he had won, and the record of the debates confirms him.

When Agassiz and Rogers clashed, the show was worth seeing. Agassiz was handsome, impetuous, and eloquent, but unguarded in speech.

---

[16] *Proceedings of the American Academy of Arts and Sciences* 4: 410–416, 424–441.
[17] Lurie, *Louis Agassiz*, pp. 295–297.
[18] Jules Marcou, *Life, Letters, and Works of Louis Agassiz*, 2 vols. (New York: Macmillan, 1896), II, 108.

Rogers had sharper features, was always alert, and possessed a keener sense of logic. He would begin speaking, according to Nathaniel Shaler, one of Agassiz's students, "with an odd gesture by which he seemed to turn his eagle eyes 'hard aport.' "[19] Shaler's respect for both men was great, and he enjoyed their tactics while he learned from their discourse.

Agassiz opened the debate by referring to Darwin as one of the ablest naturalists in Europe, whose learning and experience were being used to support "an ingenious but fanciful theory."[20] Then he mentioned the *Lingula prima*, a species of shellfish found through a great part of the geological record and still existing. The persistence of *Lingula*, he thought, was fatal to any doctrine of evolution. In reply Rogers said that some animals could resist change. On the other hand, he pointed out that another shellfish, the *Calymene blumenbachii*, extends through strata of different periods and exhibits change amounting to specific differentiation.

Agassiz returned to the floor and attacked Rogers's point that fossils change progressively. His argument was that trilobites, which he regarded as among the "highest" crustaceans, are found in the lowest fossil beds. In other divisions of the animal kingdom the highest representatives are likewise found in the earliest rocks, he declared. The earliest fishes, for example, also are among the perfect of their class. He added, in explanation, that these fishes have many reptilian characteristics.

Here Agassiz seemed to have a telling point against Darwin. Actually he was on decidedly slippery ground of his own polishing. The term "high" was popularly applied to organisms in the degree that their structures resembled man's. Agassiz himself normally used "high" in this sense. But he sometimes applied the same term to the undifferentiated forms furthest removed from man. Against Rogers he used the word in the latter sense, as his reference to perfect early fish made clear. Therefore he was really affirming a paleontological order favorable to Darwin. In this semantic jungle an audience might well go astray and think that the fossil record was against Darwin. Agassiz was unquestionably disingenuous in offering as an objection what truly was no such thing. Yet

[19] Nathaniel Shaler, *The Autobiography of Nathaniel Southgate Shaler* (Boston: Houghton Mifflin, 1909), pp. 116–117.

[20] *Proceedings of the Boston Society of Natural History* 7: 231. The first session of the debate is covered on pp. 231–236.

one can readily see how in the heat of debate the use of one word with opposed meanings might befuddle even the author of the befuddlement.

Rogers was not confused at all. He saw that he must make Agassiz explain his terminology and that the supposed argument against Darwin would then evaporate. He saw too that Agassiz was peculiarly vulnerable with his own theory of embryonic recapitulation affording a means by which to check the "progress" of fossils. This theory Agassiz had published so widely that everyone present knew about it. The good grace with which Rogers disposed of an irrelevant question indicated his confidence in the outcome. Then he pointed out that Agassiz had contradicted his own pet theory by asserting that the earliest forms are highest. Agassiz was quick to defend himself. In the process he wrecked his argument against evolution. For as he went along he made quite clear that, like Darwin, he recognized the progressive character of the fossil record. Agassiz's words ended the first round between himself and Rogers, from whom anything at this point would have been anticlimactic. No words from anyone were needed to show that Agassiz had just made a strong speech in Darwin's favor.

The issue at the next meeting was whether fossils of the same species occur in successive geological periods.[21] Agassiz maintained the negative. If he was right, animal life would not be continuous, as Darwin taught, but subject to extinction and replacement by new creation. Agassiz had satisfied himself that fossils thought to be the same are of different species. He asserted that Lyell would not have fallen into error on this point "with a sound zoologist by his side." On the same question, Agassiz also indicated certain shortcomings in the work of James Hall, author of the authoritative study of the geology of New York.

Rogers countered that a complete separation between species of separate strata must be shown before Agassiz's reasoning could be employed against Darwin. Dividing lines between species are essentially local, he continued, and many species may pass between formations in one region, while few do so in nearby places. He too cited Hall, who found that, in New York, only three of seventy fossil forms pass from the Black River limestone into the overlying Trenton limestone. But, in Canada, paleontologists found that fifty-two of seventy-five forms were

21 Ibid., pp. 236–245.

common to both series of rocks. Rogers offered emigration and return as the explanation of gaps in the fossil record. He then cited paleontologists in general as holding, contrary to Agassiz, that identical forms occur in rocks of different geological periods.

Agassiz rose in rebuttal. He did not expect the immediate acceptance of his views, though he was convinced of their truth. He believed that "after mature examination of his facts they would be generally received." As evidence that opinions change on this subject, he referred to certain fossils from the Jura, which, though once thought to be identical, were later classified as specifically different. He next showed that opinions change similarly in chemistry; chemists once did not distinguish platinum from silver. This was an unfortunate analogy for Agassiz to draw, because Rogers at once pointed out that chemistry had also reduced to one many supposedly different mineral species, for instance, diamond, plumbago, and carbon. On this point the debate ended for another evening. The session had closed, however, with Agassiz's revealing admission that his views, though indubitably true, were not generally accepted and with Rogers's adroit turning of Agassiz's argument.

The last two debates turned on the conditions under which the New York fossil-bearing rocks had been preserved. To account for the imperfection of the geological record and the absence of intermediate forms, Darwin argued that fossils would remain intact only when deposited on a subsiding sea bed. At the third meeting Rogers held that the existence of the New York fossiliferous strata was consistent with Darwin's views.[22] In response Agassiz contradicted Rogers. He maintained that there was no subsidence during the deposition of the New York strata and that, instead, the facts indicated an upheaval. If Agassiz was right and fossils could survive under such conditions, Darwin would find the absence of intermediate forms even more embarrassing. Agassiz defended his belief but soon returned to his pet subject of successive species and the method that he would use to determine their limits. Except for brief comments by President Jeffries Wyman and Rogers, there was no more to the debate at this meeting.

At the next session Agassiz made his last attack. He thought that Rogers's opinions about the New York fossil beds were "ingenious, and

[22] Ibid., pp. 245–263.

capable of explaining the facts in the case," but he saw no evidence of their truth. Then he introduced his principal topic, an exposition of his ideas on biology in opposition to Darwin's. When Rogers obtained the floor, he overlooked Agassiz's biology and supplemented his views on the geology of New York. Agassiz replied on the same issue and came to different conclusions. But he began by stating that, "during a local upheaval of the shore, the whole bottom was, in his opinion, subsiding from the shrinking caused by the cooling of the earth's crust."[23]

Rogers's eyes must really have snapped "hard aport" at this, because Agassiz had admitted the key point in their debate. Apparently he had become so involved in biology that he forgot the real question and trapped himself. Rogers had not forgotten and now pressed his advantage. He answered that he heard "with extreme surprise" Agassiz's statement disclaiming a rise in the ocean floor during the formation of the New York strata. Then he summarized the discussions so that no one could miss what Agassiz had conceded. As Rogers recalled, the discussion had begun with Agassiz's denial that strata and fossils would be extensively destroyed by a slow upheaval of the sea bed, as Darwin and others taught. Agassiz had then urged "as an insuperable objection" the extent and completeness of the New York Paleozoic series, which he maintained were deposited "during a *period of upheaval*." However, as Agassiz now recognized "the *subsidence* of the ocean-bed as essential to the theory of their formation," Rogers "thought it of no importance in this connection how that depression may have been brought about, or whether it was accompanied by a stationary or a rising condition of the ancient shore." Agassiz immediately referred to some fossil discoveries in Switzerland as disproving Darwin. But he could not escape even in his native Alps, for Rogers showed that these fossil mollusks also supported Darwin. Here their debate ended.[24]

No panel of judges chose a victor, but obviously it was Rogers. Even Jules Marcou admitted this. On the first evening Rogers forced Agassiz into defending a theory favorable to Darwin. In the second meeting he offered at least as strong an argument as Agassiz's, and this was true of the whole series. In the last debate he caught Agassiz in a flat-footed admission of the principal issue. Even more damaging to Agassiz was the

23 Ibid., p. 274.
24 Ibid., pp. 274–275.

bad impression that he made throughout. He seemed convinced that his opinions were absolute truth even when entangling himself in contradictions. His assertions, in the face of contrary fact and opinion, that he would yet be proved right, his evident feeling that a competent zoologist was one who agreed with him, and his readiness to write off as mistaken the most famous scientists could have only one effect on his listeners. Realizing that Agassiz was not objective about Darwinism, they would be disposed more favorably toward the new theory.

While nothing like these debates occurred in the other countries surveyed in this volume, nothing like them occurred in any other American scientific society either. Clearly, the reason that they took place in Boston was Agassiz's presence. The real issue at stake was whether Agassiz's or Darwin's principles would guide future scientific research. Each provided a coherent view, but both could not be right.

In this clash of world views, the success of Gray and Rogers certainly did not establish Darwin as right. Their success did, however, show that Darwinism would not collapse as the *Vestiges* had, even under the wrath of Agassiz. The debates made clear that Darwin's views must stand the trial by scientific scrutiny that he asked for. They showed too that Agassiz could challenge Darwin only by challenging the work of able scientists whose conclusions supported Darwin. He thus demonstrated the isolation to which such views as his, at this date in the nineteenth century, condemned him. Although the other papers in this volume give no indication that these debates were heeded in other countries, had Agassiz discredited Darwin, the fact would have been quickly recognized.

While the debates were progressing through 1860, the religious press was also filled with commentaries on Darwin's work. This was inevitable in view of the close relations that existed between science and religion in the United States. These early reviews of the *Origin* reflect the confidence that scientists would once again put down this theory, which seemed a resurrection of the *Vestiges*. Religious thinkers did not realize that scientists saw much more in the *Origin* than in the *Vestiges* and would ultimately present a verdict much different from their expectations. The old alliance between science and religion was thus in the initial stages of dissolution.

There were yet some common elements in the responses of scientists and religious thinkers. The most significant of these was a distrust of natural selection. Asa Gray, speaking against Francis Bowen, had conceded that natural selection probably could not accomplish all that was claimed for it.[25] Theophilus Parsons, professor of law at Harvard, also upheld evolutionary views at the American Academy but favored a saltatory version inconsistent with natural selection.[26] Most debaters, moreover, seemed principally concerned with establishing the fact of evolution before worrying very much about its cause.

Religious thinkers, on the other hand, felt more threatened by natural selection than by evolution. Evolution itself might be squared with an argument from design, but natural selection seemed to make chance the formative principle of species—a principle that could not be reconciled with divine foresight and goodness. The religious thinkers therefore summoned arguments of a scientific character against natural selection. They had an advantage in doing so, in that Darwin lacked the theory of genetics that he needed to support natural selection, as he himself knew. Critics charged, therefore, that variations under natural conditions would be diluted over the generations and that species would therefore remain constant. This charge was perfectly logical in view of the blending theory of inheritance that then prevailed, and it was a standing embarrassment to Darwin.

His critics also made much of the fact that the fossil record did not supply the sequences that might be expected in Darwin were right. He had conveniently brought this fact to the attention of his enemies by discussing the fragmentary character of the fossil record. If Darwin's critics were ungrateful in attempting to turn his argument against him, they unconsciously served his cause by educating readers in this and other scientific issues involved.

Biblical considerations formed surprisingly little part in the early controversy for two reasons. Problems like the inspiration of the Bible were secondary to the larger question of the existence of God. Furthermore Darwin's view, which he had not yet applied to mankind, seemingly would favor a single origin and thus be in agreement with Scripture. Some commentators even developed a Darwinian argument against Agassiz.

[25] *Proceedings of the American Academy of Arts and Sciences* 4: 411.
[26] Ibid., pp. 415–416.

The writer who pushed these themes furthest was a relative of Asa Gray's, Charles Loring Brace. As is well known, the separate creation of races had been offered as an argument for racial inferiority and a justification of slavery. Against this position Brace maintained the unity of man on Darwinian principles. His major blow along these lines was his book, *The Races of the Old World*, published in 1863. Though Brace did not openly admit an ulterior motive, he yet stressed one benefit from studying ethnology. "Many of our narrow prejudices and false theories in regard to Race—ideas which have been at the base of ancient abuses and long-established institutions of oppression—are removed by this study."[27]

One by one he assailed the polygenists' contentions. Their argument that insufficient time had elapsed for racial deviation from a single stock he countered with the antiquity of man. The argument that the infertility of mixed races attested to separate origins he disposed of by statistics to the contrary. On the positive side he relied heavily upon linguistics as demonstrating the unity of mankind, but he felt that philology at most made a single origin probable.

At length he indicated a source of greater hopes. "We believe that under the new lights furnished by Science during the last few years, this question can be investigated to far better advantage and with more probability of a conclusion than ever before." This statement launched him into a discussion of Darwinism and a Darwinian argument for a single human origin. Soon he was showing how natural selection could account for the black man. He came to the triumphant conclusion: "We see no difficulty on this supposition—on the Darwinian theory of an imperceptible accumulation of profitable changes through long periods of time, which few will question in regard to man—of accounting for the origin of the negro from the white man, or from the brown, or from some other race."[28]

Much of Brace's book was in direct answer to Agassiz, but he nearly committed one of the errors for which his adversary was castigated. Brace admitted that the development of human varieties and their environmental adaptation would result in "a degree of correspondence of the man

[27] Charles Loring Brace, *The Races of the Old World* (New York: C. Scribner, 1863), p. v.
[28] Ibid., pp. 448, 499.

with the *fauna* and *flora* of given latitudes." This was almost the naturalistic mixing of man and animals that Agassiz was charged with and approached his doctrine of zoological provinces. In a footnote, therefore,
Brace conceded the similarity but stated that ethnology was against Agassiz's "classification of 'realms of men,' as indicating separate origin or
local creation." His own book had been a continuous statement on the
other side, he wrote. By limiting the correspondence between men and
other organisms Brace steered between dangerous extremes.[29]

Through the years of the Civil War, studies related to Darwinism were
rare in the United States. An exception was the work of Benjamin D.
Walsh, a Philadelphia entomologist who had been introduced to biological studies by Darwin himself. Before the Boston Society of Natural
History and later in the *American Journal of Science*, Walsh stated his
belief in evolution and his doubts about natural selection. He also made
a sharp attack upon Agassiz by finding unconfirmed and unreasonable
Agassiz's insistence that no insect is native to both Europe and North
America. Agassiz attributed the presence of several butterflies in the
United States to human agency, but Walsh thought this an impossible
explanation. He noted that the larvae of some of these butterflies feed on
nettles and thistles, and he directed a prickly question toward Agassiz:
"Do men import nettles and thistles?" Walsh listed other difficulties
confronting Agassiz but at length indicated a way out of the maze. "If,
rejecting the Creative theory, we assume the Derivative Origin of Species, how simple and intelligible become the great facts of the geographical distribution of species!"[30]

Walsh knew that Agassiz would never embrace Darwinism. Indeed,
because of Agassiz's assaults, Walsh felt that he had license "to demonstrate that he [Agassiz] has totally misapprehended and misstated the
Darwinian theory, and appears never to have given himself the trouble
to read Darwin's book through." Walsh added that Agassiz had "ap-

29 Ibid., p. 503.

30 Benjamin D. Walsh, "On Certain Remarkable or Exceptional Larvae," *Proceed-
ings of the Boston Society of Natural History* 9: 286–318; idem, "Gradation from
'Individual Peculiarities,'" *American Journal of Science* 90 (1865): 282–283; idem,
"On Certain Entomological Speculations of the New England School of Naturalists,"
Proceedings of the Entomological Society of Philadelphia 1864: 207–249.

proached that book with the same feelings as many men approach a toad or a spider, viz, as something scarcely worthy his notice and disgustful to every rightly constituted mind." Walsh then showed that Agassiz was troubling himself to assail a doctrine that was not Darwinism at all.[31]

Walsh's support of Darwin, as already said, did not make him one of a crowd, for through 1865 the only other American scientist who wrote consistently on Darwin's behalf was Asa Gray. His writings appeared largely in the *American Journal of Science* and seemed calculated to keep the question alive in that periodical. Gray generally pressed two themes in this work. One was that Darwin had not destroyed the theological argument from design. Gray even went so far as to thank Darwin for having "brought teleological considerations back into botany" by his studies of orchids.[32] He admitted to a vein of "petite malice" in doing so because he knew that Darwin did not accept this view.

Gray also stressed the theme that scientists abroad were looking with increasing favor on Darwin's ideas. He noted that President George Bentham of the Linnean Society had praised Darwin, although Bentham was supposedly against him. Gray found Bentham's remarks were "better considered and more consonant with the spirit of the age than some others which eminent naturalists have thought it their duty to make."[33] Since Agassiz had recently proclaimed that God's resources could not be so meager that he must make men from monkeys, Gray's statement had an obvious application.

In spite of Gray, American scientists generally seemed content with traditional interpretations, as was shown by speculation about the *Archaeopteryx*, a strange Bavarian fossil discovered in 1861 that seemed to combine traits of birds and reptiles. Agassiz believed that this was just another "synthetic type" like the fishes with reptilian characteristics that he had been the first to point out. James Dwight Dana, who had taken no part in the Darwinian controversy so far because of illness, thought that systems of classification would now have to establish a category inter-mediate between birds and reptiles. All of these he believed must have

---

[31] Walsh, "On Certain Entomological Speculations," p. 223.

[32] Asa Gray, "Fertilization of Orchids through the Agency of Insects," *American Journal of Science* 84 (1862): 420–429.

[33] Asa Gray, "Address of George Bentham, Esq.," *American Journal of Science* 84 (1862): 286.

resulted from separate creations. Dana assessed the evolutionary significance of the fossil by stating: "We find in the facts no support for the
Darwinian hypothesis with regard to the origin of the system of life."[34]

The seeming indifference toward Darwinism was, in part at least, a
result of the war. During this time the American Association for the Advancement of Science, the one national scientific organization at the start
of hostilities, did not meet. Although the National Academy of the Sciences was formed in 1863, its membership was restricted, and it did not
foster the discussion of controversial issues, as the American Association
had. Promise that American science would take a larger role in the debate was given by the flurry of activity that followed the war. Then scientists could resume, or start, work that hostilities had precluded. Con-
gressional support of science came in the authorization of the four
surveys of the American West. These were bound to bear upon Darwin's
fortunes. He had pointed in the *Origin* to the fact that only a small part
of the surface of the earth had been explored scientifically. Now Congress was doing something to change that and to permit Americans to
play a larger part in international scientific questions.

The postwar years were also a time of advance in American higher
education. Universities were either strengthened or established as a
result of the Morrill Act, and institutions that did not benefit from this
law also began to develop their scientific departments. Even Princeton,
then called the College of New Jersey and controlled by a group of highly conservative Presbyterians, felt such stirrings. A critic called attention
to the scientific museum at Princeton to make the point that Princeton
was behind the times. He described the scientific collections as "some
forty turtles covered with the dust of thirty years, a hundred or two
faded butterflies, with their dismembered wings dangling from pins or
dropped to the bottom of the case; a few stuffed birds; perhaps ten
square feet over which shells are displayed; the most uninviting collection of minerals which we remember to have seen since the first few
months of our freshman labors in amassing a museum; and a space of
shelves, a large number of them entirely empty, over which are scattered

---

[34] *Proceedings of the Boston Society of Natural History* 9: 191; James Dwight Dana,
"On Certain Parallel Relations," *American Journal of Science* 84 (1862): 108, 315–
321.

some of those fossil shells which it would be impossible not to gather out of the green sand of New Jersey.''[35]

Princeton's future brightened in 1868 when James McCosh, a Scot from the University of Belfast, became president. McCosh was not the man to stifle Princeton's growth by foolish obstructionism. On the way to his new duties he decided to avow openly his belief that evolution and Christianity could coexist. He told the students as much shortly after arriving and from then on was a spokesman for that viewpoint.

In the years immediately after the war there were signs that scientists too were looking with increasing favor on Darwin's views. This was clear in 1867 when the American Association for the Advancement of Science met in Burlington, Vermont, the first postwar meeting on a large scale. There President J. S. Newberry summarized opinions on Darwinism in his address. Although Newberry was far from being a Darwinian, he denied that Darwin made theism impossible. He said so, he declared, in view of the possible acceptance of Darwin's theory by scientists.[36]

Newberry's statement could be supported by developments in 1866 that did indicate a growing disposition among scientists in Darwin's favor. Jeffries Wyman demonstrated the tendency when he delivered a paper blandly entitled "Notes on the Cells of the Bee" at the American Academy of Arts and Sciences.[37] Wyman was professor of anatomy at Harvard where, without complaint, he labored in Agassiz's shadow. He had, however, won the respect of scientists everywhere for his study of the gorilla. His anatomical description of it, Richard Owen said, left very little to add and nothing to correct.

Wyman's paper was essentially a report of measurements of bee cells. He had compared sections of different combs to see whether rows of ten cells were always the same length. He then filled cells with plaster of Paris and measured the resulting forms to see whether the angles were identical. To prove beyond all shadow of doubt that the angles vary, he cut away the filled cells at the point where the angles would be most regular. From the plaster form he then made a woodcut. When printed,

[35] "Dr. M'Cosh and the College of New Jersey," *Independent* 20 (November 5, 1868): 1.

[36] *Proceedings of the American Association for the Advancement of Science* 16 (1868): 1–15.

[37] *Proceedings of the American Academy of Arts and Sciences* 7: 68–83.

this showed to all the world that irregularities in cell construction are visible to the naked eye.

The neatness of Wyman's demonstration was impressive, and his audience surely realized that he was demolishing Francis Bowen's argument, delivered some six years previously against Asa Gray, that the invariability of bees' cells proved the invariability of instinct and therefore made human evolution impossible. Though Wyman never mentioned Bowen, there was no question about who was being flayed. Wyman then applied salt by stating that "those who have discussed the structure of the bee's cell" would have avoided error by making a comparative study, as Darwin had done. There seemed little question after this of Wyman's receptivity to Darwinism.

An even more startling incident occurred at the Boston Society of Natural History, also in 1866, when Alpheus Hyatt, Agassiz's student and assistant at the Museum of Comparative Zoology, applied evolutionary principles to some fossil shellfish that he was studying.[38] Hyatt was clearly not a Darwinian. He attributed the evolutionary change that he saw to a process of acceleration and retardation that was itself based upon the heritability of acquired characteristics. (Darwin did of course believe that such traits were hereditary but did not believe that they were marshaled according to Hyatt's views.) At almost the same time, Edward Drinker Cope, eventually to be recognized as one of the greatest American paleontologists, was formulating a theory nearly identical to Hyatt's. Their work is particularly significant as marking the beginning of a Neo-Lamarckian school of evolution that was soon to flourish in the United States.

Startling as Hyatt's defection from Agassiz was, Agassiz had clearly influenced the heresy, not least of all by his denunciations of Lamarck. Together, Hyatt and Cope demonstrated Darwin's wisdom in looking to "the younger naturalists" for confirmation of his theory. Although this pair were not in complete agreement with him, they yet demonstrated the appeal of evolutionary theory, as did the theological writers, like McCosh, who believed that religion and evolution could coexist.

This viewpoint became increasingly popular in the next few years and

[38] Alpheus Hyatt, "On the Parallelism between the Different Stages of Life in the Individual and Those in the Entire Group of the Molluscous Order Tetrabranchiata," *Boston Society of Natural History: Memoirs* 1, pt. 1 (1866–1867): 193–209.

was not even stifled by the *Descent of Man* (1871). This work of course was not the bolt from the blue that the *Origin* had been, because commentators had recognized from the first that the animal origin of man must be part of Darwin's vision. Neither irenic feelings nor familiarity with Darwin's theme could, however, have guaranteed the calm reception afforded the *Descent of Man*. The explanation must be found in a wave of scientific opposition that had swelled against Darwin, largely in England. This took the sting from the book and convinced religious writers that they could accept some version of human evolution. Americans were thus once more responding to the statements of European scientists.

A major figure in this opposition was William Thomson, Lord Kelvin. By cramping the age of the earth, on the basis of physics and astronomy, below Darwinian requirements, he seemed to discredit natural selection. But the most significant figures in this opposition were Richard Owen, Alfred R. Wallace, and St. George Mivart. Though they all accepted evolution, each had reservations about natural selection. Furthermore, Wallace and Mivart addressed themselves to the question of human beginnings and became the allies of those who insisted that man could not be completely explained on the basis of evolution. The comparative quiet with which the *Descent of Man* was received thus resulted from the feeling that it was not a great threat after all.

Owen's influence came first in time. One of the great English anatomists, he had opposed Darwin since the *Origin* was published. He nevertheless believed in evolution and probably had inclinations in that direction independently of Darwin. But Owen's shiftiness caused doubt as to what his opinions really were. In order to clarify them, the editors of the *American Journal of Science* printed in 1869 the final chapter of Owen's *Anatomy of the Vertebrates*.[39] The gist was that Owen accepted evolution, but not Darwin's version. Owen thought "an innate tendency to deviate from parental type, operating through periods of adequate duration, to be the most probable nature, or way of operation, of the secondary law, whereby species have been derived one from the other."[40]

[39] Richard Owen, "Derivative Hypothesis of Life and Species," *American Journal of Science* 97 (1869): 33–67.
[40] Ibid., p. 51.

That readers would find this statement very satisfactory seems doubtful, but Owen at least furnished ammunition against natural selection.

Of greater comfort to Darwin's enemies was Wallace's "Limits of Natural Selection as Applied to Man," originally published in 1868 and included in his *Contributions to the Theory of Natural Selection* (1870).[41] Naturally, any doubts of Wallace's about natural selection had peculiar force. In this work he held that natural selection seemed inadequate to account for the size of man's brain or the hairlessness of his body. Wallace also doubted that natural selection was responsible for the feet, hands, voice, and moral sense of man, though he found these difficulties less formidable. His basic premise was that a variation must either be beneficial or be destroyed in the struggle for survival. Therefore Wallace could not see why primitive peoples had such large brains: they could get along as well, he felt, with brains only slightly larger than those of gorillas. The organ in question thus seemed to have been designed in anticipation of future needs. Wallace also believed that a covering of hair would be helpful to uncivilized peoples and could not explain its absence by natural selection. He concluded that some other law was more fundamental.

Wallace's great service to orthodoxy was that he apparently justified a theistic interpretation of evolution. Even so, Mivart's *Genesis of Species* (1871) was more influential in softening the shock of the *Descent of Man*. Mivart, though a well-known English naturalist, was not of the same stature as Wallace and Owen, but he made the sharpest attack on natural selection. He argued, along with much else, that natural selection could not account for the "incipient stages" of useful structures, that species seem to develop suddenly, and that transitional forms are missing from the fossil record. He had no cause to substitute for natural selection but held that species were formed according to law and not by chance. This interpretation was in effect theistic too.[42]

Mivart devoted his final chapter to the religious implications of evolution. He concluded that this was simply God's action through natural laws and declared that Christians are free to accept it. He then considered how evolution squared with theology, which to him meant Roman Ca-

---

[41] Alfred R. Wallace, *Contributions to the Theory of Natural Selection* (New York: Macmillan, 1870).

[42] St. George Mivart, *On the Genesis of Species* (New York: D. Appleton, 1871).

tholicism. Again he concluded that no conflict existed: Saint Augustine, Saint Thomas Aquinas, and Francisco Suárez all justified an espousal of evolution. In view of the forthcoming *Descent of Man*, the most interesting part of the chapter was Mivart's treatment of human origins. While granting that man's body might have evolved, he held that the soul was directly created. He thus interpreted the infusion of the soul as the true creation of mankind and seemed to impart scientific respectability to the doctrine.

Such views from a Catholic were particularly impressive to American Protestants, who might regard Roman Catholicism as the archenemy of science. There was, besides, an additional reason that Owen, Wallace, and Mivart were influential in the United States. Their attack on natural selection coordinated beautifully with that of Alpheus Hyatt and Edward Drinker Cope. These brilliant young paleontologists had independently reached the conclusion that genera of fossil shellfish display a pattern of development that can be identified as a progression from youth to old age. Though they thus qualified as evolutionists, Hyatt strikingly so since he was the first of Agassiz's students to become one, they gave natural selection at best a secondary role in the evolutionary process. Anyone abreast of scientific thinking would perceive the similarity between their views and Mivart's. Hyatt himself saw it and thought that Mivart's silence concerning the American work did the Britisher no credit. Hyatt was the more unforgiving because reference to his work and Cope's would have strengthened Mivart's hand against Darwin.[43] Ironically Hyatt thought that Huxley, whom he suspected of counseling Mivart, was partially responsible for the lapse, though he must have changed his mind when Huxley attacked Mivart.

Mivart also drew criticism from another American, Chauncey Wright, important in the history of American philosophy. Wright felt that Mivart had not represented natural selection adequately and said so at length in the *North American Review*.[44]

From the theologians' point of view these writings, with the exception of Wright's, performed the great service of toning down the objectionable features of Darwinism. Evolution, which could be interpreted as design,

[43] *Proceedings of the Boston Society of Natural History* 4: 141–145.
[44] Chauncey Wright, "The Genesis of Species," *North American Review* 113 (1871): 63–103.

was less of a worry to them than natural selection, which they saw as an irrational process assuming divine functions. But these new works struck at natural selection and left room for God's action in the development of human beings. The upshot was that theological writers were ready for the *Descent of Man*. The English naturalists had supported Americans desirous of effecting a compromise with Darwin. The existence of such a group meant that the *Descent of Man* would be received very differently from the *Origin*.[45]

The *Descent of Man* was less likely to turn scientists against Darwin and in fact did not. The result was that Darwin's prospects brightened generally. The British opposition to him did generate assertions that Darwinism was on the wane. But those who said so usually did not point out that evolution, as separate from natural selection, was gaining adherents. Darwin himself would not willingly see natural selection discredited but had already written to Gray that he would sacrifice this more willingly than evolution.[46] On these grounds, opinion was definitely favoring Darwin. Dana, who had stated in 1863 that geology did not support evolution, summed up the state of the controversy in 1871. He stated that Darwinism, or variation according to natural selection, was losing ground in England. Evolution itself was gaining favor, he thought, though its cause was still undetermined.[47]

Other writers, and religious ones at that, realized that Darwinism, though undergoing revision, was not in decline. One of them warned against false hopes on the basis of the recent criticism. Darwinism was in reality, he said, spreading among scientists "like the measles in a school."[48] The justice of his claim would be evident to anyone who glanced through current scientific publications. The scientists seldom made formal statements of their conversion; they simply incorporated

---

[45] As evidence that criticism of the *Descent of Man* was tempered, see the following: *Independent* 23 (March 16, 1871): 6; "Darwin's Descent of Man," *Old and New* 3 (1871): 594–600; "Darwin's Descent of Man," *Monthly Religious Magazine* 45 (1871): 501–507; "Evolution and Theology," *Radical* 9 (1871): 375–385; J. B. Tyler, "Evolution in Natural History as Related to Christianity," *New Englander* 30 (1871): 464–470; John Bascom, "Darwin's Theory of the Origin of Species," *American Theological Review* 3 (1871): 349–379.

[46] Asa Gray, *Letters*, 2 vols. (Boston: Houghton Mifflin, 1893), II, 510.

[47] "The Darwinian Theory," *Independent* 23 (March 23, 1871): 1.

[48] *Independent* 23 (March 16, 1871): 6.

the new views into their thinking and tested and applied them as they went along.

Papers delivered before the Philadelphia Academy of Natural Sciences show the trend very clearly. In 1869 F. W. Meek and A. H. Worthen, who were working up the paleontological specimens of the Hayden Survey, gave an account of some fossil echinoderms. Some seemed typical of one species but varied so widely in particular characteristics as to seem not even of one genus. But the forms differing most extremely were closely connected by intermediate forms and seemed necessarily to belong to the same genus. Meek and Worthen drew no conclusions, but their facts were obviously in Darwin's favor, and they presented others of like significance. Elliot Coues, a young army surgeon who was also a brilliant ornithologist, wrote of a variety of thrush that had diverged to a point where most naturalists would rank it as a separate species. In 1870, Joseph Leidy, then the most distinguished American paleontologist, remarked that species common to both America and Europe often vary slightly, as an evolutionist would expect. Thomas Meehan, an enthusiastic Philadelphia botanist who was continually testing Darwin's theories, claimed that varieties have certainly originated in the sweet potato by evolution. In 1871, C. E. Dutton, soon to be associated with the United States Geological Survey, remarked that fossil mollusks that he had studied confirmed Cope's evolutionary theory.[49]

Such references could be multiplied from other scientific publications. The scientific publications of these years also record spectacular fossil discoveries by Cope and by Othniel C. Marsh of Yale. It was in 1871 that Marsh's attempt to bar Cope from the Wyoming fossil grounds started their feud. Marsh was successful in this attempt, but the West provided ample space for two collectors, even such energetic ones as Cope and Marsh. They now strove to outdo each other in paleontological discovery, with the result that a wealth of fossils was rapidly gathered from the West.

If Darwin was right, their labors should confirm him. Some remark-

[49] F. W. Meek and A. H. Worthen, "Descriptions of New Carboniferous Fossils from the Western United States," *Proceedings of the Philadelphia Academy of Natural Sciences* 1869: 137–172; Elliot Coues, "On Variation in the Genus Aegiothus," ibid. 1869: 180–189; Discussion, ibid. 1870: 72; Thomas Meehan, "Bud Varieties," ibid. 1870: 128–130; Discussion, ibid. 1871: 112–113.

able forms were already coming to light. By 1869, Marsh had found in Nebraska four kinds of fossil horses, the strangest of which was full grown but only about the size of a shepherd dog. Marsh was about twenty years behind Leidy in identifying the first fossil horse. But such forms were now of greater interest, since Owen and Huxley had recently traced the evolution of the horse in Europe. Marsh before long would make other discoveries that weakened the "missing-link" argument. The *Archaeopteryx* seemed an answer to the objection, but doubt about its teeth, for one reason, kept it from being as useful to the Darwinians as might have been expected. In 1872 and 1873, however, Marsh announced the discovery of other fossil birds with teeth, and there could be no doubt about these. He thereby supplied irrefutable evidence of an early connection between birds and reptiles.

Without question, evolutionary views were coming to predominate in science. Appropriately, Asa Gray rode the crest of this wave and, in 1872, became the first president of a national scientific association to proclaim his belief in evolution. Gray made his profession at Dubuque when he presided at the meeting of the American Association for the Advancement of Science. In his address he raised several questions about the California redwoods. Were they created there by the Pacific "thus local and lonely"? Were they only beginning their history? Or were they the remnants of a race approaching extinction? The audience would recognize these as questions at issue between Agassiz and Darwin. They listened as Gray reported that paleontologists placed redwoods in the Arctic during Tertiary times, when the climate was mild, and traced their later retreat southward during the ice age. When the glaciers eventually receded, Gray continued, the redwoods stayed by the Pacific. This *Sequoia* history had one meaning for Gray. He could not doubt that contemporary species were descended from ancestors in earlier geological periods. He likened "organic nature" to a river, "so vast that we can neither discern its shores nor reach its sources, whose onward flow is not less actual because too slow to be observed by the *ephemerae* which hover over its surface, or are borne upon its bosom."[50]

This figurative passage was really a public avowal of evolution, and

---

[50] "Address of Professor Asa Gray," *Proceedings of the American Association for the Advancement of Science* 21 (1873): 1–31.

Gray meant it so. He made clear, however, that things had changed since 1860, by stating that evolutionary views "have so possessed the minds of the naturalists of the present day, that hardly a discourse can be pronounced or an investigation prosecuted without reference to them."[51]

This address clearly marked Darwin's victory among American scientists. This does not mean that every scientist now accepted evolution. Indeed, at the meeting of the American Association in 1873, President J. Lawrence Smith, a chemist, attacked Darwin as more "a metaphysician with a highly wrought imagination" than a scientist.[52] Several other scientists at one of the section meetings also spoke against Darwin. But Darwin had defenders, and the meeting of 1873 was the dying flurry of his scientific opposition. There would still be debate about the processes of evolution, but evolution had clearly superseded special creation in American scientific thought.

Agassiz had thus lost his fight. Although religious thinkers still cited him against Darwin, Agassiz's stature was certainly diminished from what it had been fifteen years previously. Criticism of his stance now appeared frequently in the press, and John Fiske attacked him sharply in the *Popular Science Monthly*.[53] Of greater importance was the fact that the young biologists who studied with him had largely forsaken his views. Thus, in his own lifetime he had been unable to withhold even these young scientists, who were most subject to the force and charm that Agassiz still had in abundance, from evolutionary positions.

Reasons for Agassiz's ineffectual opposition are easily determined. For one, his own writings supplied a mass of material that could easily be interpreted according to a theory of evolution, once Darwin had pointed the way. Agassiz seemingly never perceived this and never saw until too late how some of his own statements embarrassed him. But the most important element in his downfall was the training that he gave his students. The essence of this was independent work, from which he insisted that his students come to their own conclusions on the basis of their own observations. The defection of his students therefore testified to the

[51] Ibid., p. 20.

[52] "Address of Dr. J. Lawrence Smith," *Proceedings of the American Association for the Advancement of Science* 22 (1874): 12.

[53] John Fiske, "Agassiz and Darwinism," *Popular Science Monthly* 3 (1873): 692–705.

quality of their teacher. At his death, which occurred in late 1873, critics and friends alike joined in praise of that aspect of his work and of the impetus that he had given to American science. The praise was surely justified.

There was now no scientific champion behind whom religious opponents could take shelter. Dana announced his conversion to evolution in 1874. The irony of this profession was not lost upon its author, who characterized his new position as "essentially the view taken by Professor Tayler Lewis of Schenectady, whom I once criticized on account of it."[54]

Darwin seems to have won the day quickly in the United States, especially in view of the disruption of the Civil War. His victory testifies to the independence of American scientists who found the theory confirmed by their work, despite any displeasure that they might cause. But his victory shows too that Americans were still playing the supporting role that had been theirs since colonial times.

American awareness of their culture as new and utilitarian was emphasized in 1872 when John Tyndall, the Irish physicist, visited the United States. Tyndall, like Lyell and Agassiz, was a representative of European science and a celebrity for that reason. He was additionally fascinating because of his address in 1870 on the "Scientific Use of the Imagination." In it Tyndall praised Darwin, who he admitted had "drawn heavily upon the scientific tolerance of his age." Tyndall defended Darwin, however, and before long speculated that the human mind itself might be traced ultimately to that original stuff from which the universe was formed.[55]

This materialistic inclination was enough to make him notorious in the United States, but shortly before his visit he endorsed a proposal that shocked Americans even more. This was the prayer test. Since prayers, he argued, are frequently said for a particular purpose, their efficacy could be tested. This might be done by establishing separate hospital wards, one of which would be given over to patients treated medically, while patients in the other ward would receive only the benefit of prayer. Re-

[54] Daniel Coit Gilman, *The Life of James Dwight Dana* (New York: Harper and Bros., 1899), pp. 330–331.

[55] John Tyndall, *Fragments of Science*, 5th ed. (New York: D. Appleton, 1886), pp. 423–457.

covery rates could then be established and the efficacy of prayer determined.

Religious critics were naturally upset by Tyndall's proposal, and at least one of them hinted that Christians in their charity would pray for the occupants of both wards and thus upset the experiment. The furor also led to prayer meetings for Tyndall wherever he went throughout his American journeys and did nothing to lessen the crowds that came to hear him lecture. On one of these occasions, Tyndall quoted Alexis de Tocqueville and scolded Americans for their indifference to pure science. Americans by this time were used to such admonitions and took his comments in good spirit.

Americans were seemingly in for more of the same when Huxley visited the United States in 1876. He, if anything, overshadowed Tyndall as an enemy of religion and as an outspoken champion of science. Yet Huxley made no statement like Tyndall's. He was instead a considerate guest and praised Marsh's collection of fossils from the West as unequalled in Europe. A high point of his visit came when he lectured on evolution in New York.[56] Here he stated that Marsh's horse sequence established the truth of evolution, although questions concerning its causes still remained. Huxley thus testified to the maturity of American science as well as to the accomplishments of Marsh. His testimony was particularly welcome in 1876, the centenary of the Declaration of Independence, when Americans were seeking measures of national achievement.

Huxley still saw American science in a supporting role. This was clear from his statement that Marsh's discoveries accorded perfectly with a theory announced before they were made. Huxley did not have to say who had enunciated the theory but made the acceptance and support of that theory the sign of maturity. Edward S. Morse, one of Agassiz's former students, had taken the same general position at the 1876 meeting of the American Association for the Advancement of Science in his address, "What American Scientists Have Done for Evolution."[57]

[56] *Popular Science Monthly* 10 (1876–1877): 43–56, 207–223, 285–295; these lectures are also in T. H. Huxley, *Science and Hebrew Tradition* (New York: D. Appleton, 1894), pp. 46–138.

[57] *Proceedings of the American Association for the Advancement of Science* 25 (1876): 137–176.

By 1876, however, not all American scientists docilely accepted this role. By this date the Neo-Lamarckians, who had developed significantly since their beginnings in 1866, believed that Darwin's pre-eminence as an evolutionary theorist could be conceded only if the testimony of American scientists was ignored. This does not mean that they had chosen Lamarck as their patron saint in preference to Darwin, for they became aware of their Lamarckian affinities only after developing their principles. The most prominent members of the group were Hyatt and Cope, both paleontologists, but their most ardent publicist was Alpheus S. Packard, Jr., an entomologist. There were many lesser figures who professed this same view of evolution. It was not until 1885 that Packard applied the term "Neo-Lamarckian" to them, but in 1876 they were very conscious of their dissent from Darwin. In that year Packard proclaimed that they were a distinctively American school of evolution.[58]

Neo-Lamarckism was essentially an attack upon natural selection as the main factor in evolution. Here again Darwin was in trouble because of his inadequate view of genetics. He played into the Neo-Lamarckians' hands by offering his theory of "pangenesis" in an attempt to supply the shortcoming. Since this principle allowed the heredity of acquired characteristics, which the Neo-Lamarckians built upon too, they were pleased to note this drift in his opinions. Ironically, when the revival of Mendelian genetics supplied the necessary theory, it was Neo-Lamarckism, not Darwinism, that failed.

This was in the future, however. By 1876 the Neo-Lamarckians had developed a theory of evolution that, they were confident, supplied answers where Darwin could not. To begin with they believed that evolutionary change was not always minute and creeping. They felt instead that it was sometimes rapid and proceeded by large steps. They held this position on the basis of their own work, largely in paleontology, but they felt that they thereby satisfied two of the main objections against Darwin: the absence of intermediate forms and Lord Kelvin's argument that the world had not existed for the length of time that natural selection required. Huxley had considered Kelvin's objection in the New York lectures and maintained that biologists would accept the time frame

58 Alpheus S. Packard, "A Century's Progress in American Zoology," *Independent* 28 (July 1876): 9.

established by other sciences, confident that evolution had worked within it. To Packard this was subterfuge.

Neo-Lamarckism explained evolutionary change as resulting largely from the relationship between organism and environment. The effort of organisms, like that of Lamarck's giraffe to reach fruit on trees, would concentrate growth forces and produce physical change that the Neo-Lamarckians believed was hereditary. Such views, of course, are not ac cepted today, but the Neo-Lamarckians' environmentalism should not be dismissed as oversimplified. Generally the environmental relationship, as they perceived it, was a complicated one involving action and reaction, habits and instincts, use and disuse of particular organs, the struggle for food, and other conditions of existence. George Gaylord Simpson has stated that the Neo-Lamarckian emphasis upon environmental factors in evolution was a major contribution to evolutionary theory.

The Neo-Lamarckians also believed that evolution moved according to discernible patterns. Hyatt identified the pattern as one of youth, maturity, and senescence. This simply meant that evolutionary change, as traced through a fossil series, would be healthful until the vitality of the group was depleted. Then the change would be erratic and a portent of coming extinction. Species in youthful stages demonstrated a process of acceleration, by which Hyatt meant that growth speeded up so that at reproduction an organism had more advanced traits to pass along than its ancestors had. Later generations would make further advances, which were also transmitted, until the group passed into its period of decline. Then acceleration would reverse itself, with increasingly degraded traits coming to the fore and racial death the ultimate result.

By the end of the century there were probably more Neo-Lamarckians than Darwinians in American science. This is the less surprising when one notes how well Neo-Lamarckism agreed with ideas and attitudes long prevalent among Americans. It certainly accorded with the distrust of natural selection that had been noticeable since 1860 in the American response to Darwin. Clarence King, of the United States Geological Survey, in 1877 combined his dislike for natural selection with a glorification of the American environment in a well-publicized statement of Neo-Lamarckism.[59] King did so by addressing himself to the very

---

[59] Clarence King, "Catastrophism and Evolution," *American Naturalist* 11 (1877): 449–470.

fossils that Huxley had cited—"that strange procession of fossil horse skeletons among whose captivating splint-bones and general anatomy may be descried the profiles of Huxley and Marsh." King went on to say that "a mere Malthusian struggle was not the author and finisher of evolution." He maintained instead that vast geological change occurring between the members of the sequence explained the evolution of the horse.[60]

King's Neo-Catastrophism in fact glorified the American West, which he felt contradicted Darwin. An editorial writer for the *Nation* emphasized King's criticism and gave reasons why King's view should be more authoritative than that of any European. King, the writer noted, had worked for twelve years upon a geological section which included "nine hundred miles of mountainous country." "When it is remembered," the writer continued, "that the longest possible European section of continuous mountain elevations is not only far shorter than this, but far inferior for purposes of generalization, it will be seen that the Americans have a great advantage in the breadth of their view." He went on to say that "It is therefore a matter of some consequence that Mr. King, whose field experience is greater than that of any European, and whose abilities are certainly entitled to high respect, should now assure us that the facts of our Western geology show the clearest proof of violent catastrophes recurring at intervals between periods of permanence, and in each case accompanied by a complete change in existing species." In other words, Lyell and Darwin were both wrong in the light of western geology.[61]

King's address, delivered at Yale, placed Marsh in the position of denying Darwin or of being un-American. He escaped the dilemma, in an address before the American Association for the Advancement of Science, by saying that he regarded natural selection as the "most potent" cause of evolutionary change. He used the term "natural selection," he said, "in the broad sense in which that term is now used by American evolutionists." "Under this head" he included "not merely a Malthusian struggle for life among the animals themselves, but the equally important contest with the elements and all surrounding Nature."[62] Thus, by

60 Ibid., p. 470.

61 "The American Naturalist," *Nation* 25 (August 30, 1877): 137.

62 Othniel C. Marsh, "Introduction and Succession of Vertebrate Life in America," *Popular Science Monthly* 12 (1878): 695.

enlarging the conception of natural selection beyond the Neo-Lamarck-ian definition, he escaped offense to Darwinians and American environ-mentalists both.

The assertion of American scientific accomplishment also accorded with Emerson's plea for independence in our cultural activities. Emerson had espoused the principles of Lamarck when it was unfashionable to do so and had stated in the "Poet" (1844) that "within the form of every creature is a force impelling it to ascend into a higher form." In the same essay he had stated that "America is a poem in our eyes; its ample geography dazzles the imagination, and it will not wait long for metres."[63] The Neo-Lamarckians thus drew upon strains present in our culture before the *Origin of Species*.

Neo-Lamarckism appealed to religious thinkers because of the com-parative ease with which an argument from design could be built upon it. Since the Neo-Lamarckians saw a pattern in evolution, they seemed to declare that evolution was a more orderly process than Darwin al-lowed. The saltatory aspect of the theory allowed the argument that something was guiding these steps toward a foreseen end. The Neo-Lamarckians knew that their work lent itself to these interpretations and did not discourage them. Some of them, indeed, were anxious to foster just such a reconciliation. Packard was one of these. In 1880 he prophe-sied that "a second Paley, in the light of the law of evolution, will write a new natural theology."[64]

Finally, Neo-Lamarckism was attractive because it could be related to American political and social development. Just as a split was develop-ing among biologists concerning the true theory of evolution, so there was a division among political thinkers concerning the implications of evolution. One group drew upon the principles of Herbert Spencer and William Graham Sumner to argue for "laissez faire," while Henry George pressed Neo-Lamarckism into the service of reform.

Joseph Le Conte, a California geologist whose views resembled Clar-ence King's, was possibly George's mentor in Neo-Lamarckism. At any rate he showed how readily Neo-Lamarckian ideas could be translated

[63] *The Complete Works of Ralph Waldo Emerson*, ed. Edward Waldo Emerson, 12 vols. (Boston: Houghton Mifflin, 1903–1904), III, 20, 38.

[64] Alpheus S. Packard, Jr., "The Law of Evolution," *Independent* 32 (February 5, 1880): 10.

into social concepts. He referred to the principle of acceleration as "a sort of a young-Americanism in the animal kingdom."[65] And he related embryonic development to the early stages of a civilization, which he described as follows:

Commencing with a condition in which each individual performs all necessary social functions, but very imperfectly; in which each individual is his own shoemaker and tailor, and house-builder and farmer, and therefore all persons are socially alike; as society advances, the constituent members begin to diverge, some taking on one social function and some another, until in the highest stages of social organization this diversification or division and subdivision of labor reaches its highest point, and each member of the aggregate can do perfectly but one thing. Thus, the social organism becomes more and more strongly bound together by mutual dependence, and separation becomes mutilation. I do not mean to say that this extreme is desirable, but only that an approach to this is a natural law of social development. *Is not this the law of differentiation?*[66]

Le Conte's statement seems an anticipation of Frederick Jackson Turner, who certainly explained American development by Neo-Lamarckian principles.[67] The first sentence of the second paragraph of the "Significance of the Frontier in American History" is genuine Neo-Lamarckian rhetoric: "Behind institutions, behind constitutional forms and modifications, lie the vital forces that call these organs into life and shape them to meet changing conditions."[68] Turner's emphasis upon the influence of geography is Neo-Lamarckian too. His insistence that civilization begins anew on each frontier is in accord with the principle of acceleration. And his statement that the closing of the frontier ended the first phase of American history can be related to the Neo-Lamarckian division of fossil sequences into youth, maturity, and senescence.

Turner, of course, came well after the day when King and Marsh were exchanging definitions of natural selection. Although they disagreed

[65] Joseph Le Conte, *Evolution and Its Relation to Religious Thought* (New York: D. Appleton, 1888), p. 161.

[66] Ibid., p. 25.

[67] William Coleman, "Science and Symbol in the Turner Frontier Hypothesis," *American Historical Review* 72 (1966): 22–49.

[68] Frederick Jackson Turner, "Significance of the Frontier in American History," in *Annual Report of the American Historical Association for the Year 1893* (Washington, D.C.: Government Printing Office, 1894), pp. 197–227. The quote is from p. 197.

over the factor that guided evolution, they both accepted the fact. Darwin in this sense had triumphed. And there was no question that by 1880 evolutionary views prevailed among American scientists, as was clear from a controversy that broke out in that year between the editors of the New York *Independent* and the *Presbyterian Observer*.

The *Independent* had long maintained that to insist upon an irreconcilable conflict between evolution and religion was theologically dangerous. Consistent with this opinion, the editors printed the following statement: "We are all taught in our best schools, by our scientific authorities almost without exception, and we laymen in science are, therefore, compelled to believe, that man was, at least so far as his physical structure is concerned, evolved from irrational animals. We, therefore, cannot help doubting, as every thinking and scholarly young man in these schools must and does doubt, whether the story of the fall in the first Adam is historical."[69] The Presbyterian editor sent the above sentences to a number of college presidents, who were requested to state whether the quotation reflected conditions in their institutions.

The presidents all answered that such doctrines were not taught in their classrooms. President Noah Porter of Yale stated: "The enclosed does not give a correct representation of the teaching in this college by our scientific authorities." President Ezekiel Robinson of Brown was even blunter: "We do not teach the doctrine stated in the enclosed slip." President Eliphalet Nott Potter of Union required only one word to answer: "No." The reply most favorable to the *Independent* naturally enough was McCosh's, and even that gave little support. He wrote: "In answer to your inquiries I have to state that we do not teach in this college that man is 'evolved from irrational animals.' I teach that man's soul was made in the image of God, and his body out of the dust of the ground. I do not oppose development, but an atheistic development."[70]

The replies indicated that the controversy was going against the *Independent*, which then contrasted the presidents' statements with the known views of scientists on their staffs. Thus, after quoting Porter, the *Independent* noted that Marsh, Dana, Addison Emery Verrill, and other

[69] "Do Our Colleges Teach Evolution?" *Independent* 31 (December 18, 1879): 14–15.

[70] "Scientific Teaching in the Colleges," *Popular Science Monthly* 16 (1880): 556–559.

evolutionists all taught at Yale—a tactic that pretty well robbed Porter's statement of its force. The same device was less successful when applied to schools without scientists of such standing. At Williams, for instance, President Paul A. Chadbourne himself taught biology and held that evolution was not proved but would be harmless to religion if it were. The *Observer*, for obvious reasons, had not polled the presidents of Harvard, Johns Hopkins, and Michigan, as the *Independent* pointed out. Eventually the *Popular Science Monthly* got into the dispute and challenged the *Observer* to name two working naturalists in the United States who were not evolutionists.[71] John W. Dawson, who was a Canadian, and Arnold Guyot, then a very old man, were the only two who could be settled upon, clear evidence of prevailing opinion among American scientists.

Certainly not everyone in the United States followed the lead of these scientists, and the particular grounds on which individuals came to terms with Darwin is a fascinating study. Asa Gray, who all along had labored to assist in such personal crises, renewed his efforts in 1880 when as a patriarch he spoke on religion and evolution at the Yale Theological School. He repeated there, he said, only what he had said or fore-shadowed in his early writing on Darwinism. He did however make a statement at Yale that provided both a means of reconciliation with Darwin and an insight into a little-recognized effect of Darwin's thought.

Gray emphasized the evolutionary connection between human beings and the lower animals by stating: "We are sharers not only of animal but of vegetable life, sharers with the higher brute animals in common instincts and feelings and affections. It seems to me that there is a sort of meanness in the wish to ignore the tie. I fancy that human beings may be more humane when they realize that, as their dependent associates live a life in which man has a share, so they have rights which man is bound to respect."[72]

Gray was not the first to express such sentiments on Darwinian grounds. About a year before, an author in the *Unitarian Review* published a similar passage. He had stated that the doctrine of the transmigration of souls had led to considerate treatment of animals. He then continued:

71 Ibid.
72 Asa Gray, *Natural Science and Religion: Two Lectures Delivered to the Theological School of Yale College* (New York: C. Scribner's Sons, 1880), p. 54.

A like wholesome influence is also exerted by the general diffusion and acceptance of the Darwinian doctrine of evolution. The establishment of an original affinity between man and beast, however remote the kinship may be, necessarily creates a current of sympathy extending even to the most insignificant members of the great and widely diversified family, and rendering it impossible to neglect or maltreat the "poor relations" to whom we are united by the warm and living ties of blood. It is said that,—

> "Aspiring to be man, the worm
> Mounts through all the spires of form."

A clear perception of this truth would cause even the most heedless wayfarer to take heed to his feet and step aside rather than tread upon the humble embodiment of such lofty aspirations.[73]

Rather ironically the quotation about the aspiration of worms was from Emerson and Lamarckian, but Darwin unquestionably fostered a sentimental attitude toward animals. This may seem surprising, since his thought is usually considered hostile to such views. Yet he promoted a flood of animal stories in the journals of his day by raising such questions as the variability of instinct and the existence of intelligence in animals. Sometimes these were printed with reference to Darwin, sometimes not. Anyone familiar with scientific issues would nevertheless see them in an evolutionary context.

Some of the creatures whose doings were reported seemed to have intelligence of a high order and other admirable qualities besides. Thus, dogs protected escaped canaries from cats, avenged canine friends against other dogs, and one in Texas even went for the doctor after his mistress suffered a sudden attack of nausea. Naturally this last dog fetched a basin first.[74] A grateful pike was also reported. He swam to the edge of a pond and placed his battered head on the feet of a benefactor who had bandaged it some days before. Probably the most ludicrous of all these stories concerned a pet boa constrictor whose master was seriously ill. The boa slithered into the sickroom, but the master took no notice. This so crushed the snake that he slithered out again, refused to eat, and pined away.[75]

---

[73] E. P. Evans, "Mythical and Symbolical Zoology," *Unitarian Review* 12 (1879): 117–138. The quote is from p. 120.

[74] "Notes," *Popular Science Monthly* 5 (1874): 766; "Canine Sagacity," ibid. 10 (1876): 202–206; "A Pretty Big Dog Story," ibid. 11 (1877): 488–489.

[75] "Pet Snakes," *Popular Science Monthly* 10 (1876): 119.

The kind of thinking that these stories reflected had a scientific aftermath in the Darwinian argument that the antivivisectionists developed. Darwin himself, despite his reputation as the avenger of horses beaten in his vicinity, opposed the antivivisectionists. Their sentimentalism worked in his favor, however, by providing a basis for accepting evolution. Henry Ward Beecher demonstrated the point by stating that he was not worried by the possibility of man's having evolved from a lower form. He was not even worried by Darwin's attempt to trace conscience to animal instinct. Grant the connection, and what would it prove? "It would only prove that the higher forms of animals also participate in some of the rudimentary forms of our moral sense. As an argument for the immortality of a virtuous dog, or a heroic horse, it might have some force."[76]

No doubt Beecher's argument swayed some of his readers. How his reasoning would go over with those who had opposed Darwin in the early, bitter controversy over the *Origin* is another question. They must have felt uncomfortable indeed to learn that they had really been against the admission of Fido and Old Bess to the Kingdom of Heaven.

Beecher nevertheless demonstrated the trend in the United States. He followed the example of the scientists who had supported Darwin and, by doing so, had asserted their own importance in the international scientific community. While other Americans might or might not take pride in these developments, at least one British writer deplored their effects in the United States. According to the story that he told, a lecturer here advertised that he would commit suicide at the end of his next talk. On the scheduled evening he presented "a most interesting discourse," which he concluded by drawing a pistol and shooting himself. The English reporter set the sequel in italics: "*At his lodging was found a will, leaving all his property to purchase the works of Darwin, Tyndall, and Huxley for the public library of the district.*" Then the writer demanded: "After that, can any rational being doubt that Mr. Darwin has much to answer for?"[77]

American scientists would answer, to somewhat different effect from what the writer expected, that Darwin had indeed.

76 "A Word on Evolution," *Christian Union* 19 (1879): 50.
77 "The Dangers of Darwinism," *Popular Science Monthly* 15 (1879): 68–71.

UNITED STATES

Bibliographical Essay

MICHELE L. ALDRICH

Darwinism in America has been investigated not only by historians of science but also by historians with a wide range of other interests. Consequently, a respectable number of books and articles exist on the impact of the evolution controversy on American religion, philosophy, social thought, literature, and education. Enough material now exists for comparing the American reaction with that in other countries. When historians begin such evaluations, however, they will discover lacunae in the treatment of the American experience that make such comparisons somewhat tentative in nature. For example, there is no equivalent on American scientific institutions to Frederick Burkhardt's paper in this book. Another problem is carelessness in the American literature about whether a response was triggered by a writing of Darwin (and, if so, which book or edition) or by another thinker, particularly Herbert Spencer.[1] This essay, then, has two purposes: to gather citations for persons wishing to compare the American scene to other countries, and to warn about what seem to be inadequacies in the secondary literature.

SURVEYS

George Daniels reviews the impact of evolution on American thought, but, consistently with the nature of his book as an overview that depends heavily on other writers, he says nothing new about the subject. His

---

[1] James Allen Rogers, "Darwin and Social Darwinism," *Journal of the History of Ideas* 33 (1972): 265–280.

bibliography is useful for citations to the secondary literature.[2] Edward Pfeifer's dissertation on Darwinism in the United States to 1880 concentrates on the interaction between scientists and theologians in the debate over the validity and the applications of natural selection.[3] The dissertation is more detailed than Pfeifer's essay in this volume but does not discuss the social, political, and literary impact of Darwinism. Shorter treatments of the general topic of Darwinism in America by Sidney Ratner, William Edenstein, Joseph Boromé, Bert Loewenberg, and Vincent Hopkins are marred by the tendency to make pro-Darwin forces the good guys and anti-Darwinians the incarnation of evil, but each essayist offers some insight on the controversy.[4]

Ratner differentiates between the impact of *Origin* and that of *Descent*, using the theologian James McCosh to illustrate his point. Boromé also credits *Descent* for a resurgence of the debate, in an essay which ties the evolution controversy to such other developments in the American past as the Civil War. Hopkins suggests that Darwin's and Spencer's relevance for conservative legal thought might be worth investigating (Philip Wiener has already written on evolutionary ideas among liberal lawyers).[5] Loewenberg argues that the history of Darwinism in America is divisible into two phases, a period to 1880 of probation and polemics, with gradual acceptance among scientists, and a period to 1900, when popular acceptance came and disciplines other than science adopted evolution.

Paperback anthologies of source material on Darwin in America were

[2] George Daniels, *Science in American Society: A Social History* (New York: Knopf, 1971), pp. 223–264.

[3] Edward J. Pfeifer, "The Reception of Darwinism in the United States, 1859–1880" (Ph.D. dissertation, Brown University, 1957; *Dissertation Abstracts* 18: 1024).

[4] Sidney Ratner, "Evolution and the Rise of the Scientific Spirit in America," *Philosophy of Science* 3 (1936): 104–123; William Edenstein, "The Early Reception of the Doctrine of Evolution in the United States," *Annals of Science* 4 (1939): 306–318; Joseph Boromé, "The Evolution Controversy," in *Essays in American Historiography: Papers Presented in Honor of Allan Nevins*, ed. Donald Sheehan and Harold Syrett (New York: Columbia University Press, 1960), pp. 169–192; Bert Loewenberg, "Darwinism Comes to America, 1859–1900," *Mississippi Valley Historical Review* 28 (1941): 339–369; Vincent Hopkins, "Darwinism in America," *Thought* 34 (1959): 256–268. (I owe the citation on Boromé to Edward Pfeifer.)

[5] Philip Wiener, *Evolution and the Founders of Pragmatism* (Cambridge: Harvard University Press, 1949).

published by Raymond Wilson in 1967 and by Daniels in 1968. Daniels is concerned largely with the debate over evolution in science and religion. Wilson covers the controversy in philosophy and social thought as well as in theology and science. Where possible, Wilson points out which ideas were derived directly from Darwin, which indirectly, and which were not Darwinian at all but misascribed.[6]

One kind of anthology on Darwin frequently encountered is the collection of essays by scientists on the present standing of Darwin's ideas. These are useful for learning the current version of the truth of Darwinian ideas, but the authors so rarely give enough historical background on the subject and are so selective as to the "true" predecessors of their own ideas that their essays are not of much use for studying the past impact of Darwin. Eventually, such books themselves become sources for the history of Darwinism in a country at the time they were published. The American Association for the Advancement of Science's *Fifty Years of Darwinism* is now primary rather than secondary material.[7]

SCIENCE

Bernard Barber has proposed a model of scientists' resistance to new scientific ideas that might help structure the early American response to Darwin's *Origin.* Barber contends that the stereotype of scientists as open-minded men and women has stopped historians from investigating the resistance of scientists to new ideas as thoroughly as they have investigated that of other thinkers, particularly theologians. Barber believes that scientists initially reject new ideas because the ideas threaten older systems of thought that have proved functional to the scientific community and because new theories may pose a psychological or personal threat in their very novelty. A scientific hierarchy based on age, professional standing, and educational background might also account for a

---

[6] George Daniels, ed., *Darwinism Comes to America* (Waltham, Mass.: Blaisdell, 1968); Raymond Wilson, ed., *Darwinism and the American Intellectual* (Homewood, Ill.: Dorsey, 1967). Rogers believes that Wilson still underestimates the importance of Spencer (see note 1 above).

[7] American Association for the Advancement of Science, *Fifty Years of Darwinism* (New York: Henry Holt, 1909). Other examples of this kind of literature are commented on by Bert Loewenberg, "Darwin and Darwin Studies, 1959–1963," *History of Science* 4 (1965): 15–54.

slowness in adoption of a new thesis.[8] In testing Barber's model, historians of American science must go beyond the Gray-Agassiz-Dana confrontations to study the reaction of large numbers of lesser-known scientists, such as Moses Ashley Curtis.

Most of the historical work on Darwinism in American science has concentrated on the acceptance rather than the rejection of natural selection. In 1876 and again in 1887, Edward S. Morse reviewed American zoologists' work on variation, embryology, geographical distribution, and so on.[9] He was descriptive rather than critical, and what is now needed is a historian to note his citations, reread the papers, and impose some order on the subject. Andrew Rodgers, in his book on American botany from 1873 to 1892, reports that Darwin's work encouraged the study of paleobotany in America, relatively neglected before the Civil War, and gave a theoretical reason to investigate plant distribution.[10] Nathan Reingold shows that Darwinian evolution changed paleontology from a largely empirical science to an organized search for evidence to test the Darwinian synthesis.[11]

There were a large number of American scientists who did not find the natural-selection hypothesis adequate for explaining life in the past —the group of scientists that Edward Pfeifer discusses in his article on American Neo-Lamarckism. These men, mostly paleontologists, reacted constructively to problems in the Darwinian model (notably the failure to account for the cause of variation) by building a new theory rather than by sniping at Darwin as did others abroad. They replaced Darwin's gradualism with saltatory changes in the history of species to answer Kelvin's objections to the length of time required for Darwinian evolution. Pfeifer discovered that a large number of Agassiz's students became Neo-Lamarckians: they were encouraged to reject Darwin by their teach-

[8] Bernard Barber, "Resistance by Scientists to Scientific Discovery," *Science* 134 (1961): 596–602; excerpted in *Darwin*, ed. Philip Appleman (New York: Norton, 1970), pp. 288–292.

[9] Edward S. Morse, "Address [on the Contributions of American Zoologists to the Darwinian Theory of Evolution]," *Proceedings of the American Association for the Advancement of Science* 25 (1876): 137–176; and 36 (1887): 1–43.

[10] Andrew Rodgers, *American Botany, 1873–1892: Decades of Transition* (Princeton: Princeton University Press, 1944).

[11] Nathan Reingold, *Science in Nineteenth-Century America: A Documentary History* (New York: Hill and Wang, 1964), pp. 181–199, 236–250.

er, while he exposed them to Lamarck's ideas (albeit in a derogatory way) when other American naturalists were ignoring Lamarck completely.[12]

Detailed studies of individual scientists' responses to Darwin exist so far only for Asa Gray, Louis Agassiz, and James Dwight Dana.[13] William Sanford analyzes Dana's gradual adoption of Darwinism, showing the accommodation of the theory required by his religious ideas and by the importance of Dana's pre Darwinian geological uniformitarianism and biological catastrophism. Sanford does not discuss Dana's personal motivations in his change of ideas—his run-in with Agassiz over Jules Marcou and his friendship with Darwin inspired by their cooperation on theories of coral-reef formation.[14] Anderson Hunter Dupree mentions the personal side of Darwin and Asa Gray's relationship as a factor in Gray's advocacy of natural selection. Like Dana, Gray interpreted Darwinism theistically, but he accepted the scientific ideas of the *Origin* far more quickly than Dana because they fit so well with his studies of plant distribution. There are a few minor faults in Dupree's book—he overstates the affection of European scientists for Spencer and underestimates the importance of Gray's friendship with Charles Loring Brace—but the book remains a good place to begin reading on the relevance of Darwin for American science.[15]

Historians of science have studied Louis Agassiz more than other American opponents of Darwin. Ernst Mayr details Agassiz's scientific objections to Darwinian evolution and finds four underlying concepts —typology; distinctiveness of species; a planned, rational universe; and "an ontological concept of evolution." He shows that, before 1859, Agassiz had encountered evolutionary ideas that were philosophical or meta-

[12] Edward J. Pfeifer, "The Genesis of American Neo-Lamarckism," *Isis* 56 (1965): 156–167.

[13] For a short early treatment of all three scientists, see Bert Loewenberg, "The Reaction of American Scientists to Darwinism," *American Historical Review* 38 (1932–1933): 687–701. Also, Walter Hendrickson has written about the St. Louis Academy's debate in "An Illinois Scientist Defends Darwinism," *Transactions of the Illinois State Academy of Sciences* 65 (1972): 25–29.

[14] William Sanford, "Dana and Darwinism," *Journal of the History of Ideas* 26 (1965): 531–546.

[15] A. Hunter Dupree, *Asa Gray* (Cambridge: Harvard University Press, 1959), pp. 233–306, 355–383.

physical and that ran counter to correct methods of scientific investigation. Mayr is skeptical of the importance of Agassiz's religious beliefs
in his rejection of Darwinism.[16]

While Mayr concentrates on Agassiz's early (1859–1861) response
to *Origin*, Edward Lurie adds a fuller assessment of the later phases of
his reaction, from 1861 to 1868, when Agassiz fought the theory of evolution before popular audiences, and from 1868 to 1873, when he returned to evaluate it on scientific grounds and began to admit some merit
to the idea of natural selection. Lurie presents a longer and more complex
interpretation of Agassiz's response in his book-length biography of the
man. Here he agrees with Mayr on the importance of Agassiz's past encounters with evolution as a metaphysical challenge to Cuvierian science.[17]

Apart from Dana, Gray, and Agassiz, American naturalists during the
second half of the nineteenth century have been inadequately studied.
For instance, the older biographies of Edward Drinker Cope and Othniel C. Marsh spend more time on the flamboyant personal feud between
the two than on their attitudes toward Darwinism. No studies have appeared of Joseph Leidy, Ferdinand Hayden, Fielding Meek, Joseph Le
Conte, or scores of other scientists who faced the problem of dealing
with natural selection. Until these other thinkers are written about, generalizations and tests of such models as Barber's will remain tentative and
open to debate.

RELIGION

One problem with the literature on this subject is the habit of historians of science of keeping the internal history of religion fixed and of
making theological thought vary solely in response to science during the
period after the Civil War. For an important exception to this method,
see John C. Greene's book, *Darwin and the Modern World View*, espe-

---

[16] Ernst Mayr, "Agassiz, Darwin, and Evolution," *Harvard Library Bulletin* 13
(1959): 165–194.

[17] Edward Lurie, "Louis Agassiz and the Idea of Evolution," *Victorian Studies* 3
(1959): 87–108; idem, *Louis Agassiz: A Life in Science* (Chicago: University of Chicago Press, 1960), pp. 252–387 (since the footnotes are omitted from the paperback
version of this book, the hardcover edition is preferable).

cially the sections on Catholicism and Darwinism.[18] Greene is not particularly concerned with Americans as apart from general Western thought, but his method is useful for the American scene nevertheless.

Andrew D. White's analysis of Darwinism and religion in *A History of the Warfare of Science with Theology* is an extended argument for a theistic interpretation of evolution, with footnotes to the theological disputants' writings.[19] Edward A. White connects Andrew D. White's work with Social Darwinism and considers also the religious thought of William James, David Starr Jordan, John Dewey, and John Fiske as related to the new theories of evolution.[20]

Essays by Herbert Schneider and Stow Persons survey evolution and religion in America from 1859 to the twentieth century. Both are concerned with theologians rather than the popular mind. Schneider distinguishes two main variants of evolutionary theology, the Presbyterian adaptation of natural selection to Calvinistic doctrines and the Unitarian blend of rationalist theology with a Spencerian belief in progress through nature. He is impatient with those who rejected Darwinism as a threat to faith and spends little time on a fourth group who separated religion from science.[21] Persons dissects the work of anti-Darwinian theologians as well as that of atheistic evolutionists but restricts his sample to a few men who he asserts but does not prove are "characteristic."[22] These essays should be supplemented with the excellent bibliographical essay by Gail Kennedy on the debate among the intellectuals in his otherwise undistinguished anthology on evolution and religion.[23]

John Campbell's dissertation outlines three phases of the religious re-

[18] John C. Greene, *Darwin and the Modern World View* (Baton Rouge: Louisiana State University Press, 1961).

[19] Andrew D. White, *A History of the Warfare of Science with Theology in Christendom*, 2 vols. (New York: Appleton, 1896), I, 70–88 (ch. 1, part iv); excerpted in Appleman, *Darwin*, pp. 423–428.

[20] Edward A. White, *Science and Religion in American Thought: The Impact of Naturalism* (Palo Alto: Stanford University Press, 1952).

[21] Herbert Schneider, "The Influence of Darwin and Spencer on American Philosophical Thought," *Journal of the History of Ideas* 6 (1945): 1–18.

[22] Stow Persons, "Evolution and American Theology," in *Evolutionary Thought in America*, ed. idem (New Haven: Yale University Press, 1950), pp. 422–453.

[23] Gail Kennedy, ed., *Evolution and Religion: The Conflict between Science and Theology in Modern America* (Boston: Heath, 1957), pp. 110–114.

action to *Origin*: a period of relative calm, when reviewers cited scientists' objections to Darwinian ideas; an outburst of criticism against Darwin's defenders and paraphrasers (especially T. H. Huxley), during which the arguments of *Origin* were found guilty by association with biblical criticism, materialism, and atheism; and a time of accommodation, when Darwin's work was given various theistic interpretations.[24] Another dissertation, by Windsor Roberts, documents these same three phases and also analyzes the residue of "irreconcilables."[25]

Bert Loewenberg and Dirk Struik restrict their interpretations of Darwin and religion to the New England scene. Loewenberg asserts in his man-by-man review of theologians who rejected natural selection that no supporters appeared, a position that requires that he ignore the pro-evolutionary Unitarians.[26] Struik's chapter on "Science and Religion" simply indicates who was on which side in the early debates on the theory of natural selection.[27]

A few detailed accounts of different sects' reactions have appeared thus far, but most of the Protestant churches and the Judaic responses have yet to be treated separately. Deryl Johnson traces "The Attitudes of Princeton Theologians toward Darwinism from 1859 to 1929" and provides a useful index for Calvinistic accommodations to evolution.[28] Reginald Deitz has examined eastern Lutheranism and discovered that its thinkers were inclined to accept a theistic form of evolutionary science, a tendency that prevented the "explosive conflicts which rent other Protestant churches."[29] Catholicism's reaction is briefly sketched by

[24] John Campbell, "A Rhetorical Analysis of the *Origin of Species* and of American Christianity's Response to Darwinism" (Ph.D. dissertation, University of Pittsburgh, 1968; *Dissertation Abstracts* 30A: 849).

[25] Windsor Roberts, "The Reaction of the American Protestant Churches to the Darwinian Philosophy, 1860–1900" (Ph.D. dissertation, University of Chicago, 1936; microfilm copy, Divinity School Library, Yale University).

[26] Bert Loewenberg, "The Controversy over Evolution in New England, 1859–1893," *New England Quarterly* 8 (1935): 232–257.

[27] Dirk Struik, *Yankee Science in the Making*, new rev. ed. (New York: Collier, 1962), pp. 373–397.

[28] Deryl Johnson, "The Attitudes of Princeton Theologians toward Darwinism from 1859 to 1929" (Ph.D. dissertation, University of Iowa, 1968; *Dissertation Abstracts* 29: 4092).

[29] Reginald Deitz, "Eastern Lutheranism in American Society and American Christianity, 1870–1914: Darwinism, Biblical Criticism, and the Social Gospel" (Ph.D. dissertation, University of Pennsylvania, 1958; *Dissertation Abstracts* 19: 784).

John R. Betts and limned with great detail by John L. Morrison. Betts's article is more accessible than Morrison's dissertation but is too short to present the conflicts within the church over the issue of evolution, making the Catholic response seem more unified than it was.[30] Morrison shows the dissension among theologians, lay thinkers, and Catholic scientists, connecting the disputants' stands to the political quarrel over liberalizing the church.[31] The Catholic reaction is also the subject of Harry W. Paul's essay in this book.

PHILOSOPHY

No book-length survey of the impact of Darwinism on all branches of American philosophy has yet appeared. Three shorter treatments are available, however. Max Fisch concentrates mainly on the pragmatists and, except for its compactness and inclusion of later pragmatists, his treatment is superseded by Wiener's book.[32] William Quillian contrasts deterministic philosophers who emphasized instinctual bases of behavior and biological justification of actions with a second, later set of theorists who stressed social factors and had room for idealism. Both groups claimed to have derived their ideas from the concept of evolution, making them the Conservative and Reform Darwinists of philosophy.[33] In a long chapter on evolution in his *History of American Philosophy*, Herbert Schneider ranges over geology, poetry, theology, and sociology to show the complex interaction of these disciplines with philosophy in America after the Civil War. He distinguishes where possible between thinkers whose main source for evolutionary ideas is Spencer and thinkers who derive directly from Darwin but is careless about what edition of *Origin* or other books a thinker is reacting to.[34]

[30] John R. Betts, "Darwinism, Evolution and American Catholic Thought, 1860–1900," *Catholic Historical Review* 45 (July 1959): 161–185. (Edward Pfeifer supplied me with this reference.)

[31] John L. Morrison, "A History of American Catholic Opinion on the Theory of Evolution, 1859–1950" (Ph.D. dissertation, University of Minnesota, 1951). (I am grateful to Thomas Glick for telling me of Morrison's work and lending me a copy.)

[32] Max Fisch, "Evolution in American Philosophy," *Philosophical Review* 56 (1947): 357–373; Wiener, *Evolution and the Founders of Pragmatism*.

[33] William Quillian, "Evolution and Moral Theory in America," in *Evolutionary Thought*, ed. Persons, pp. 398–419.

[34] Herbert Schneider, *History of American Philosophy*, 2d ed. (New York: Columbia University Press, 1963), pp. 277–371.

Treatments of specific philosophers have also appeared. James Collins has a few pages on Josiah Royce and George Holmes Howison in an essay that is otherwise concerned with contemporary uses of evolution in philosophy.[35] J. H. Kultgen critically examines the evolutionary elements of Fiske's and Dewey's metaphysics.[36] When discussing Whitehead in his general review of the "Philosophical Adventures of the Ideas of Evolution, 1859–1959," James Fulton points out Whitehead's departures from and uses of evolutionary theories.[37] Philip Wiener's *Evolution and the Founders of Pragmatism* has separate chapters on the thought of Chauncey Wright, Charles S. Peirce, William James, John Fiske, Nicholas St. John, and Oliver Wendell Holmes, in which he discusses the uniqueness of each man's work. His closing chapter summarizes the common beliefs of the group that are attributable to evolution—a theory of the natural selection of ideas in which truth evolves through testing ideas in practice, the notion that new ideas are useful variations that help man survive in the natural and social world, and faith in biological justifications for individualism.[38]

SOCIAL THOUGHT

Surveys

This is the area in which American intellectual historians have done the most work on Darwin in America. At least in part their concentration has been inspired by the early appearance of Richard Hofstadter's influential book, *Social Darwinism in American Thought.* Hofstadter tries to distinguish between Spencer's influence in America and the ideas directly inspired by Darwin. He then sets William Graham Sumner and Lester Frank Ward as counterpoints, showing that evolutionary ideas could be used to shore up the status quo or to call for change, depending on the way society fit into the process. If society itself was believed to be controlled by biological forces, then reform was not only futile but poten-

---

35 James Collins, "Darwin's Impact on Philosophy," *Thought* 34 (1959): 185–248.

36 J. H. Kultgen, "Biological Evolution and American Metaphysics," in *The Impact of Darwinian Thought on American Life and Culture*, ed. American Studies Association of Texas (Austin: University of Texas, 1959), pp. 84–92.

37 James Fulton, "Philosophical Adventures of the Ideas of Evolution, 1859–1959," *Rice Institute Pamphlet* 46 (1959): 1–31.

38 Wiener, *Evolution and the Founders of Pragmatism* (see note 5 above).

tially dangerous; if it was thought that social institutions directed the operation of biology to select the socially fit, then reform was justified and healthy in a civilization. Hofstadter briefly traces the implications of both sets of ideas for ethics, the social sciences as disciplines, pragmatism, and racism and imperialism.[39]

Three essays treat the themes in Hofstadter in more compact fashion. Eric Goldman's *Rendezvous with Destiny* has a chapter on the Reform Darwinists and the Conservative Darwinists (apparently he coined these terms);[40] Goldman is partial to the first, whereas Hofstadter tries to be fair to both. Merle Curti reviews "The Impact of Evolutionary Thought on Society" with respect to Reform Darwinism, politics and history, psychology, and philosophy.[41] Charles Rosenberg sketches the scientific theories that influenced American social thinkers after 1859, suggesting the importance of Spencer and Darwin relative to other scientists.[42]

A fourth paper, by Robert Bannister, rather strange for its ability to make Sumner disappear, argues that Conservative Darwinists were straw men invented by Reform Darwinists to react against. Bannister believes that the reform thinkers were not inspired by Darwin's own work but went back to it after rejecting the conservatives in order to strengthen their case.[43]

In 1956 Sidney Fine published a new interpretation of the two varieties of Social Darwinists. He believes even more firmly than Hofstadter that Spencer was the central figure in American work on the problem. Fine analyzes the laissez-faire theorists and defenders in the first half of his book and examines the rival concept of the general welfare state, its popularizers, philosophers, and practitioners, in the second half.[44] He

[39] Richard Hofstadter, *Social Darwinism in American Thought*, rev. ed. (Boston: Beacon, 1955; first published, 1944).

[40] Eric Goldman, *Rendezvous with Destiny* (New York: Knopf, 1952), pp. 84–104.

[41] Merle Curti, *The Growth of American Thought*, 3d ed. (New York: Harper and Row, 1964), pp. 540–563.

[42] Charles Rosenberg, "Science and American Social Thought," in *Science and Society in the United States*, ed. David Van Tassel and Michael Hall (Homewood, Ill.: Dorsey, 1966), pp. 135–162.

[43] Robert Bannister, "Survival of the Fittest Is Our Doctrine: History or Histrionics?" *Journal of the History of Ideas* 31 (1970): 377–398.

[44] Sidney Fine, *Laissez-Faire and the General Welfare State* (Ann Arbor: University of Michigan Press, 1956).

uses a massive number of sources but presents his results in a readable style.

The Separate Professions

According to Edwin Boring, American psychology encountered Darwinian ideas from three sources other than directly from *Origin*: German psychologists who transmitted evolutionary ideas to their American students, Darwin's *Expression of the Emotions in Man and Animals* (1872) and Francis Galton's *Hereditary Genius* (1896), and John Dewey's functionalism.[45] Boring's essay is flawed by his stilted writing style and his fondness for such undefined phrases as "American character" and "American faith." Edward Corwin's essay on Darwinism in American political thought and Robert Faris's on the idea of evolution in American sociology add nothing new to Hofstader.[46]

Irving Goldman wrote about American anthropologists in a general survey of "Evolution and Anthropology" in 1959. He has a useful brief section on Franz Boas's criticism of evolutionary anthropology, but his section on Lewis Henry Morgan is disappointing. Goldman does not show that Darwinian ideas inspired Morgan's stage theory; Comte could be as likely a source. Goldman also contradicts himself when he asserts that "professional field work" did not begin until *Origin* inspired anthropologists and then says that Morgan's *League of the Iroquois* (1851) was the "first scientific study of an American Indian tribe."[47] Ralph Gabriel's section on Morgan and John Wesley Powell is more satisfactory. He argues that an older strand of thought, which he calls the democratic faith and which is a central theme in his book, is present as the base on which Morgan and Powell built a Reform Darwinist social philosophy.[48]

In an article that reviews all the literature on the social thought of American businessmen, Irwin Wyllie sharply criticizes Hofstadter for his

[45] Edwin Boring, "The Influence of Evolutionary Thought upon American Psychological Thought," in *Evolutionary Thought*, ed. Persons, pp. 268–298.

[46] Edward Corwin, "The Impact of the Idea of Evolution on the American Political and Constitutional Tradition," in *Evolutionary Thought*, ed. Persons, pp. 182–199; Robert Faris, "Evolution and American Sociology," in ibid., pp. 60–80.

[47] Irving Goldman, "Evolution and Anthropology," *Victorian Studies* 3: 55–75.

[48] Ralph Gabriel, *The Course of American Democratic Thought*, 2d ed. (New York: Ronald Press, 1956), pp. 170–182.

argument that Conservative Darwinism underpinned their ideas. Wyllie points out that the remarks of the few Gilded Age businessmen who did use Darwinist analogies were usually ex post facto apologetics rather than a philosophy that guided operations. Far more popular were justifications drawn from the pre-Darwinian theories of laissez-faire competition.[49] Joseph Spengler finds that economists were prone to adopt evolutionary ideas if they were working with economic phenomena that required long time dimensions and historical explanations. The professions' development of a theory of competition well before 1859 had already made economists' thinking compatible with the part of natural selection that posited a struggle for existence. Spengler does find a lag in the creation of a systematic evolutionary economics, which he attributes to the dominance of the field by theologians who were suspicious of Darwinian theories. When secular, professional economics emerged later in the nineteenth century, its adherents used evolutionary ideas to mark their departure from the older group that had previously dominated the field. Spengler concludes that evolutionary economics made two contributions to the discipline: it made economic dynamics more important than old laissez-faire statics as a subject of study, and it used the new psychology that grew in part from Darwinism to build a theory of consumer choices.[50]

Individual Thinkers

Robert McCloskey has a long section on William Graham Sumner that is no improvement on Hofstadter.[51] Conway Zirkle's sharp criticism of Lester Frank Ward as the American carrier of the Marxist disease into sociology well demonstrates Zirkle's irritating habit of lumping American Neo-Lamarckism with Russian Lysenkoism under the rubric "Marxian biology."[52] His book is less a piece of intellectual history than a cold-

[49] Irwin Wyllie, "Social Darwinism and the American Businessman," *Proceedings of the American Philosophical Society* 103 (1959): 629–638.

[50] Joseph Spengler, "Evolution in American Economics, 1800–1940," in *Evolutionary Thought*, ed. Persons, pp. 202–266.

[51] Robert McCloskey, *American Conservatism in the Age of Enterprise: A Study of William Graham Sumner, Stephen J. Field and Andrew Carnegie* (Cambridge: Harvard University Press, 1951).

[52] Conway Zirkle, *Evolution, Marxian Biology, and the Social Scene* (Philadelphia: University of Pennsylvania Press, 1959), pp. 169–176.

war polemic. Theodore Roosevelt's Social Darwinism and his views on imperialism have been reanalyzed by D. H. Burton and found to be more complex than Hofstadter had said. Burton makes Roosevelt less deterministic (but still paternalistic) in his race thinking, an interpretation explained by influences that softened Roosevelt's Social Darwinism: belief in the Social Gospel, pragmatism in foreign policy decision, and older ideas of ethics and politics absorbed from Christian traditions. Burton traces a fading of Roosevelt's Conservative Darwinism from his Spanish-American War years through the years of the presidency and in his later political life.[53]

Specific Concepts

Although generally superseded by Hofstadter, Julius Pratt's account of the connection between Darwinism and imperialism in *Expansionists of 1898* is still intriguing for its accusation that intellectuals rather than politicians and journalists promoted hawkish ideas right up to May, 1898.[54] However, the specifically Darwinian content of these ideas may be overdrawn by Pratt, as the thought of imperialists seems to have drawn on a general view of the natural basis of violent racial warfare that could have come from many sources.

John Haller's book on the concept of scientific racism from 1859 to 1900 is more careful in this regard. Haller finds that Darwin (particularly in *Descent*) quieted the debate over a single or diverse origin for the human race. Evolution, though not necessarily the Darwinian brand (Neo-Lamarckism was important as well), led scientific racists to emphasize the historical development of races after their origins, an especially keen problem because of the alleged bad effects of emancipation upon black Americans. Haller argues that when biology was insufficient for "proving inferiority," as with Chinese Americans, sociology took over the function of buttressing racism. However, Haller suggests, American sociology was only marginally connected to Darwin through Spencer and Fiske.[55]

[53] D. H. Burton, "Theodore Roosevelt's Social Darwinism and Views on Imperialism," *Journal of the History of Ideas* 26 (1965): 103–118.

[54] Julius Pratt, *Expansionists of 1898* (Baltimore: Johns Hopkins Press, 1936), pp. 1–33.

[55] John Haller, *Outcasts from Evolution: Scientific Attitudes of Racial Inferiority, 1859–1900* (Urbana: University of Illinois Press, 1971).

Cynthia Russett introduces her book with a comparison of the process of adopting evolution in the social sciences to the introduction of equilibrium into the same disciplines. She finds that evolutionary concepts dynamized equilibrium and made the concept organic rather than mechanistic in its social application. However, Russett has also discovered that the idea of equilibrium was borrowed from physics (indirectly from Josiah Willard Gibbs) and from physiology (via Claude Bernard) rather than from Darwin's description of nature in *Origin*.[56]

George Stocking's careful paper on Lamarckism among American scientists concentrates on the work of Powell, Ward, and G. Stanley Hall. He finds that the concept had two important functions—it explained racial differences unilinearly by stages, and it explained the evolution of the mind. Stocking argues that the initial resistance to new genetical ideas among social scientists was due to the extreme hereditary basis it gave to social traits, a theory that they feared would stifle reform because of its determinism. Stocking attributes the demise of Lamarckism to the movement among social sciences to become autonomous disciplines rather than remain subordinate to biology.[57]

Literature

Hunting for evidence on the influence of Darwin in literature is not easy, since poets and novelists rarely interrupt their works to acknowledge sources of ideas explicitly. Except in such rare cases as Jack London's *Martin Eden* (1908), with its clearly expressed debt to Herbert Spencer, literary historians have but two indirect ways of suggesting the impact of Darwin. First, they analyze the portrayal of nature and the history, personality, behavior, and interaction of characters to learn the extent of an author's reliance on evolution. Unfortunately, there is a tendency to accept Darwinian-sounding phrases or ideas that Darwin shared with other scientists (for example, the importance of violence in nature) as adequate proof of the relevance of the *Origin* in a piece. Second, literary historians read an author's diaries, library lists, marginal notes in scientific books, reading notes, critical essays and reviews, inter-

---

[56] Cynthia Russett, *The Concept of Equilibrium in American Social Thought* (New Haven: Yale University Press, 1966).

[57] George Stocking, "Lamarckianism in American Social Science, 1890–1915," *Journal of the History of Ideas* 23 (1962): 239–256.

views with journalists, and autobiographies. This is hard work and occasionally fruitless, but the reward can be sufficient evidence to pin down the scientific sources with some confidence.

In a poorly written monograph full of useful citations, Harry Clark shows how revealing writers' criticism of their contemporaries can be of Darwinian influence in their own thought. He found Hippolyte Taine to be an important inspiration for reviewers who adopted the new scientific concepts in literature. Clark is careless about interchanging the terms *evolution, natural selection, heredity, determinism, struggle for existence,* and the like, but a more careful person might profitably build on his research and write the work this should have been.[58]

There are two essays that evaluate the general relevance of Darwinism for naturalism in American literature. Arthur E. Jones's *Darwinism and Its Relation to Realism and Naturalism in American Fiction, 1860–1900* is outshone by Malcolm Cowley's more detailed and incisive "Naturalism in American Literature."[59] Cowley reviews the factors that led the naturalists to use Darwinian or Spencerian ideas—their newspaper experience, admiration for local-color novelists, and distaste for the pieties and literary limitations of the dominant "genteel tradition." Cowley finds that the naturalists preferred Spencer as a scientific source to Darwin and even to Huxley, because "he gave them a unified world picture to replace the Christian synthesis" that underpinned the genteel tradition.[60]

Some work on Darwinism in separate genres in American literature has appeared. C. N. Stavrou looked for Darwinian ideas in drama and, apart from the plays of Eugene O'Neill, didn't find many. He believes that playwrights would have a hard time making the predetermined nature of man usually associated with Darwin (correctly or not) a viable basis for their work. Elton Miles summarizes the concern with evolution in the literary criticism of William Dean Howells and Thomas Sargent Perry as "pseudo-Darwinian"—their understanding of evolution had been

[58] Harry Clark, "The Influence of Science on American Literary Criticism, 1860–1910, including the Vogue of Taine," *Transactions of the Wisconsin Academy of Science, Arts, and Letters* 44 (1955): 109–144.

[59] Arthur E. Jones, *Darwinism and Its Relation to Realism and Naturalism in American Fiction, 1860–1900, Drew University Studies* 1 (1950); Malcolm Cowley, "Naturalism in American Literature," in *Evolutionary Thought,* ed. Persons, pp. 300–333.

[60] Cowley, "Naturalism in American Literature," p. 304.

filtered through Emile Zola in the first case and Hippolyte Taine in the second, and was quite different from a direct application of Darwin's ideas.[61] In his massive study (about one-half of which consists of quotations from poems and essays) of evolutionary concepts in American poetry, Frederick Conner reports that the ideas entered the poets' thoughts through teleological evolution, scientific or religious criticism of Darwin, Social Darwinism like that promulgated by Spencer, Neo-Lamarckism, pragmatism, Taine's work, and, most important, through pre-Darwinian evolutionary writers. *Origin* and *Descent* were rarely of direct influence.[62]

So far, separate monographs have managed to sift evolutionary ideas from the writings of Stephen Crane, Henry Adams, Ellen Glasgow, Walt Whitman, Theodore Dreiser, Jack London, and Frank Norris.[63] Taken with the surveys cited above, these works constitute a basis for a book-length evaluation of the literary uses of Darwinism and evolutionary thought.

POPULARIZATION

This area of the history of Darwinism in America has been relatively neglected, probably for two reasons. First, historians think of themselves as intellectuals and prefer to write about people they identify with, not

[61] C. N. Stavrou, "Darwinism in American Drama," in *Impact of Darwinian Thought*, ed. American Studies Association of Texas, pp. 37–50; Elton Miles, "The Influence of Darwinism on American Literary Criticism," in ibid., pp. 27–36.

[62] Frederick Conner, *Cosmic Optimism: A Study of the Interpretation of Evolution by American Poets from Emerson to Robinson* (Gainesville: University of Florida Press, 1949).

[63] David Fitelson, "Stephen Crane's *Maggie* and Darwinism," *American Quarterly* 16 (1964): 182–193; William Jordy, *Henry Adams: Scientific Historian* (New Haven: Yale University Press, 1952), pp. 172–219; Julius Raper, "Ellen Glasgow and Darwinism, 1873–1906" (Ph.D. dissertation, Northwestern University, 1966; *Dissertation Abstracts* 27A: 2541); James Tanner, "The Lamarckian Theory of Progress in *Leaves of Grass*," *Walt Whitman Review* 9 (March 1963): 3–11; Randall Stewart, "Dreiser and the Naturalistic Heresy," *Virginia Quarterly Review* 34 (1958): 100–116; J. D. Thomas, "The Natural Supernaturalism of Dreiser's Novels," *Rice Institute Pamphlet* 44 (1957): 112–125; idem, "The Supernatural Naturalism of Dreiser's Novels," *Rice Institute Pamphlet* 45 (1959): 53–69. (Marion Boutelle supplied the reference on Whitman.) For Jack London, see Zirkle, *Evolution*, pp. 318–337, and the citations in Hensley Woodbridge, *Jack London: A Bibliography* (Georgetown, Calif.: Talisman Press, 1966), pp. 290–297, 354–357. For citations on Norris, consult John S. Hill, *The*

about The Masses. Second, the sources are scattered and hard to sift through, with small amounts of material generated after a large amount of reading. Tabloids, cheap books, family journals, farmers' magazines, and printed versions of itinerant lecturers' speeches and of sermons exist in large numbers; looking for Darwinian traces among them is a long process. Thomas Glick illustrates the variety of evidence available from such studies in his work on Darwinism in Texas. An American equivalent of Leo Henkins' work on popular English novels has yet to be written. There is also no American equivalent for Alvar Ellegård's monograph on the treatment of Darwin by English newspapers and magazines. Hofstadter wrote very briefly about popular exposure to Darwinism in America. He found that Spencer's and Huxley's visits to the United States resulted in a burst of publicity about Darwinism; perhaps these are good dates to tap the popular press. Finally, children's literature both here and abroad needs to be examined for the introduction of evolutionary ideas.[64]

Two American thinkers responsible for popularizing evolutionary ideas have already been written on. Charles Haar is more concerned with the means by which E. I. Youmans spread scientific ideas (lectures, teaching aids, and the *Popular Science Monthly*) than with the content of these ideas, although he does mention that Youmans diffused ideas more Spencerian in cast than Darwinian.[65] A scholarly, critical biography on John Fiske by Milton Berman has sections of Fiske's interpretations of Darwin and of Spencer throughout the book.[66]

Merrill Checklist of Frank Norris (Columbus, Ohio: Merrill, 1970), pp. 4–25, and Kenneth Lohf and Eugene Sheehy, *Frank Norris: A Bibliography* (Los Angeles: Talisman Press, 1959), pp. 69–90.

[64] Thomas F. Glick, *Darwinism in Texas: An Exhibition in the Texas History Center, April 1972* (Austin: Humanities Research Center of the University of Texas, 1972), pp. 7–8, 13–15; Leo Henkins, *Darwinism in the English Novel 1860–1910* (New York: Corporate Press, 1940); Alvar Ellegård, *Darwin and the General Reader: The Reception of Darwin's Theory of Evolution in the British Periodical Press, 1859–1872* (Göteborg: Elanders Boktryckeri Aktiebolag, 1958). Hofstadter, *Social Darwinism in American Thought*, pp. 22–24. The suggestion about children's literature came from Barbara Greenberg.

[65] Charles Haar, "E. I. Youmans: A Chapter in the Diffusion of Science in America," *Journal of the History of Ideas* 9 (1948): 193–213.

[66] Milton Berman, *John Fiske: The Evolution of a Popularizer* (Cambridge: Harvard University Press, 1961).

The history of the teaching of Darwinian evolution has been confined to higher education. Walter Metzger's chapter on Darwinism in controversies within the educational profession is an ingenious examination of the conflict as part of the attempt to secularize colleges and universities that had been under clerical control in the first half of the nineteenth century. Professors advocating Darwinism used evolution as an issue to wrest control from trustees and presidents who were church-oriented. As Alexander Winchell's 1878 experience at Vanderbilt in Tennessee suggests, this tactic martyred a professor or two along the way. Metzger also shows that scientists who argued for evolution could be as arrogant and dogmatic as the theologian-educators who opposed them.[67] Clifford Peterson has found that college biology courses did not include Darwinian ideas until the mid-1870's, and that traditional colleges, which resisted new ideas in nearly all fields, held out even longer against evolution. Peterson attributes the introduction of scientific Darwinism to the adoption of *Elementary Biology* (1875), by T. H. Huxley and H. N. Martin, as a college text, to the indirect influence of German university biologists, and to the inclusion of new biological subdisciplines inspired by Darwin's ideas (such as eugenics) in the curriculum.[68]

Conclusion

It seems as if we are ready for a book-length synthesis on the influence of Herbert Spencer on American thought. Such a book would bring together the strands of Spencerism that are evident in nearly every part of American intellectual history and, in doing so, would encourage care among historians about what exactly has been Darwin's impact. Otherwise, we must continue to use loosely defined phrases to describe evolution in America—in itself a vague phrase. A monograph on Spencer would also attempt to answer the question of why Spencer was so much more popular in America than in his own country.

So far, there seems to be little connection between the literature by

[67] Walter Metzger, "Darwinism and the New Regime," in *The Development of Academic Freedom in the United States*, ed. Richard Hofstadter and Walter Metzger (New York: Columbia University Press, 1955), pp. 320–366.

[68] Clifford Peterson, "The Incorporation of the Basic Evolutionary Concepts of Charles Darwin in Selected American Biology Programs in the Nineteenth Century" (Ph.D. dissertation, Columbia University, 1970; *Dissertation Abstracts* 31A: 6431). (Thomas Glick lent me his copy of this dissertation.)

historians on Darwin in America and the major schools of historiography. This lack of correlation between historiographical schools and historical work on Darwin in America may indicate that we have not been asking very imaginative questions about the topic. Apart from Barber and Peterson, no one dealing with the subject has proposed a model for testing. Apart from some economists who have done cost-benefit analyses of artificial breeding, no one has asked questions that require quantification to answer. Finally, since Hofstader (and he was not that inclusive either), no one has attempted a survey of the whole question in a detailed, ambitious book. There is certainly enough secondary literature to warrant the undertaking.

RUSSIA

Biological Sciences

ALEXANDER VUCINICH

Darwin's theory could not have come to Russia at a more propitious time. After the Crimean defeat, the country was in the midst of a national awakening that called for a critical reassessment of dominant values and engendered a strong sentiment in favor of fundamental reforms. The writer Ivan Turgenev heralded the coming of the new era when he wrote in 1855 that the fall of Sevastopol should serve the same purpose for Russia that the fall of Jena served for the Prussians—that it should be the main catalyst of a new era of national emancipation from both the shackles of political oppression and the debilitating burden of the lingering vestiges of the feudal system.[1] The unrelenting clamor for a critical national "self-examination" went hand in hand with the search for new principles, cultural values, and intellectual standards. The most concrete and monumental products of the collective search for a way out of the fetters of feudalism and stagnation were the Great Reforms, which emancipated the serfs, liberalized the judicial system and university administration, and inaugurated the *zemstvos* as territorial units granted an unusual degree of independent initiative in the civic affairs of local importance. According to Boris Chicherin, the 1860's marked the first appearance of "many publics" in Russia.[2] A. I. Herzen noted that the 1860's marked a turning point in Russian literary history: while the

---

[1] Cited by A. A. Kizevetter, *Istoricheskie otkliki* [Historical comments] (Moscow, 1915), p. 193.

[2] Boris N. Chicherin, *Neskol'ko sovremennykh voprosov* [Some modern questions] (Moscow, 1862), p. 26.

heroes of earlier literary works were "superfluous men," now the time had come for a literary portrayal of the "man of action."[3] Still another writer saw in the 1860's an accelerated process of the secularization of wisdom; this was the time, he said, of the search for "a substitution of anthropology for religion, inductive method for deductive method, materialistic monism for idealistic dualism, empirical aesthetics for abstract aesthetics, and the theory of rational egotism for morality based on super-sensory principles."[4] In a reference to university education a famous embryologist, educated during the 1860's, noted that the students were particularly attracted to scientific articles published in general journals and to translations of popular scientific works. They were especially interested in writings presenting new theories of nature. On their reading lists, an important place was occupied by the books of Ludwig Büchner, Carl Vogt, and Jacob Moleschott which had a wide underground circulation; the fact that these books were banned by the censors made them so much the more popular. When the banned foreign books began to appear in hectographed Russian translations, "the positivist and materialistic world view"—which "answered all the questions asked by young people"—acquired new strength and popularity.[5]

Darwinism reinforced three guiding principles of the ideology of the rebellious intelligentsia: it placed the weight of scientific evidence behind the notion of historical relativism; it supported a materialistic view totally opposed to the spiritualistic and irrationalistic bent of autocratic ideology; and it introduced causality, as the fundamental principle of scientific explanation, into a domain of inquiry that had previously been heavily laden with teleological presuppositions and allowances for providential interference. Darwinism reinforced not only contemporary science, safeguarded by the academic community, but also the scientific world view that dominated the thinking of the intelligentsia, who regarded science as the most trustworthy index of social progress. M. A.

[3] A. I. Herzen, *Izbrannye stati iz "Kolokola"* [Selected articles from *The Bell*] (Geneva, 1887), p. 248.

[4] Nestor Kotliarevsky, "Ocherki iz istorii obshchestvennogo nastroeniia shestidesiatykh godov" [Essays on the history of social orientations in the sixties], *Vestnik Evropy* [The messenger of Europe] 11 (1912): 275. All translations are mine.

[5] I. I. Mechnikov, *Akademicheskoe sobranie sochinenii* [Collected works: academic edition], 14 vols. (Moscow, 1959), XIV, 10.

Antonovich, who published a systematic survey of Darwin's fundamental ideas in 1864, wrote many years later that the educated people of the 1860's favored the theory of evolution because it had a "broad philosophical base," it explained a general biological phenomenon simply and naturally, it placed emphasis on change at the time when change was uppermost in the minds of "the men of the sixties," and it was widely supported by many leaders of contemporary science.[6]

Most contemporaries agreed that the Darwinian evolutionary theory found an enthusiastic reception in Russia and that negative criticism came from isolated quarters and could muster only widely scattered and feeble support. The weak voice of the critics was paralleled by a rapid decline in idealistic philosophy. Eastern Orthodox theology, entangled in spiritualism and ethicism, did not have alert and able spokesmen to fight the new heresy; however, once the church recognized the danger, the theological journals began to carry anti-Darwinian articles, in most cases translations from Western religious journals. I. Krasovsky, in an article published in the "Theological herald" in 1865, was one of the Russian clerics who tried to refute the new theory. Relying mostly on ethical judgments, he treated Darwin's theory as "natural-philosophical fables" doomed to be short-lived.[7] Krasovsky did recognize, however, that the evolutionary conception was not a radically new development in science but rather the crowning point in the long history of a scientific idea.

The leading biologists who helped strengthen the empirical base of the evolutionary theory and who started their scientific careers during the 1860's were fully aware of the favorable reception of Darwin's ideas in Russia. According to Alexander O. Kovalevsky: "Darwin's theory was received in Russia with profound sympathy. While in Western Europe it met firmly established old traditions which it had first to overcome, in Russia its appearance coincided with the awakening of our society after

---

[6] M. A. Antonovich, *Charlz Darvin i ego teoriia* [Charles Darwin and his theory] (St. Petersburg, 1896), p. 234.

[7] As cited by V. P. Alekseev, "Evoliutsionnaia ideia proiskhozhdeniia cheloveka v russkoi nauke do Darvina i proniknovenie v nee darvinizma" [The evolutionary idea of the origin of man in Russian science before Darwin and the diffusion of Darwinism], in *Ocherki istorii russkoi etnografii, fol'kloristiki i antropologii* [Essays on the history of Russian ethnography, folklore and anthropology], ed. R. S. Lipets (Moscow, 1971), V, 223.

the Crimean War and here it immediately received the status of full citizenship and ever since has enjoyed widespread popularity."[8] I. I. Mechnikov contended that the evolutionary theory was readily accepted, not only because the general conditions favored it, but also because it did not encounter a strong antievolutionary tradition.[9] Although Mechnikov did not give a detailed explanation of his statements, it is an incontrovertible fact that Russia had a strong tradition in pre-Darwinian evolutionism. For example, Caspar Wolff, a member of the St. Petersburg Academy of Sciences in the eighteenth century, was the first embryologist to attack the theory of preformation, and Karl von Baer, a member of the same academy in the nineteenth century, contributed much more to the study of evolutionary processes in nature than he was willing to admit in his old age. In the *Origin of Species*, Darwin fully endorsed von Baer's view that "the embryos of mammalia, of birds, lizards, and snakes, probably also of chelonia are in their earliest states exceedingly like one another, both as a whole and in the mode of development of their parts; so much so, in fact, that we can often distinguish the embryos only by their size."[10]

In 1860, the "Russian messenger" published an article by the youthful A. N. Beketov entitled "Harmony in nature" and written before Darwin's ideas had reached Russia. Obviously impressed with Lamarck's transformist ideas, Beketov argued that the law of a double adaptation of species—the functional adaptation of an organism's parts to the organism as an indivisible whole and the further adaptation of the whole to the surrounding environment—is the law that governs the gradual transformation of the living world and determines the survival or extinction of individual species.[11] Implicit in Beketov's argument was the general idea that nature is subject to constant irreversible change,

[8] As cited by N. Umov, "Po povodu sbornika" [On the occasion of the symposium], in *Pamiati Darvina* [In memory of Darwin], by M. M. Kovalevsky et al. (Moscow, 1910), p. 1.

[9] Mechnikov, *Akademicheskoe sobranie sochinenii*, p. 52.

[10] Charles Darwin, *The Origin of Species by Means of Natural Selection* (New York: Fowle, n.d.), II, 241.

[11] A. N. Beketov, "Garmoniia v prirode" [Harmony in nature], in *Izbrannye proizvedeniia russkikh estestvoispytatelei pervoi poloviny XIX veka* [Selected works of Russian naturalists from the first half of the nineteenth century], ed. G. S. Vasetsky and S. R. Mikulinsky (Moscow, 1959), pp. 545–582.

which takes place according to regular and universal natural laws. Beketov showed conclusively that Russia was intellectually ready to enter the age of modern evolutionary biology.

Sir Charles Lyell was the person who introduced Darwin's theory to the Russian reading public. In January, 1860, two months after the publication of the *Origin of Species*, the "Journal of the Ministry of National Education" published a Russian translation of Lyell's report to the twenty-ninth meeting of the British Association for the Advancement of Science, held in September, 1859, in which a mention was made of the revolutionary significance of Darwin's forthcoming book. Darwin, according to Lyell, "appears to me to have succeeded, by his investigations and reasonings, in throwing a flood of light on many classes of phenomena connected with the affinities, geographical distribution, and geological succession of organic beings, for which no other hypothesis has been able, or has even attempted, to account."[12] Lyell gave full support to Darwin's conclusion that the powers of nature that give rise to varieties in plants and animals are the same as those that produce new species over longer periods of time. In his introductory remarks to Lyell's report, the Russian writer noted that the reason for the antitransformist views in biology, most categorically expressed by Cuvier, must be sought in the too narrow empiricist orientation of naturalists, who were much more careful in gathering reliable facts than in drawing theoretical conclusions. The creationist orientation of an overwhelming number of pre-Darwinian biologists was becoming rapidly incompatible with the mounting empirical evidence in favor of transformism. The Russian author was particularly critical of H. G. Bronn, the German zoologist who in 1858 received a prize from the Paris Academy of Sciences for a work that categorically denied the possibility of the origin of species by means of "the transformation of a small number of primeval forms of life."[13]

In 1860, S. S. Kutorga, a professor at St. Petersburg University, pre-

[12] Sir Charles Lyell, "On the Occurrence of Works of Human Art in Post-Pliocene Deposits," *Report of the Twenty-Ninth Meeting of the British Association for the Advancement of Science* (London: John Murray, 1860), p. 95.

[13] S. L. Sobol', "Iz istorii borby za darvinizm v Rossii" [From the history of the struggle for Darwinism in Russia], *Trudy Instituta Istorii estestvoznaniia i tekhniki* [Studies of the Institute of the History of Natural Science and Technology] 14 (1957): 200.

sented his students with a general review of the basic ideas contained in the *Origin of Species*. Soon after this, a series of articles published in two journals, the "Messenger of the natural sciences" and the "Library for reading," took Darwin's ideas far beyond the university community. The "Library for reading" published the first detailed survey of Darwin's ideas in the Russian language. The anonymous writer commended Darwin for his courageous effort to "reconstruct" the time-honored but outdated picture of the organic world. He admired Darwin for the precision of his logic and the richness of his argumentation and was convinced that no previous theory of the origin of species was so close to the truth. The reader was treated to a bonanza of well-digested information on the leading forerunners of Darwin's theory, particularly Goethe, Lamarck, Etienne and Isidore Geoffroy Saint-Hilaire, and A. R. Wallace.[14] Another journal informed the readers that Darwin's evolutionary idea was not a wild guess but "an inevitable result of the basic postulates of modern science," that it was a product of a long evolutionary process in the history of scientific ideas, and that its influence would be felt far beyond the boundaries of biology.[15]

In 1864, S. A. Rachinsky, professor of plant physiology at St. Petersburg University, produced the first Russian translation of the *Origin*. Although not a masterpiece of translation art, the book sold out so quickly that in 1865 it went through a second printing. By this time Darwin's ideas were discussed not only by scientists but also by such popular writers as Dmitri Pisarev and M. A. Antonovich. The dramatic tone of his lengthy article on the wide implications of Darwin's theory helped establish Pisarev as one of the most respected and influential popularizers of natural science during the early 1860's. The famous neurophysiologist Ivan Pavlov noted many years later that Pisarev's popular essays helped attract many young Russians to natural-science studies in the institutions of higher education. Pavlov himself was one of these students.[16]

---

[14] Cited by B. E. Raikov, "Iz istorii darvinizma v Rossii" [From the history of Darwinism in Russia], *Trudy Instituta Istorii estestvoznaniia i tekhniki* 31 (1960): 27–28.

[15] I. Sheneman, "Teoriia Darvina" [Darwin's theory], *Zagranichnii Vestnik* [Foreign messenger] 2 (1864): 210.

[16] I. P. Pavlov, *Polnoe sobranie sochinenii* [Complete works] (Moscow and Leningrad, 1952), VI, 441; see also K. N. Davydov, "A. O. Kovalevskii kak chelovek i

Among the men trained in science, two were particularly important as popularizers of Darwin's scientific ideas: N. N. Strakhov and K. A. Timiriazev.

In his graduation paper at St. Petersburg University, Strakhov presented three original algebraic theories giving solutions for inequalities of the first degree.[17] This paper, as well as his master's thesis dealing with the wristbones of mammals, was published in the "Journal of the Ministry of National Education." After his efforts to obtain a university position had failed, Strakhov became a free-lance writer contributing to several popular journals. It is generally believed that he was the translator of Lyell's report, published in the "Journal of the Ministry of National Education," which introduced Darwin's evolutionary theory to the Russian reading public. In an article published in the journal "Time" in 1862, he surveyed Darwin's ideas and criticized the efforts of Clémence Royer, the French translator of the *Origin*, to give Darwinian theory a broader sociological meaning. Strakhov may be rightfully called the first Russian scholar to stand firmly against Social Darwinism, a sociological orientation that did not have a single supporter among leading Russian naturalists and social thinkers. Strakhov greeted Darwin's theory as an addition to science and as a modern world view. He credited Darwin for having been the first scientist to formulate an "immanent law" of the development of organisms—that is, a law that had no room for supernatural interference with the evolution of living forms. Although Darwin's theory did not answer all the intricate questions of biological evolution, it was built on sound foundations.[18] Despite his deep and enthusiastic involvement in the popularization of Darwin's ideas during the years immediately after the publication of the *Origin*, Strakhov was not a typical Russian Darwinist of the 1860's: he did not allow Darwin's natural-science "materialism" to displace his own inclination for idealis-

uchenyi" [A. O. Kovalevsky as a man and a scholar], *Trudy Instituta Istorii estestvoznaniia i tekhniki* 31 (1960): 333.

[17] B. V. Nikol'skii, *Nikolai Nikolaevich Strakhov: Kritiko-biograficheskii ocherk* [Nikolai Nikolaevich Strakhov: an essay in critical biography] (St. Petersburg, 1896), pp. 22–24.

[18] N. N. Strakhov, *O metode estestvennykh nauk i znachenie ikh v obshchem obrazovanii* [On the method of the natural sciences and the role of these sciences in general education] (St. Petersburg, 1865), pp. 178–182. (This book is a collection of essays published by Strakhov in various journals during the late 1850's and early 1860's.)

tic philosophy. He was ready to admit that Darwin's ideas came to Russia at the time when philosophical materialism was a reigning philosophy; but he was equally ready to emphasize that the demise of idealism was a transitory phenomenon and to fight for a revitalization of anti-materialistic philosophy. He endorsed Rudolf Virchow's dictum that "sick idealism" should be transformed into "healthy idealism" rather than abolished altogether as advocated by the Nihilist intelligentsia.[19] A decade later, the idealistic philosophical bias predisposed Strakhov to undertake a vicious attack on the *Descent of Man* as a caricature of empirical science and a flagrant attack on man's moral code.[20] He became one of the most belligerent Russian anti-Darwinists of the nineteenth century.

K. A. Timiriazev, a pioneer in plant physiology, was the first scholar educated in the atmosphere of the Great Reforms to help spread the basic ideas of the new evolutionary orientation. Like many scientists of the 1860's who came from noble families of rapidly declining fortunes, he combined an insatiable thirst for scientific knowledge with a profound dedication to democratic ideals and to the philosophy of "realism" advanced by the leaders of the Nihilist movement. He enrolled in St. Petersburg University in 1861, a year marked by growing student unrest that led to the frequent closing of the university. Before graduating in 1866, he had written "Garibaldi in Caprera," "The hunger in Lancashire," and "Darwin's book: its critics and commentators," all published in the influential journal "Fatherland notes" in 1863–1864. The article on Darwin appeared in 1865 as a book entitled "Charles Darwin and his theory." This book provided several generations of Russian students with the most thorough and lucid exposition of Darwinian evolutionary thought.[21]

Although Timiriazev was the first link between the interpreters of evolution in the academic community and the Nihilist intelligentsia, his book on Darwin contained no statements of direct ideological import. However, in prefacing the book with a statement by Auguste Comte on

---

[19] N. N. Strakhov, *Iz istorii literaturnogo nigilizma, 1861–1865* [From the history of literary nihilism, 1861–1865] (St. Petersburg, 1890), pp. 123–124.

[20] N. N. Strakhov, *Bor'ba s Zapadom v nashei literature* [The struggle with the West as reflected in our literature], 3d ed., 2 vols. (Kiev, 1897), II, 250–280.

[21] K. A. Timiriazev, *Kratkii ocherk teorii Darvina* (Moscow, 1865).

selection as a source of harmony between organisms and the changing environment, he was eager to show that the scientific theory of evolution and the philosophical orientation of the Nihilist intelligentsia, imbued with the spirit of Comtian philosophy, had much in common. The book endorsed Darwin's theory in its entirety. For the next forty years Timiriazev was Russia's staunchest defender of Darwin's legacy; he fought the Neo-Darwinists and Neo-Lamarckians with the same fury that he fought the anti-Darwinists.

Darwin's works that followed the *Origin* were received with equal enthusiasm. In August, 1867, Darwin wrote to Lyell that he was visited by a young Russian "who is translating my new book into Russian."[22] The book was the *Variation of Animals and Plants under Domestication,* and the young Russian was Vladimir Kovalevsky, who subsequently became a well-known evolutionary paleontologist. At that time the *Variation* was not yet published, and it seems most probable that the translation was made from a set of proofs given to Kovalevsky by Darwin. Thanks to Kovalevsky's rapid work, the first section of the Russian translation of the *Variation* was published several months prior to the publication of the English original.[23] *The Descent of Man,* Darwin's major work dealing with the evolution of man, was published in 1871: in 1871–1872 not less than three translations of this work were published in Russia, two under the editorship of the famous neurophysiologist I. M. Sechenov. Darwin's *Expression of the Emotions in Man and Animals* was published in England in November, 1872, and in Russia in December, 1872.

The rendering of Darwin's works into Russian was only partly responsible for the rapid diffusion of modern transformist ideas in Russia. Equally important were the translations of Western works analyzing the nature of the new ideas and the revolutionary changes in the general scientific orientation expected to come in their wake. Lyell's *Antiquity of Man,* T. H. Huxley's *Evidence as to Man's Place in Nature,* Carl Vogt's *Lectures on Man and His Place in the Creation and the History of the*

[22] Francis Darwin, ed., *The Life and Letters of Charles Darwin,* 2 vols. (New York: Basic Books, 1959), II, 256.

[23] While Darwin's book was delayed by the printer's work on the plates of illustrative material, the publishers of the Russian translation used the illustrations prepared for A. E. Brehm's *The Life of Animals.*

Earth, and many articles on evolution written by leading Western scientists were translated into Russian in quick succession. While Lyell's work placed the up-to-date paleontological information behind the evolutionary theory, Huxley made challenging excursions into the taxonomy of anthropoids and the major achievements of comparative embryology. Huxley performed a function that Darwin—in the *Origin*—had painstakingly avoided: he made the evolution of man the central theme of his study. He gave a scientific backing to the Nihilist ideology based on the notions of the historical relativity of moral values and of natural science as the safest path to a virtuous life and the general betterment of human existence. He gave the evolutionary theory a much broader basis: he carried it from its scientific moorings into social thought. Like Huxley, Carl Vogt made excursions into all the basic sciences, providing illustrative material for the grand law of organic evolution; however, unlike Lyell and Huxley, he favored a polygenic theory of the evolution of the basic human races—a view that did not find a single supporter among the early Russian evolutionists.

In addition to these works, all addressing themselves to the general public, there were also translations of technical studies in organic evolution appealing almost exclusively to a narrow circle of specialists. One of these studies was Fritz Müller's *Für Darwin*, based on the first intensive use of the Darwinian theory in embryology. In undertaking an empirical study of the Crustacea, Müller was inspired by Darwin's assertion that embryology stands to make a substantial contribution to the transformist view of nature. Darwin thought that embryology was of particular importance to the study of evolution because it was in an ideal position to throw light upon primeval history of species, which was of great help in the search for a natural system of classification. He said: "As the embryo often shows us more or less plainly the structure of the less modified and ancient progenitor of the group, we can see why ancient and extinct forms so often resemble in their adult state the embryos of existing species of the same class." "Embryology," he added, "rises greatly in interest, when we look at the embryo as a picture, more or less obscured, of the progenitor, either in its adult or larval state, all of the members of the same great class."[24]

---

[24] Charles Darwin, *Origin*, II, 254, 255.

In 1865, Lyell's *Principles of Geology* was published in a Russian translation. This work, the first volume of which was published in 1830, had inspired the young Darwin to dedicate his life to the study of evolutionary processes in organic nature. It should also be noted that during the early 1860's Russian readers were given an opportunity to read in their own language many Western biological classics that, while not directly identified with Darwin's ideas, presented the modern theoretical and methodological advances in all the branches of biology. Included in this list were Matthias Schleiden's study of plants, Isidore Geoffroy Saint-Hilaire's *General Biology*, and Claude Bernard's *An Introduction to the Experimental Method in Medicine.*

The reception of Darwin's theory by the wide spectrum of Russian scientific institutions was uneven: while at one extreme there were institutions that completely ignored biological evolution, at the other extreme were institutions that not only played a major role in the speedy diffusion of evolutionary ideas but also made Darwin's theory the point of departure of mushrooming scientific research.

During the 1860's and 1870's, the Academy of Sciences, with a membership consisting primarily of old scholars whose most productive years belonged to the pre-Darwinian past, did not produce a single Darwinian scientist of consequence. The biological sciences in the Academy were dominated by foreigners with distinguished records of scholarly achievement but of a decidedly conservative political cast—as guests of the Russian government, foreign academicians were noted for their inclination to avoid the great theoretical ideas of modern science that could be interpreted as inimical to the sacred values of the autocratic ideology. In his annual report on the activities of the Academy's staff in 1869, K. S. Veselovsky, permanent secretary of the Academy, made a direct reference to the contribution of Darwin's theory to the widening of the research base of natural science. He noted the growing interest in the evolution of animals and emphasized the limitless promises of the recent removal of the traditional boundaries separating botany and zoology from paleontology. "A comparison of presently existing forms of organic life with fossils buried in various strata of the earth—and belonging to various geological periods—will pave the way for an ultimate understanding of the general laws which have governed the transformation of life from its first appearance on earth to its present diversity and richness."[25] However, all the scientists mentioned by Veselovsky, with the exception of one, worked outside the Academy and were in their twenties. In 1867 the Academy elected Darwin a corresponding member. The members of the nominating committee recommended Darwin for the high quality of his scholarship, but they also noted that none of his works contained more errors than the *Origin of Species*.[26]

The learned societies founded before the 1860's and typified by the Moscow Society of Naturalists and the Russian Geographical Society greeted the new evolutionary ideas with complete silence. These societies were closed corporations with a clear aristocratic bias in the selection of new members. They discriminated against younger scholars, as well as against persons who did not speak French or German. They were dedicated to enriching descriptive natural history and had little use for theoretical natural philosophy—for scientific theory. In 1864 a group of Moscow University professors, dissatisfied with and openly critical of the caste exclusiveness and scientific myopia of the Moscow Society of Naturalists, which preferred to use the German language in scientific communication, founded a new scholarly association—the Society of the Admirers of Natural Science, Anthropology, and Ethnography—dedicated to attracting a large membership, using the Russian language in all its proceedings and publications, and keeping abreast of the most modern scientific ideas. It did not come as a surprise to its members when, in 1864, G. E. Shchurovsky opened the first session of the anthropological section of the new society with a long paper on anthropoid fossils and their significance for the understanding of the evolution of man.[27]

During the 1860's the universities became the major centers of scien-

---

[25] K. S. Veselovsky, "Otchet Imperatorskoi Akademii nauk po Fiziko-matematicheskomu i Istoriko-filologicheskomu otdeleniiam za 1869 god" [Report on the Physical-Mathematical and Historical-Philological Departments of the Imperial Academy of Sciences for 1869], *Zapiski Akademii nauk* [Journal of the Academy of Sciences] 17 (1870): 2–3, 16.

[26] For details on these elections, see G. A. Kniazev, "Izbranye Ch. Darvina chlenom-korrespondentom Peterburgskoi Akademii nauk (po materialam Arkhiva Akademii nauk SSSR)" [The election of Darwin as a corresponding member of the St. Petersburg Academy of Sciences (based on the materials of the archives of the Academy of Sciences of the U.S.S.R.)], *Priroda* [Nature] 11 (1931): 117–120.

[27] Alekseev, "Evoliutsionnaia ideia," p. 231.

tific research, a position previously occupied by the Academy of Sciences. The popular interest in the natural sciences was recognized by the university charter of 1863, which considerably expanded the curriculum coverage of these fields. University education and research benefited from the rapid growth and expansion of laboratories and libraries, the organization of modern naturalist societies anchored in individual universities, the frequent visits of professors to Western institutions of higher learning, and the expanded postgraduate studies of young Russians in the leading German and French universities. While the Academy of Sciences and most older naturalist societies were insulated from the social and ideological fermentation that swept the country after the Crimean War, the universities—to use a phrase of the eminent surgeon and anatomist N. I. Pirogov—were the true barometers of social pressure generated by new ideological currents and political strivings. For all these reasons, it is small wonder that the universities became the centers of evolutionary research. Only a few years after the publication of the *Origin of Species*, Russian universities produced a number of scholars ready to make the evolutionary conception the starting point of their research and to lay the groundwork for a rich scientific tradition in Russian Darwiniana.

Alexander O. Kovalevsky brought to an end the period of exclusive concern with the diffusion and popularization of Darwin's ideas and opened the period of original scientific research generated by—and directly related to—the new ideas of organic evolution.[28] A founder of modern comparative embryology, Kovalevsky never doubted the fundamental correctness of Darwin's theory; he was guided, not by the question of whether Darwin was correct or not, but by the question of why he must be correct. Darwin was interested in embryology as a source of

[28] For biographical information on A. O. Kovalevsky, see L. Ia. Bliakher, "Aleksandr Onufrievich Kovalevskii," in *Liudi russkoi nauki: Biologiia, meditsina, sel'skokhoziaistvennye nauki* [Men of Russian science: biology, medicine, agricultural sciences], ed. I. V. Kuznetsov (Moscow, 1963), pp. 157–172; K. N. Davydov, "A. O. Kovalevskii i ego rol' v sozdanii sravnitel'noi embriologii" [A. O. Kovalevsky and his role in the development of comparative embryology], *Priroda*, no. 4 (1916): 579–598; A. D. Nekrasov and N. M. Artemov, "Aleksandr Onufrievich Kovalevskii," postscript to *Izbrannye raboty* [Selected works], by A. O. Kovalevsky (Moscow, 1951), pp. 536–621.

substantive information confirming the evolutionary point of view; Kovalevsky was interested in the evolutionary point of view as an integrating and interpretive principle of embryological research. Kovalevsky was recorded in the annals of science as a thorough empiricist and one who wrote careful and remarkably precise summaries of his personal research ventures without showing the least inclination toward far-reaching abstractions and complex schemes of logical constructions. However, he made it abundantly clear that this preoccupation with the minutiae of empirical embryology derived a broad scientific meaning from the evolutionary idea of the morphological unity of animal types: the central concern of all Kovalevsky's empirical studies was the correlation of homologies with the various stages in the embryonic development of animals. The implicit purpose of his research was to contribute to a universal theory of embryonic growth—to erase the prevalent pre-Darwinian notion of a morphological abyss separating the vertebrates from the invertebrates.

Mechnikov made a pertinent comparison between Kovalevsky and Ernst Haeckel, as opposite types of evolutionary biologists. Haeckel was so preoccupied with the formulation of the universal laws of biogenetic relevance that he was willing to overlook the products of empirical research that could not be fitted into his schemes. He was not essentially a research scholar but a combination of a grand synthesizer and a formulator of hypotheses—basing much of his theory of gastraea on empirical data supplied by Kovalevsky. Kovalevsky, on the other hand, was convinced that in more delicate areas of evolutionary research—such as embryology—extreme care should be exercised not to overlook a single empirical detail.[29]

Kovalevsky's first study—which brought him a magister's degree at St. Petersburg University in 1865—analyzed the growth of the lancelet (*Amphioxus lanceolatus*), which at that time was usually classified as a vertebrate. The morphological simplicity of the lancelet led Kovalevsky to believe that it might represent a species occupying a transitional position between the vertebrates and invertebrates. His belief was not disappointing: the embryonic development of the lancelet fell into two

[29] I. I. Mechnikov, *Stranitsy vospominanii* [Pages from memory] (Moscow, 1946), pp. 31–32.

clearly distinct phases—an earlier phase that followed the pattern of growth common to invertebrates and a later phase that repeated the stages in the growth of vertebrates. Kovalevsky's research produced two results: it showed, first, that the lancelet should be classified as an invertebrate of the highest development and, second, that the embryological growth of vertebrates and invertebrates is basically similar.

Kovalevsky then undertook to study the ascidians, which at that time were classified by some as mollusks and by others as worms. His research showed that the embryonic development of the larvae of the ascidians displayed features characteristic for lancelets and lower vertebrates. He concluded that the ascidians were a degenerated vertebrate form. His discovery made a great impression on his contemporaries, more so in the West than in Russia. In the *Descent of Man*, Darwin took serious note of Kovalevsky's interpretation of the embryonic development of ascidians. He stated:

M. Kovalevsky has lately observed that the larvae of the Ascidians are related to the Vertebrata in their manner of development, in the relative position of the nervous system and in possessing a structure closely like the *chorda dorsalis* of vertebrate animals; and in this he has since been confirmed by Prof. Kupffer. M. Kovalevsky writes to me from Naples, that he has now carried these observations yet further; and, should his results be well established, the whole will form a discovery of the greatest value. Thus if we may rely on embryology, ever the safest guide in classification, it seems that we have at last gained a clew in the source whence the vertebrates were derived. I should then be justified in believing that at an extremely remote period a group of animals existed, resembling in many respects the larvae of our present ascidians, which diverged into two great branches—the one retrograding in development and producing the present class of Ascidians, the other rising to the crown and summit of the animal kingdom by giving birth to the Vertebrata.[30]

Kovalevsky returned several times to the study of lancelets and ascidians, but he also expanded his research interests to cover many additional species of invertebrates, represented mostly by the marine microfauna. All these studies added essential information in support of the homologies of the embryos of vertebrates and widely scattered representatives of invertebrates. The evolutionary basis of his theoretical orientation, the pre-

[30] Charles Darwin, *The Descent of Man* (New York: Collier, 1900), pp. 175–176.

cision and remarkable skill of his research techniques, and the general significance of his findings assured Kovalevsky of a notable place in the mainstream of the biological thought of the early post-Darwinian era. He not only helped strengthen the hold of Darwin's theory on modern scientific thought but also made a solid contribution to the accelerated growth of modern biology in Russia. In a paper delivered at the Eleventh Congress of Russian Naturalists, held in St. Petersburg in 1901, V. V. Zalensky, a distinguished member of the St. Petersburg Academy, noted that the evolutionary theory was "the main stimulus for the entire range of Kovalevsky's research." He added: "This was not a mere involvement in theory but a conscious and clear awareness of the great importance of evolution for the study of animals as the surest way to solve the grand questions of organisms, a problem which preoccupied the human mind since time immemorial. By the entire orientation of his work, Kovalevsky was an evolutionist, and he contributed to the theory of evolution more than any of his contemporaries."[31] In 1890, the St. Petersburg Academy of Sciences honored Kovalevsky by electing him into its full membership; the election was a particularly important event in the history of the Academy, for it marked its first major step toward the recognition of the limitless potentialities of Darwin's theoretical legacy.

Kovalevsky was not the only Russian scientist to help lay the foundations of modern comparative embryology; the contributions of I. I. Mechnikov belong to the same category of distinguished achievement.[32] Mechnikov's name is usually associated with the phagocytic theory, built upon a study of intracellular digestion among invertebrates, which helped explain the origin of multicellular animals and lay the foundations for comparative pathology. It was his work in the latter field that

31 Bliakher, "Aleksandr Onufrievich Kovalevskii," p. 172.
32 For details on Mechnikov's life and work, see G. K. Khrushchov, "Il'ia Il'ich Mechnikov," in *Liudi russkoi nauki*, ed. Kuznetsov, pp. 192–200; V. A. Dogel' and A. E. Gaisinovich, "Osnovnye cherty tvorchestva I. I. Mechnikova kak biologa" [Basic characteristics of I. I. Mechnikov's original work in biology], introduction to *Izbrannye biologicheskie proizvedeniia* [Selected biological works], by I. I. Mechnikov (Moscow, 1950), pp. 677–725; A. E. Gaisinovich, "Velikii russkii biolog I. I. Mechnikov" [The great Russian biologist I. I. Mechnikov], introduction to *Izbrannye proizvedeniia* [Selected works], by I. I. Mechnikov (Moscow, 1956), pp. 6–27.

earned him a Nobel Prize in 1908, which he shared with the distinguished German pathologist Paul Ehrlich.

In 1863, as an eighteen-year-old student at Khar'kov University, Mechnikov wrote an essay on the *Origin of Species* in which he criticized the main ideas of Darwin's thesis, particularly its derivation of the struggle for existence from Malthus's law of the discrepancy in the growth rates of population and food resources. This, however, did not prevent him from concluding that the theory presented in the *Origin of Species* was destined to have a great future and from considering himself one of its "most ardent followers." For unknown reasons, he left the manuscript unpublished.[33] After intensive study under several leading German biologists, he settled on embryological research, concentrating on the Mediterranean marine invertebrates. But, even in this research, he resisted a close adherence to the Darwinian theory; he clung steadfastly to his original idea that Darwin had advanced too many general ideas of purely hypothetical nature. To Kovalevsky, Darwin's theory was the incontrovertible basis of comparative embryology; to Mechnikov it was, at best, a challenging hypothesis requiring careful empirical testing. Mechnikov did not hesitate to criticize the broad evolutionary conclusions reached by Kovalevsky in his studies of lancelets and ascidians as devoid of a solid empirical basis. However, the deeper he became immersed in embryological research, the more closely he became identified with the basic theoretical principles of Darwin's legacy.

From 1865 to 1869, Mechnikov wrote about thirty papers on the embryonic growth of lower animals ranging from tapeworms to the *Balanoglossus*. He established that all animals investigated had two basic germ layers—ectoderm and endoderm—thus giving an embryological explanation of the unity of the animal kingdom.[34] The evolutionary commitment of his research led him to concentrate on comparative embryological

[33] The manuscript was published in 1950; see Mechnikov, *Izbrannye biologicheskie proizvedeniia*, pp. 655–672. Mechnikov was also critical of Fritz Müller's use of the biogenetic law as a proof of the Darwinian conception of evolution, rather than as a hypothesis in need of empirical testing and verification (I. I. Mechnikov, "Sovremennoe sostoianie nauki o razvitii zhivotnykh" [The current status of the scientific study of the development of animals], *Zhurnal Ministerstva Narodnogo prosveshcheniia* [Journal of the Ministry of National Education] 142 [1869]: 160–164).

[34] For details, see S. Zalkind, *Ilya Mechnikov: His Life and Work* (Moscow, 1959), pp. 47–49.

studies of animals whose morphological affinity had not yet been established. Some of his conclusions were not upheld by subsequent research; yet his evidence in favor of the genetic relationship of the Echinodermata, the Enteropneusta, and the Chordata has gone unchallenged.

In the early 1870's, Mechnikov shifted his research emphasis from the homologous features of the germ layers of all animals to the incipient steps in the development of multicellular animals from an ovum, a problem that brought him into direct conflict with Haeckel's gastraea theory and produced challenging ideas on the evolution of specialized digestive cells and on the role of nutrition in metabolism. In the improvement of intracellular digestion he saw the primary force behind the emergence of multicellular organisms. The study of digestion, in turn, led him to the formulation of the phagocytic theory and the theory of immunity. At the time when comparative pathology became his major concern, he left Russia to become a distinguished associate of the Pasteur Institute in Paris.[35]

Despite his persistent criticism of certain aspects of Darwin's theory, it would not be an exaggeration to say that Mechnikov's entire scientific work and all his theories were parts of a brilliant search for the deeper meanings of the scientific legacy of the great English naturalist.[36] To Mechnikov, Darwinism was a theoretical orientation of a very broad nature. He saw its basic source of power in the combination of a historical view of nature, a comparative approach to biological phenomena, and an identification of purposiveness in the organic world with the processes of adaptation as a means of survival. If his scientific work had weaknesses, they did not stem from a lack of scholarly dedication and experimental skill, but from his temperament: unsettled and excitable, he moved too swiftly—particularly in his embryological work—from one research un-

[35] Mechnikov was among the scientists who gathered in Cambridge in 1909 to commemorate the centenary of the birth of Charles Darwin and the fiftieth anniversary of the publication of the *Origin of Species*. The event prompted him to write an article on the relationship of Darwinism to medicine, in which he cited his own work in inflammation as an example of the fruitfulness of the evolutionary approach in scientific research directly related to medical problems. The world of microorganisms, he argued, is not immune to changes through adaptation to the changed external environment (I. I. Mechnikov, 'Darvinizm i meditsina" [Darwinism and medicine], in *Pamiati Darvina*, by Kovalevsky et al., pp. 112–116).

[36] Gaisinovich, "Velikii russkii biolog I. I. Mechnikov," p. 14.

dertaking to another to do justice to all of them; his embryological research, for example, covered representatives of almost all the major groups of invertebrates and also some of the vertebrates.[37]

The third towering figure among the pioneers of Russian Darwinism was Vladimir Kovalevsky, who helped inaugurate evolutionary paleontology, a field in which he achieved results comparable to the most signal accomplishments of Alexander Kovalevsky (his older brother) and Mechnikov.[38] As a young man—before he chose paleontology as an academic specialty—he helped Darwin's ideas reach the ever-widening circles of the Russian enlightened public. He was the moving force behind the publication of the Russian translation of Darwin's *Variation*, and he was one of the chief translators of Darwin's *Expression of the Emotions in Man and Animals*. This time, too, the translation was rendered from galley proofs sent by Darwin.[39] Kovalevsky translated some and edited the rest of an entire series of scientific books that, while not necessarily dedicated to the problems of organic evolution, contributed to the ad-

---

[37] N. M. Kulagin, "I. I. Mechnikov kak zoolog" [I. I. Mechnikov as a zoologist], *Priroda*, no. 5 (1915): 703.

[38] For details on V. O. Kovalevsky's life and work, see L. Sh. Davitashvili, *V. O. Kovalevskii*, 2d ed. (Moscow, 1951); A. A. Borisiak, *V. O. Kovalevskii: Ego zhizn' i nauchnye trudy* [V. O. Kovalevsky: his life and scientific work] (Leningrad, 1926); L. Sh. Davitashvili, "V. O. Kovalevskii, ego nauchnaia deiatel'nost' i znachenie ego trudov po paleontologicheskoi istorii semeistva loshadinykh" [V. O. Kovalevsky, his scientific work and importance of his studies in the paleontological history of the family of equines], postscript to *Paleontologiia loshadei* [The paleontology of horses], by V. O. Kovalevsky (Moscow, 1948), pp. 258–314; D. N. Anuchin, "Kovalevskii, Vladimir Onufrievich," *Biograficheskii slovar' professorov i prepodavatelei Imperatorskogo S.-Peterburgskogo universiteta za istekuiushchuiu tret'iu chetvert' veka ego su shchestvovaniia, 1869–1894* [Biographical dictionary of professors and lecturers at the Imperial St. Petersburg University during the third quarter of the first century of its existence, 1869–1894] (St. Petersburg, 1896), pp. 324–326. For Kovalevsky's place in the development of evolutionary paleontology, see Iu. I. Poliansky, "Osnovnye puti razvitiia darvinizma" [Main currents in the development of Darwinism], in *Istoriia evoliutsionnykh uchenii v biologii* [The history of evolutionary theories in biology], ed. V. I. Poliansky and I. I. Poliansky (Moscow and Leningrad, 1966), pp. 311–312; V. A. Alekseev, *Osnovy darvinizma* [The foundations of Darwinism] (Moscow, 1964), pp. 291–295; L. Sh. Davitashvili, *Istoriia evoliutsionnoi paleontologii ot Darvina to nashikh dnei* [History of evolutionary paleontology from Darwin to the present time] (Moscow and Leningrad, 1948), pp. 74–127.

[39] Davitashvili, *V. O. Kovalevskii*, pp. 152–154.

vancement of the philosophy of science in which the *Origin of Species* was firmly anchored. Some translations, typified by T. H. Huxley's *On Our Knowledge of the Causes of the Phenomena of Organic Nature* (1864), were dedicated to an expansion and popularization of Darwin's theory.

Kovalevsky dedicated only four years—from 1869 to 1873—of his eventful and mercurial life to original scientific research. After intensive preparatory work in all the major earth sciences, geological field work in the Permian formation in Thuringia and the Cretaceous formations in England and southern France, and careful studies of ungulate fossils deposited in leading Western European museums, he undertook to reconstruct the main trunk in the evolution of the horse from the late Eocene period to the present time—from ancient palaeotheres and paloplotheres to modern *Equus*. The same problem attracted the attention of T. H. Huxley, who in 1872 pointed out that the rich collections of ungulate fossils made it possible to reconstruct the evolution of the horse through a long period of geological time. While the list of Kovalevsky's publications was not long, the results of his findings marked the turning point in the development of paleontology from a closed science encumbered by the strong tradition of Georges Cuvier's and Louis Agassiz's antitransformism to an open science guided by the evolutionary principle.

In his major studies, Kovalevsky concentrated on the modifications in the skeletal structure of horses as a result of fundamental adaptations of various organs to changing environmental conditions. In the opinion of a modern expert, Kovalevsky "made out an essentially correct story of the mechanical evolution of the horse's foot and dentition" even though he studied the fossils representing genera that were not directly ancestral to *Equus*.[40] Particularly in the evolution of the feet of the ungulates— characterized by increasing morphological simplification and reduction —Kovalevsky was able to illustrate the role of natural selection as a mechanism of variation leading to the emergence of new species. He found out that the pressure of changing environmental conditions led to the emergence of two types of skeletal modifications in feet: "adaptive"

[40] D. M. S. Watson, "The Evidence Afforded by Fossil Vertebrates on the Nature of Evolution," in *Genetics, Paleontology, and Evolution*, ed. Glenn L. Jepsen et al. (New York: Atheneum, 1963), p. 48. Also see George Gaylord Simpson, *Horses* (New York: Oxford University Press, 1951), pp. 87–88.

and "inadaptive." "Adaptive" modifications represented more drastic deviations from established forms and were considerably less frequent than "inadaptive" modifications. He viewed evolution as a process of many phases and many lines: evolution means the survival of animal forms with adaptive morphological features and the extinction of forms with inadaptive features. There is no direct and one-sided relationship between morphological specialization and progress; in some cases specialization and progress are synonymous; in other cases a trend opposite to specialization is equal to progress. Among the ungulates the evolutionary simplification of the skeleton was a sign of progress, that is, of better adaptation to the rigors of environment.[41] Kovalevsky was also firmly convinced that without the close scrutiny of aberrant forms there could be no meaningful understanding of the full scope of the evolutionary process.

The American paleontologist Henry Fairfield Osborn gave the following appraisal of the quality, historical meaning, and methodological bent of Kovalevsky's work: "The remarkable memoirs of Vladimir Onufrievich Kovalevskii (1842–1883) . . . are monuments of exact observation of the details of evolutionary change in the skull, teeth, and feet, and of the appreciation of Darwinism. In the most important of these memoirs, entitled *Versuch einer natürlichen Classification der Fossilen Hufthiere* (1875), we find a model union of detailed inductive study with theory and working hypothesis. These works swept aside the dry traditional fossil lore which had been accumulating in France and Germany. They breathed the new spirit of recognition of the struggle for existence, of adaptation and descent."[42]

No Russian scientist of the early phase of evolutionary science had had more personal contact with Darwin than Kovalevsky. He made several visits to Darwin's home in Down, and there are tangible indications that the two exchanged occasional letters. In the only letter that has been preserved, Darwin expressed high hopes for the promising scientific contributions of the Kovalevsky brothers and informed Vladimir that his paleontological ideas had found influential supporters in England. He was pleased to hear from Vladimir that he had decided to dedicate his

[41] Davitashvili, *Istoriia evoliutsionnoi paleontologii*, p. 108.
[42] H. F. Osborn, *The Age of Mammals in Europe, Asia and North America* (New York: Macmillan, 1910), p. 6.

forthcoming monograph on *Anthracotherium* to him.[43] On at least two occasions Vladimir transmitted to his brother Darwin's requests for clarifications of specific problems in invertebrate morphology.[44]

The scientific work of the Kovalevsky brothers and Mechnikov represented the crowning point in the reception and early application of Darwin's ideas by Russian natural scientists. But how were these scientists accepted by the Russian scientific community? While it was true that the St. Petersburg Academy of Sciences published their papers, it was also true that all three had difficulty in finding suitable academic employment. At the time when the most coveted academic positions were concentrated in St. Petersburg and Moscow, A. O. Kovalevsky and Mechnikov had no choice but to accept teaching positions in provincial universities, equipped with poor libraries and laboratories and wanting in intellectual stimulus for sustained scientific work. Mechnikov's candidacy for a position in the Medical and Surgical Academy in St. Petersburg proved futile. He served fifteen years on the faculty of the newly founded Odessa University, resigning in 1882 to avoid the grueling pressure of academic intrigue. In 1887, before he had reached the peak of his scientific career, he left Russia to join the Pasteur Institute in Paris, where he remained until his death in 1916. A. O. Kovalevsky spent twenty-two years of his academic career in Kazan, Kiev, and Odessa universities fighting the depressing monotony of provincial isolation by extensive correspondence with Western European embryologists, occasional scientific trips to the Mediterranean Sea, and cooperative research ventures with eminent foreign scientists.

---

[43] The dedication reads in part: "It gives me great joy to dedicate this work to you not because I consider my work particularly worthy of such a dedication but because it gives me an opportunity to express my profound respect for you personally. From the very beginning of my scientific work you have been my best teacher and friend: you have paid full attention to my entire work and, during my prolonged stay in England, you have created favorable conditions for my research. Your intercession opened many collections and libraries to me which otherwise might have remained closed. Your name and your friendship were the best recommendations which opened all the doors to me" (V. O. Kovalevsky, *Sobranie nauchnykh trudov* [Collection of scientific works] [Moscow, 1960], III, 99).

[44] For details on personal relations between Darwin and Kovalevsky, see Davitashvili, *V. O. Kovalevskii*, pp. 144–171.

Vladimir Kovalevsky fared much worse: during the 1870's all his efforts to secure a teaching position fell on deaf ears. It is difficult to identify the main reason for this misfortune. The authorities did not forget that he was the publisher of an entire series of translations of Western scientific or semiscientific works with "materialistic" leanings. His undergraduate training was in law rather than in natural science. His study abroad was neither sponsored nor supervised by the Ministry of National Education. He echoed the Nihilist motto that the function of science is not only to understand nature but also to usher in a better society free of the vestiges of feudal institutions. He was a paleontologist at a time when the university curriculum did not have a clearly defined slot for paleontology. Like his brother and Mechnikov, he encountered formidable opposition from men, occupying powerful positions in the scientific establishment, who had achieved scholarly eminence by their work in nontransformist biology and paleontology of the pre-Darwinian epoch. It is most likely that the professional hardships of the three evolutionists did not result solely—and perhaps not even primarily—from their firm identification with Darwin's theory. The signal characteristic of the early phase of Russian Darwinism was that the most original and innovative studies were carried out outside St. Petersburg and Moscow, the country's centers of science.

Vladimir Kovalevsky was at first a dedicated Nihilist fighting against the sacred culture of autocratic Russia by helping accelerate the diffusion of modern scientific knowledge, and later he became a dedicated scientist. N. D. Nozhin followed an opposite course: he began as a searching scientist of great ambition and talent and ended as an ideologist dedicated primarily to emancipating his country from both the decaying feudal law and the burgeoning capitalist relations. After having studied chemistry under Robert Bunsen at Heidelberg University, he moved to Tübingen University, where he studied zoology under H. G. Bronn, the German translator of the *Origin of Species*, but not a Darwinist. In 1863 he went to Italy to conduct research on the embryonic growth of selected species of Mediterranean microfauna for the purpose of establishing morphological links between vertebrates and invertebrates. It was in Italy that he established close working relations with A. O. Kovalevsky. In his spare time he translated Fritz Müller's *Für Darwin* into Russian;

this work was generally acclaimed as the first successful effort to combine meticulous embryological research with Darwinian transformism.

During his sojourn in Germany, Nozhin participated in several *kruzhoks* ("circles") of Russian students eager to find ways of bringing modern political and social ideas to their native land. It was at this time that Nozhin began to think of Darwin's theory as a source of a unitary picture of the evolution of the universe, both natural and social. To him Darwin's work was the culminating point of nineteenth-century science, and science was the only source of sound guideposts for purposive action in social development. Like the champions of Nihilism, he preached "a visionary faith in science" as the true power of reason and declared that "all scientific knowledge in the hands of its honest servants stands in direct opposition to the existing order" and that "in the world there is only one evil—ignorance—and only one way to salvation—science."[45] He argued that only by knowing the laws of nature could man infuse more humanity into his social existence. However, he rejected the struggle for existence as the moving force of evolution; he called it an aberration, a pathological force exercising a negative influence on both natural and social evolution. He called Darwin a "bourgeois-naturalist" for his emphasis on competition—rather than on cooperation—as the mainspring of biological and social development.[46]

Nozhin gave the Populist sociology—particularly as articulated by N. K. Mikhailovsky—its guiding idea. While A. O. Kovalevsky saw in Darwin's theory a fruitful method of scientific analysis, Nozhin saw in it the culminating point in the evolution of the modern scientific world view—a triumphant victory of reason over metaphysical mysticism and religious dogma. Nozhin was the first Russian scientist to express two thoughts that found strong following during the subsequent decades: first, not the struggle for existence but the elimination of transcendental causation in the development of nature was Darwin's major contribu-

---

45 N. D. Nozhin, "Nasha nauka i uchenye" [Our science and scholars], *Knizhnyi Vestnik* [Book messenger], no. 1 (1866): 21. See also Davydov, "A. O. Kovalevskii kak chelovek i uchenyi," pp. 332–333; Viktor Chernov, "Gde kliuch ponimaniiu N. K. Mikhailovskogo?" [The key for the understanding of N. K. Mikhailovsky], *Zavety* [Legacy], no. 3, sec. 2 (1913): 114.

46 N. D. Nozhin, "Nasha nauka" [Our science], *Knizhnyi Vestnik*, no. 7 (1866): 175.

tion to science; and, second, sociology was much in debt to Darwin's theory—not to the notion of the struggle for existence, but to the unitary developmental scheme and rational models for social analysis.

Karl von Baer, an eminent member of the St. Petersburg Academy of Sciences and the greatest embryologist of the pre-Darwinian era, occupied the leading and most influential position among the early Russian anti-Darwinists. He belonged to the group of scientists who prepared the ground for the emergence of Darwin's evolutionary theory. His naturalist historicism had philosophical roots in epigenetic embryology, to which he had made contributions of lasting value. He thought that the structural unity of more advanced animals made transformation within the world of vertebrates a distinct possibility. He was an evolutionist, for he was firmly convinced of a continuous and causally linked change in natural phenomena; but he was only a partial transformist, for he categorically rejected the idea of the genetic unity of the animal kingdom. His "theory of types," which he advanced independently of Cuvier, allowed for transformation only within firmly established animal groups. He could not help but recognize that some of his own most formidable contributions had gone into the making of the theory propounded in the *Origin of Species*. It was no surprise that in 1860 he wrote to Huxley: "J'ai énoncé les mêmes idées sur la transformation des types ou origine d'espèces que M. Darwin."[47] Nor was it a surprise that Darwin greeted von Baer's approval of his theory as "magnificent news." Soon, however, Darwin learned that von Baer had changed his mind and had allied himself with the leaders of the anti-Darwinian movement.

Huxley and Darwin had good reason to expect von Baer's endorsement of the evolutionary idea; but they should not have been surprised when he suddenly turned his back on it. Von Baer's evolutionary thought was subject to constant shifts in interpretation and emphasis; at some times it assumed very broad proportions, while at other times it was drastically limited in both scope and meaning. Von Baer's scientific orientation was not immune to heavy ideological influences; in attacking Darwin's theory during the 1860's, he was not merely defending an old

---

[47] Darwin, *Life and Letters*, II, 122. "I have stated the same ideas on the transformation of types or origin of species as Mr. Darwin."

scientific legacy but was also fighting the scourge of Nihilist materialism and the widespread current attack on traditional values. To make the situation even more complicated, von Baer oscillated between two notions of evolution. At one extreme was the notion of the universal development of nature, embodied in the Schellingian *Naturphilosophie*, which viewed evolution as a gradual expansion of the domination of spirit over matter and favored Aristotelian teleology over Newtonian causation, at least in the interpretation of the organic world. At the other extreme were the new scientific ideas coming from embryology, paleontology, and several other natural sciences, which invited broad causal-mechanistic interpretations. Von Baer did not make an effort to blend these two orientations into a logically coherent and functional theory; moreover, after he became acquainted with the *Origin of Species*, his speeches and papers showed an increased reliance on the antimechanistic interpretation of evolution embodied in the *Naturphilosophie*.

In his attacks on Darwin's theory, von Baer did not limit himself to the idea of transformism. He bitterly opposed the "materialistic" orientation of modern natural science, which received powerful support from Darwin's theory. In a paper read in 1861, on the occasion of the opening of the Entomological Society in St. Petersburg, von Baer lamented the current popularity of scientific materialism and argued that a science grounded in idealistic ontology would give a much more complete picture of the universe.[48] The basic weakness of materialistic science, he argued, was that it did not—and could not—account for a teleological element in the processes of nature. During the early 1860's, von Baer was not a consistent antitransformist. For example, in the article "Anthropology," written for the "Encyclopedic dictionary," published in 1862, he viewed the brain of anthropoids as occupying an intermediate position between the cerebral cortex of man and that of the higher animals. The entire article was written in the spirit of transformism, allowing no room for creationist ideas.[49] Since the essay made no reference to

---

48 K. S. Veselovsky, "Otchet po Fiziko-matematicheskomu i Istoriko-filologicheskomu otdeleniiam Imperatorskoi Akademii nauk za 1861 god" [Report on the Physical-Mathematical and Historical-Philological Departments of the Imperial Academy of Sciences for 1861], *Zapiski Akademii nauk* [Journal of the Academy of Sciences] 1 (1862): 26.
49 Alekseev, "Evoliutsionnaia ideia," p. 221.

Darwinian transformism, it was most probably written before von Baer had read the *Origin of Species*. In all his subsequent writings that treated the problems of evolution directly or indirectly, he was adamant in his condemnation of Darwin's theory on both moral and scientific grounds. Darwin's theory, according to von Baer, was inadequate because its major concept—natural selection—was a hypothesis without foundations in empirical data, it ignored the role of heterogenesis (as formulated by Albert von Kölliker in 1864) in biological transformation, it disregarded purposiveness in the life of nature, and it advanced the unsupportable thesis of the anthropoid origin of man. It was the latter component of Darwinian theory that von Baer found most irritating and that prompted him to lecture and write extensively against the new evolutionary idea.

In 1865–1867 the journal "Naturalist" carried in numerous installments a Russian translation of von Baer's anthropological essay "Man and Ape," directed mainly against Huxley's extension of Darwin's evolutionary theory to include man. The translation was too cumbersome to appeal to the general reader and contained many naïve and sweeping statements inserted by the translator.[50] In 1868, von Baer repeated the same arguments in a popular lecture at Dorpat University. In 1873, angered by the ideas advanced by Darwin in the *Descent of Man*, he published two articles critical of the modern evolutionary theory. In the first article, published in the *Augsburger allgemeine Zeitung* and immediately reprinted as a pamphlet, he rejected the antiteleological orientation of Darwin's followers and reaffirmed his allegiance to the romantic *Naturphilosophie*, which regarded purposiveness as a universal attribute of living nature.[51] He thought that Darwinian theory was popular not because of its scientific merits but because it was in full harmony with the materialistic bent of modern ideologies. Von Baer objected particularly to the Darwinian emphasis on chance variation in the emergence of new species, for it denied supernatural causation in the development of nature. In the second article, published by the St. Petersburg Academy of

---

[50] B. E. Raikov, *Karl Ber: Ego zhizn' i trudy* [Karl von Baer: his life and work] (Moscow and Leningrad, 1961), p. 399.

[51] Karl von Baer, "Zum Streit über den Darwinismus," *Augsburger allgemeine Zeitung*, no. 130 (1873): 1986–1988.

Sciences, he attacked Darwin's claim that Alexander Kovalevsky's study of ascidians might lead to the discovery of transitional forms between invertebrates and vertebrates; his wrath centered on the "dilettantes" who saw the ancestors of man in ascidians.[52] However, he admitted that he did not reject the idea of "the transmutation of living forms" and said that he merely wanted to point out that there was no scientific information corroborating the theory of the evolutionary unity of the animal world. While not excluding the possibility of "transmutation" processes in organic nature, he was eager to show that the Darwinian notion of "transformism" was a hypothesis without an empirical basis. He also wanted to show that Darwin's notion of the evolution of man was a threat to the ethical ideals of humanity.

Von Baer elaborated both of these arguments in a long essay presented in the second volume of his *Reden*, which was published in 1876, the year of his death. Criticizing those who claimed that Darwin's law of evolution, as a fundamental explanation of living nature, was comparable to Newton's law of gravitation, as a fundamental explanation of physical nature, he pointed out that Darwin's theory was not a general evolutionary theory in the first place, but a particular explanation of the work of selection in nature. He advanced detailed arguments against natural selection and the struggle for existence, relying on information selected from morphological disciplines and zoogeography. The basic weakness of Darwin's theory, according to him, was that it violated the rigor of scientific methodology, the inductive-empirical orientation of modern natural science, and the verification standards and procedures. He ended the essay with this advice to scientists: "I want to offer only one thought to the scientists: a hypothesis may be necessary and valuable only if it is treated as a hypothesis, that is, if one takes its basic premises as topics of special inquiry. But a hypothesis may be unnecessary and harmful if, by disregarding proofs, we treat it as an end-product of our search for knowledge. Our knowledge is fragmentary. Some persons may find satisfaction in filling in the gaps in scientific knowledge by relying on presuppositions, but that is not science."[53]

---

52 Karl von Baer, "Entwickelt sich die Larve der einfachen Ascidien in der ersten Zeit nach dem Typus der Wirbelthiere?" *Memoirs de l'Académie des Sciences de St.-Petersburg*, 7th ser. 19, no. 8 (1873): 1–36.

53 Karl von Baer, *Reden, gehalten in wissenschaftlichen Versammlungen und kleinere*

Although von Baer's essay was wordy, repetitious, inconsistent in its basic theoretical propositions, and saturated with moral pathos, it was a work of notable historical value. As has been pointed out by B. E. Raikov, it is a basic document for the understanding of von Baer's complex and extensively ramified world view, which reflected a dedicated search for a middle ground between the new ideas in biology and the echoes of the old science born at the beginning of the nineteenth century.[54] It is also the most thorough synthesis of early anti-Darwinian arguments, as advanced by the representatives of various branches of biology. Even though it opened the doors for a systematic and thorough criticism of Darwin's transformist ideas, its contemporaries received it with inexplicable silence; a fact that caused much grief to Strakhov.[55] It was not until 1885 that N. Ia. Danilevsky, an ichthyologist and philosopher of history, published the first volume of his *Darwinism*, which, much indebted to von Baer's work, was the first comprehensive critique of Darwin's theory published in the Russian language.[56] But neither the work of von Baer nor that of Danilevsky could prevent the Darwinian theory from finding in the Russian scientific community a most enthusiastic, gracious, and appreciative host.

Aufsätze vermischten Inhalts, 2 vols. (St. Petersburg, 1876), II, 473; also in Raikov, *Karl Ber*, p. 412.

[54] Raikov, *Karl Ber*, p. 412.

[55] Strakhov, *Bor'ba s Zapadom*, II, 279.

[56] K. M. Ber, *Avtobiografiia* [Autobiography] (Moscow, 1950), p. 525.

RUSSIA
Social Sciences

JAMES ALLEN ROGERS

The natural sciences in general and Darwinism in particular became intimately linked with Russian social and political thought from the 1860's. This situation was not, of course, unique to Russia. Darwinism also possessed social and political significance in Western Europe. But there Darwinism acquired adherents of all hues on the political spectrum who read Darwinism in the light of their own experience. Their relative freedom to separate political and religious beliefs, to express those political beliefs, and to engage in political activity, however limited, had no corollary in Tsarist Russia where "all politics was by definition revolutionary."[1] No "conservative" school of thought in Russia could even conceive of using Darwinism to rationalize a social structure resting on an autocracy that found its ultimate justification in divine sanction. The conservative and religious rejection of Darwinism meant that Darwinism acquired among Russian social thinkers a social and political context reflecting only the radical side of the political spectrum.

These radical thinkers were primarily publicists for various journals rather than philosophers or social scientists. The controversy over the social significance of Darwinism in Russia took place in these journals rather than in academic institutions or learned journals. A study of the reception of Darwinism by Russian social thinkers from the 1860's to the 1880's is consequently a study primarily of the impact of Darwinism

[1] Richard Pipes, *Struve, Liberal on the Left, 1870–1905* (Cambridge: Harvard University Press, 1970), p. 275.

on Russian radical thought. The response to Darwinism by Russian social thinkers helped lay the foundation for the school of Russian subjective sociology, a unique contribution of Tsarist Russia to the social sciences.

After the 1880's the response to Darwinism by Russian social thinkers reflected European rather than specifically Russian controversies over Darwinism.[2] Since these European controversies are less relevant to a study of the response of Russian social thinkers to Darwinism, this paper will cover only the period through the 1880's.

Russian social thinkers became acquainted with Darwinism in the early 1860's against the background of a strong Russian scientific interest in the possibility of biological evolution. From the eighteenth century until the appearance of Darwin's works in Russia in the 1860's, many Russian naturalists had speculated on the feasibility of organic and inorganic evolution.[3]

The strong interest of Russian social thinkers in Darwinism was not the result, however, of the history of Russian scientific interest in the possibility of biological evolution. It was instead the result of the role that the natural sciences had come to play in Russian radical thought in the late 1850's and the 1860's. Russian radical thinkers hoped to undermine belief in the social and political foundations of the Russian autocracy by a widespread diffusion of the leading theories of the natural sciences. Their attitude was reinforced by the increased repression of the Russian autocracy and by the apathy of the Russian masses. This led many of the young Russian radicals to lose faith not only in reform from above but also in revolution from below. They believed that they could depend only upon the "truths" of the natural sciences, and they began to call themselves "critical realists."

During this period the young Russian radicals became acquainted with Darwin's *Origin of Species* in Russian translation (1864).[4] They found

---

[2] These controversies, particularly as they related to sociology, have been well covered in Chapter 14 of Alexander Vucinich's *Science in Russian Culture, 1861–1917* (Stanford: Stanford University Press, 1970), pp. 424–473.

[3] These naturalists and their theories are covered in detail in B. E. Raikov's *Russkie biologi-evoliutsionisty do Darvina* [Russian biological-evolutionists before Darwin], 4 vols. (Moscow and Leningrad: Academy of Sciences, 1951–1959).

[4] The translator was Sergei A. Rachinsky (1833–1902), professor of botany at the University of Moscow. He began the translation in 1862 and published an article on

in Darwinism the keystone to the structure of their materialistic philosophy. For them, Darwin became the Newton of biology, who had explained the development of the organic world without recourse to creator
or purpose by a comprehensive "mechanistic" law that made possible a
consistent philosophical materialism. The Russian radicals welcomed
Darwinism with extraordinary enthusiasm, and they soon made acceptance of Darwinism the badge of progressive thought.

Because the Russian translation of 1864 offered neither preface nor
explanation to the reader unfamiliar with the significance of Darwin's
discovery, the young critical realist and popularizer of science Dmitri
Pisarev (1840–1868) wrote a book-length review to acquaint the Russian public with Darwinism. Pisarev's major contribution to the social
interpretation of Darwinism in Russia was to read into Darwin's theory
the primary tenet of critical realism: rational egoism. "The conclusion is
that every species constantly operates only for its own sake," Pisarev
wrote, "and the fullest egoism constitutes the fundamental law of life
for the entire organic world."[5] Pisarev exaggerated the Malthusian struggle for existence without understanding that Darwin had used the concept only in a metaphorical sense. Moreover, Pisarev made no distinction between biological and social evolution and did not see the social
implications of his theory of history based primarily on the Malthusian
struggle for existence. (The equation of Darwinism with the Malthusian struggle for existence, which Pisarev introduced into the discussion
of the social significance of Darwinism, was to confuse that discussion
among radical and conservative Russian thinkers from that time.)

Pisarev's interpretation of Darwinism was carried to a more extreme
position later in 1864 by his friend Varthalomew Zaitsev (1841–1882),
a radical publicist. Advocating a theory of polygenesis, Zaitsev believed
that the various human races originated from different species of the
higher animals. This led him to believe that the black race was inferior to

---

Darwinism in 1863: "Tsvety i nasekomye" [Flowers and insects], *Russkii vestnik*
[The Russian herald] (January 1863).

[5] Dmitri Pisarev, "Progress v mire zhivotnykh i rastenii" [Progress in the world of
animals and plants], in *Sochineniia* [Works], by idem, 6 vols. (St. Petersburg, 1897),
III, 360. Pisarev took the idea of rational egoism from Nikolai Chernyshevsky (1828–
1889), the leading critical realist, but he emphasized only the individualistic or
"nihilistic" side of it. All translations are mine.

the white and was therefore "naturally" enslaved.[6] Zaitsev's racism shocked the Russian revolutionary intelligentsia. It was incomprehensible to them that anyone on their side of the political spectrum should condone slavery or racism. Maxim A. Antonovich (1835–1918) spoke for the revolutionary intelligentsia when he declared in a leading radical journal: "Whatever zoology says, common sense and the general welfare must be respected."[7]

The controversy over Zaitsev's racism led the radicals to question the relationship of Darwinism to the idea of progress and the underlying assumption that the natural sciences were the repository of an impersonal and objective truth. Reflections on the human implications of Darwinism gave rise to an anthropomorphic criticism of Darwinism guided by the Proudhonian concept of *mutualité*.[8] The idea of mutual aid furnished the young biologist and radical, Nikolai D. Nozhin (1841–1866), with the basic concept for the biological and social interpretation of Darwinism.[9]

Nozhin agreed that there was a struggle for existence in human society, but he found its origin not in biology but in the division of labor favored by economists for the accumulation of capital. Nozhin insisted that, in the absence of a division of labor, biological organisms of the same species strive to help one another and struggle only with the forces of nature. In concluding his polemic against Zaitsev and the Malthusian aspect of natural selection, Nozhin declared that only mutual aid among

[6] V. A. Zaitsev, *Izbrannye sochineniia* [Selected works], ed. B. P. Koz'min, 2 vols. (Moscow: All-Union Society of Former Political Convicts and Exiles, 1934), I, 228–231, 428–442. Zaitsev's interpretation of biological evolution came almost entirely from the theory of polygenesis advocated by Carl Vogt in *Vorlesungen über den Menschen, seine Stellung in der Schöpfung und in der Geschichte der Erde* (Geissen: J. Ricker, 1863).

[7] *Sovremennik* [The contemporary], January 1865: 163. Antonovich did, however, favorably review the Russian translation of the *Origin of Species* in *Sovremennik*, March 1864: 167.

[8] The concept of mutual aid was developed in Pierre Joseph Proudhon's posthumously published work, *De la capacité politique des classes ouvrières* (Paris: Librairie internationale, 1865). Proudhon's influence on the more radical critical realists is discussed in James A. Rogers, "Proudhon and the Transformation of Russian Nihilism," *Cahiers du Monde Russe et Soviétique* 13 (October 1972): 514–523.

[9] Nozhin joined the Proudhonian idea of mutual aid to the ideas on the variation of species through growth and adaptation that he had already gained from the reading of Fritz Müller's *Für Darwin* (Leipzig: Wilhelm Engelmann, 1864), which Nozhin translated into Russian as *V zashchitu Darvina* [In defense of Darwin].

men would produce fully integrated individuals enjoying health and freedom.[10]

From the time of Nozhin's unexpected death in 1866, Russian radicals began to accept Darwinism largely within the biological and sociological interpretation first espoused by Nozhin. For them, it clarified the relationship between the natural sciences and human progress. Nozhin had transformed the critical-realist doctrine of objective science and automatic progress into a historical theory finding its fullest expression in mutual aid leading to human solidarity. The concept of mutual aid furnished a criterion to relate Darwinism to the idea of historical progress by bringing purpose back into evolution. That purpose was the fullest development of the freedom of the integral personality, derived from a transformation of society based on mutual aid rather than on a division of labor.

Nozhin's concept of mutual aid and human progress left, however, the most important question unanswered for the radicals: was mutual aid implicit in the process of biological evolution itself, or was it the result of man's desire for human solidarity once he advanced to a certain stage of evolution? The Russian radical thinkers gave three different answers to this question. The socialist revolutionaries, following Nikolai Mikhailovsky, believed that mutual aid was a necessary, although "subjective," choice of man. The anarchists, after Peter Kropotkin, tried to prove that mutual aid was inherent in all biological evolution. The economic determinists and Marxists insisted that mutual aid could only be the product of economic, rather than biological, determinism.

Nikolai Mikhailovsky (1842–1904) had been Nozhin's roommate in St. Petersburg and began his interpretation of Darwinism under the influence of Nozhin's ideas. While Nozhin had attacked the interpretation of Darwinism that emphasized the Malthusian struggle for existence, Mikhailovsky focused on the deficiencies he saw in Herbert Spencer's "survival of the fittest," a phrase that Darwin had equated with natural selection beginning with the fifth edition of the *Origin of Species*.[11]

---

[10] N. D. Nozhin, "Po povodu statei *Russkogo slova* o nevol'nichestve" [In connection with the articles in *Russkoe slovo* about slavery], *Iskra* [The spark] 1865, no. 8: 114–117; idem, "Nasha nauka i uchenye" [Our science and scientists], *Knizhnyi vestnik* [The book herald] 1866, nos. 1–3, 7, esp. no. 7, p. 175.

[11] "The expression often used by Mr. Herbert Spencer of the Survival of the Fittest

Mikhailovsky opposed Herbert Spencer's attempt to replace an anthropocentric view of human progress with an impersonal and objective law of evolution. He agreed with Spencer that it was possible to describe objectively the elements of human history. But he felt that, to evaluate these elements with regard to the idea of progress, it was necessary to understand the ideals of the human participants. These ideals, reflecting the subjective desires of individuals, were the teleological element of sociological investigation. But Spencer had no place for teleology in his approach. He could only describe goals objectively or attempt to determine whether such goals were attainable. He could not answer the question of what is progress because his positivistic approach was unable to decide whether progress should be judged from the viewpoint of the individual or of society. Spencer could only propose that the progress of the individual and the progress of society were parallel, thus sacrificing the heterogeneity of the individual to the heterogeneity of society.[12]

Mikhailovsky believed that men must become conscious of the necessity of an anthropocentric point of view. They must believe, however, not that they were the goal of nature, but rather that their own welfare must be their goal. Instead of engaging in a self-destroying division of labor, they must engage in forms of cooperative labor that protected rather than destroyed the integrity of the individual. Mikhailovsky suggested that the application of the Spencerian "survival of the fittest" to human society in the name of Darwinism confused biological and social progress.[13]

Mikhailovsky's subjective view of human progress was his answer to the equation of Spencer's survival of the fittest with Darwin's theory of natural selection. It marked the beginning of the school of Russian subjective sociology and the incorporation of Darwinism, without its Malthusian or Spencerian metaphors, into the subjective socialism of Russian

---

is more accurate, and is sometimes equally convenient" (Morse Peckham, ed., *The Origin of Species by Charles Darwin: A Variorum Text* [Philadelphia: University of Pennsylvania Press, 1959], III, 15.1:e). (Professor Peckham's complex system of references to the six editions is explained in his preface.)

[12] N. K. Mikhailovsky, "Chto takoe progress?" [What is progress?], in *Sochineniia* [Works], by idem, 4th ed., 10 vols. (St. Petersburg: M. M. Stasiulevich, 1906–1913), I, 90–116.

[13] Ibid., I, 150–165.

revolutionary thought. Mikhailovsky's "subjective method" was extraordinarily important in the development of the socialist revolutionary interpretation of Darwinism because it brought together a belief in the natural sciences with the belief that man could consciously set his own goals and use the natural sciences to help achieve them. Mikhailovsky thus liberated the moral conscience of the socialist revolutionaries from the fetters of an impersonal scientific determinism at the same time that he made it possible for them to continue to believe in the existence of universal scientific laws.[14]

Russian anarchists, who followed Peter Kropotkin (1842–1921), believed, by contrast with Mikhailovsky, that mutual aid was not a subjective choice of man. For them, mutual aid was inherent in the biological evolution of all species, including man. While traveling in Siberia in the 1860's, Kropotkin observed the severe struggle for existence waged by animals, not against one another, but together against the adverse environment. He had read the *Origin of Species* in 1863, and it led him to speculate that cooperation, rather than a "struggle for existence" within the species, was more important for "progressive" biological evolution.[15]

Kropotkin admitted that there was a struggle for existence in nature, but he insisted that it was not a struggle within the species but rather a collective struggle against an adverse environment, which created society among animals and men: "Under *any* circumstance sociability is the greatest advantage in the struggle for life. . . . Language, imitation, and accumulated experience are so many elements of growing intelligence of which the unsocial animal is deprived."[16] Kropotkin believed that Malthus exaggerated the role of struggle within the species because he failed to study the natural checks on overpopulation, which were far more widespread and effective than was simple competition within the species

14 Mikhailovsky overlooked the problems inherent in his "subjective method." While he derived moral values from his quasi-positivistic system, his "subjective method" was not a unique guide to ethics because the values were dependent upon the subjective interpretation of what constituted progress.

15 Peter Kropotkin, *Memoirs of a Revolutionist*, ed. James A. Rogers (New York: Doubleday/Anchor, 1962), pp. 299–300; first published, 1899. See also *Petr i Aleksandr Kropotkiny, Perepiska* [Peter and Alexander Kropotkin, correspondence], ed. N. K. Lebedev, 2 vols. (Moscow: Academia, 1932–1933), II, 115–125.

16 Peter Kropotkin, *Mutual Aid: A Factor of Evolution* (New York: McClure Phillips & Co., 1903), pp. 57–58.

in keeping organisms and their subsistence in balance. Darwinian natural selection, in Kropotkin's view, was not dependent upon the Malthusian struggle and, indeed, functioned to avoid competition as much as possible.[17]

The reception of Darwinism by Russian advocates of economic determinism differed from that by the followers of Mikhailovsky and Kropotkin. It was conditioned by the attitude of Karl Marx and Friedrich Engels toward Darwinism as well as by the Russian tradition of socialist revolutionary thought. The difference between the Marxist and the Russian approach to Darwinism first became apparent to Engels in 1875 when Peter Lavrov (1823–1900) sent him his article of 1875, "Socialism and the Struggle for Existence." Lavrov had written his article against European thinkers who were using Darwinism as a scientific argument against the possibility of socialism. Lavrov insisted that the idea of struggle in Darwinism must itself be viewed from an evolutionary point of view. One of the first weapons to develop in the struggle for existence was an inheritable social instinct that led animals and men to struggle collectively against the harsh conditions of existence. In the higher stage of man's evolution, the unconscious instinct of mutual aid yielded to a conscious recognition of the human value of solidarity. The higher stage would have culminated in a recognition of the brotherhood of all men if it had been able to transcend the boundaries of national and religious chauvinism.[18]

Lavrov believed that socialism arose when the further evolution of mankind was blocked by the claims of the state, religion, and the bourgeoisie. He concluded that natural selection must lead to the victory of socialist societies rather than to their extermination, and only in this sense did socialists agree that natural selection was the best means to resolve the social problem.[19]

Engels replied to Lavrov that he would condemn the bourgeois use of Darwinism as bad economics rather than as bad moral philosophy. He

[17] Ibid., pp. 73–74.

[18] Peter Lavrov, "Sotsializm i bor'ba za sushchestvovanie" [Socialism and the struggle for existence], *Izbrannye sochineniia* [Selected works], ed. I. A. Teodorovich, 4 vols. (Moscow: All-Union Society of Former Political Convicts and Exiles, 1934–1935), IV, 99–104.

[19] Ibid., p. 109.

explained that the essential difference between human and animal society
is that animals only collect subsistence, while men produce it. This makes
it impossible to transfer laws from nature to human society. In addition,
men produce, at a more advanced stage of evolution, not only necessities
but also luxuries for a minority. The struggle among men is not so much
for existence as for the socially produced means of development. Men
produce far more subsistence and means of development under capitalism than can be consumed, because the workers are artificially separated
from the product of their labor. Capitalism finds it necessary by its inner
logic to increase its production, which is already too large for consumption. The result is periodic cycles of depression in which not only the
products but the productive forces as well are destroyed. Engels concluded that it is nonsense to talk about a struggle for existence in human
society as though it were a natural law.[20]

The first Russian social thinker to serve as a direct link between the
views on Darwinism of Marx and Engels and the pre-Marxist Russian
radicals was Peter Tkachev (1844–1885), a publicist and Jacobin revolutionary. He became acquainted with Marxism only after he had formulated a theory of historical progress based on economic determinism.[21]
In that theory he made a distinction between social laws of a given period
and eternal laws of nature. The only connection between these two types
of laws is the persistent triumph of the strong over the weak. Tkachev
believed that in nature the progress of the stronger organism led to its
domination over the weak. But in human society such domination reflected, not physiological improvement, but only the domination of
wealth. This did not further human progress and led only to the exploitation of the poor. The problem was not that force prevailed in both
nature and human society but that force in society was the artificial creation of a transient economic order. It was necessary to transform the content and form of that social force to further human progress.[22]

---

[20] M. V. Osinov, ed., *Perepiska K. Marksa i F. Engel'sa s russkimi politicheskimi
deiateliami* [Correspondence of K. Marx and F. Engels with Russian political figures]
(Moscow: Association of State Publishing Houses, 1947), pp. 171–172.

[21] Peter Tkachev, *Izbrannye sochineniia* [Selected works], ed. B. P. Koz'min, 6 vols.
(Moscow: All-Union Society of Former Political Convicts and Exiles, 1932–1937), I,
69–70; V, 24.

[22] Ibid., II, 27–51, 112–114.

The controversy over Darwinism convinced Tkachev that men must find their goals not in nature or in science but in a concrete analysis of their own historical evolution. Men must not merely adapt themselves to their surrounding conditions but must also adapt those conditions by a revolutionary transformation of the economic forces of society into a form compatible with their own goals. Although Tkachev's economic determinism had points in common with that of Marx and Engels, Tkachev used Marxism only selectively to support his Jacobin argument: the necessity for a preventive revolution in Russia to forestall the development of capitalism and preserve the peasant commune as the focal point of an indigenous socialism based on human solidarity. This Jacobin policy based on a selective use of Marxism brought Tkachev into conflict both with Engels and with socialist revolutionaries, such as Lavrov.

If Tkachev introduced the first economic-determinist interpretation of Darwinism into Russian social thought, George Plekhanov (1857–1918) introduced the first Marxist interpretation of Darwinism into Russia. His attempt to prove that Marxism and Darwinism were entirely compatible resulted in a subtle but constant oscillation between a Marxist economic determinism and a non-Marxist biological and environmental determinism. Plekhanov learned from reading Darwin that a sense of beauty plays a fairly important role in the life of animals and is to be explained biologically. But Plekhanov believed that the differing opinion among men regarding what is beautiful shifts the explanation from biology to sociology and the materialist conception of history. Plekhanov explained that man's nature makes possible aesthetic concepts, but environmental conditions determine what these concepts are. He carried this line of reasoning to the point of implying that the geographical environment was the determining cause of social evolution.[23]

Plekhanov thought it possible to be a Darwinist in biology and a Marxist in sociology because the investigations of Marx began where the investigations of Darwin ended and the methodology was the same in both thinkers. In Plekhanov's interpretation, Darwin explained the origin of species not by the nature of the organism but by its adaptation

---

[23] G. V. Plekhanov, *Izbrannye filosofskie proizvedeniia* [Selected philosophical works], 5 vols. (Moscow: State Publishing House of Political Literature, 1956–1958), I, 612–614; V, 294.

to the influence of external conditions. Marx explained the historical development of man not by the nature of man but by the character of those social relations between men that arose when men acted on external nature. "That is why," Plekhanov concluded, "one can say that Marxism is Darwinism in its application to social science."[24]

If the Russian social thinkers on the radical side of the political spectrum disagreed in their specific interpretations of Darwinism, they were in agreement in accepting Darwinism as a scientific theory of biological evolution and in rejecting the application of Darwin's Malthusian and Spencerian metaphors to human society.[25]

Among the staunch supporters of the Russian state and religion, Darwinism found, by contrast, little favor. The major attack on Darwinism from the conservative side of the political spectrum did not come, however, for more than two decades after the 1864 Russian translation of the *Origin of Species*.

Nikolai Danilevsky (1822–1885), Pan-Slavist and scientist, completed in 1885 a massive scientific and philosophical work against Darwinism, to which he had given the last fifteen years of his life. His belief in teleological evolution led him to see in Darwin's idea of random variations only a meaningless universe of chance. Moreover, Danilevsky found the Malthusian aspect of Darwin's theory as indigestible as had his predecessors on the other side of the political spectrum. He insisted that the Malthusian struggle was the essential part of Darwin's theory but

---

[24] Ibid., I, 690 n–691 n. Plekhanov followed Engels in the attempt to make Darwinism and Marxism compatible in all senses. Lenin, by contrast, emphasized that "the application of biological ideas *in general* to the domain of the social sciences is *meaningless*" (V. I. Lenin, *Sochineniia* [Works], 4th ed., 40 vols. [Moscow: Association of State Publishing Houses, 1941–1942], XIV, 315). Plekhanov and the later Russian Marxists were able to have contradictory views on the relation of Marxism to Darwinism because Marx and Engels never developed a consistent doctrine on that point. See James A. Rogers, "Marxist and Russian Darwinism," *Jahrbücher für Geschichte Osteuropas* 13 (June 1965): 199–211.

[25] The only prominent exception to this acceptance of Darwinism by Russian radical social thinkers was Nikolai Chernyshevsky. His opposition to Darwinism was based on the moral and scientific arguments of the socialist revolutionaries who repudiated Darwin's Malthusian and Spencerian metaphors as applied to human society, but not Darwinism as a scientific theory of biological evolution. See James A. Rogers, "The Russian Populists' Response to Darwin," *Slavic Review* 22 (September 1963): 456–468.

that it, in fact, had no effect upon the differentiation of species. Moreover, the existence of design in biological evolution made such a struggle unnecessary. Danilevsky believed that he had proven both empirically and logically that natural selection was not an operative factor in the evolutionary process.[26]

The attack on Darwinism begun in 1885 by Danilevsky was followed by other conservative and religious thinkers. Konstantin Pobedonostsev (1827–1907), the conservative spokesman for the autocracy and Orthodoxy, believed that Darwinism had become a new religion and was subversive of Christianity: "It is plain that to him [Darwin] the fundamental law of life is *the preservation of the strong and the extirpation of the weak.*"[27] Pobedonostsev followed his predecessors in confusing Darwinism with the Malthusian struggle for existence and the Spencerian survival of the fittest, neither of which he could accept within the framework of his religious view.

In Tsarist Russia the politically conservative and religious criticisms of Darwinism were often inseparable. The result was that a reconciliation between Darwinism and religion was as unwelcome to conservative and religious thinkers as it was to revolutionary thinkers. When Sergei Rachinsky, the first Russian translator of the *Origin of Species,* tried to publish in the Holy Synod's official journal an article demonstrating the essential harmony of Darwinism and Christianity, Pobedonostsev, as Procurator of the Holy Synod, refused to allow its publication.[28]

The major effect of Darwinism on Russian conservative and religious thought, as it had been on revolutionary thought, was to support old loyalties rather than to develop new ones. There were striking similarities in those loyalties, which arose from the special conditions of Russian social and intellectual life. Although Russian thinkers of all persuasions

---

[26] N. Ia. Danilevsky, *Darvinizm: Kriticheskoe izsledovanie* [Darwinism: a critical investigation], 2 vols. (St. Petersburg: M. E. Komarov, 1885–1889), I, pt. 2, p. 496. Danilevsky received strong support in his work against Darwinism from a close friend, Nikolai Strakhov (1828–1896), an eminent literary critic and philosopher.

[27] K. P. Pobedonostsev, *Reflections of a Russian Statesman*, trans. R. C. Long (Ann Arbor: University of Michigan Press, 1965), p. 183.

[28] George L. Kline, "Darwinism and the Russian Orthodox Church," in *Continuity and Change in Russian and Soviet Thought*, ed. E. J. Simmons (Cambridge: Harvard University Press, 1955), pp. 318–319, n. 44.

borrowed heavily from Western European thinkers, they did not work within a political and economic structure that reinforced the same aspects of those theories. Russia had no tradition of political economy comparable to that of Western Europe. It had experienced neither a laissez-faire ideology in economics nor a political structure allowing various groups to strive for power and a secular rationalization of their success. Consequently, no group in Russia looked to analogies from the natural sciences, and particularly from Darwinism, to justify its position in society. The structure of that society in the last half of the nineteenth century gave little reinforcement to an ideology, supposedly based on Darwinism,[29] which claimed that the key to human progress was an uncontrolled and individual struggle for existence in human society. To Russian radical and conservative thinkers, that ideology appeared not merely irrelevant but actually dangerous to the moral idea of human solidarity, whether they were defending the contemporary social structure or the desired socialist society of the future.

[29] What is commonly referred to as "Social Darwinism" generally turns out on closer examination to be either "Social Spencerianism" or "Social Malthusianism." See James A. Rogers, "Darwinism and Social Darwinism," *Journal of the History of Ideas* 33 (April–June 1972): 265–280.

THE NETHERLANDS

ILSE N. BULHOF

In 1899 a dismayed Abraham Kuyper, professor at the Protestant Free University of Amsterdam, wrote that "Our nineteenth century fades away under the hypnotic spell of the dogma of Evolution."[1] By the century's end, Darwinism had indeed become an established feature of the intellectual landscape of the Netherlands. Had the acceptance of Darwinism, with its novel concepts of "natural selection" and "struggle for life" and its startling consequences for the origin of man, been a difficult process in the Netherlands? In this essay I will try to give an impression of the various responses to Darwin's ideas in the Netherlands. In the first part, I will sketch the reception of Darwinism by the world of science, centered around the universities. The natural scientists, I found, had remarkably little difficulty in accepting Darwin.

Science, however, is practiced in a social context, and scientific ideas, in addition to a purely scientific value, also have a social function. What that function is depends on whose theories they are and who supports them. Scientific insights can be considered as the mental products through which a social group accounts to itself for the ways in which it holds the natural world to be organized and for its own place in it. From society's point of view, scientific ideas fulfill a function comparable to that of religious, philosophical, or political convictions: they articulate the social identity of the group in terms of its ideational beliefs. Conflicts between social groups are often fought out in the guise of intellectual quarrels,

---

[1] Abraham Kuyper, *Evolutie: Rede bij de overdracht van het rectoraat aan de Vrije Universiteit, 20 Oct. 1899* [Evolution: address delivered at the transference of the rectorship at the Free University, Oct. 20, 1899] (Amsterdam and Pretoria: Boekhandel vrhn Höveker en Wormser, 1899).

the more so when the groups do not have a well-defined profile, as was the case in nineteenth-century Holland.

The history of science—and the history of ideas in general—tends to focus on the emergence and spread of ideas in the dominant and usually articulate and well-documented groups of society. The older and partly superseded beliefs as a rule do not figure in the picture. In the second part of this essay I will try to indicate the responses to Darwin's theories by the other social groups existing in the Netherlands at the time, in an attempt to outline a sociology of Darwinism in the country. I will draw attention especially to the harmony between the scientific theories and the social position of the persons holding them.

In order to understand the various responses to Darwinism, we have to know something about Dutch society. The poet Heinrich Heine once said that, if the world were coming to its end, he would go to Holland, because there everything happened fifty years later. The saying contains some truth. During the nineteenth century the Netherlands was predominantly an agricultural society with some manufacturing towns in the west: Amsterdam, Haarlem, Leiden, to name a few. Other towns were engaged in commercial activities: Amsterdam again, Rotterdam, and Dordrecht. The industrial revolution had not yet reached the country. Labor movements started only after 1865, social democracy after 1887. Before 1880 there were no real entrepreneurs operating in Dutch business, and the laborers lacked class-consciousness. The workingman was just a poor man, and his boss was his benefactor. International trade, traditionally the backbone of the country's economy, was floundering until about 1870.

In 1848 the progressive liberal bourgeoisie, led by Jan Rudolf Thorbecke (1798–1872), succeeded in revising the constitution. From then on, the king was obliged to rule with a parliament that was chosen by the wealthy citizens. During the years 1850–1870, the liberal bourgeoisie initiated important measures that brought the country back into the mainstream of European life: the digging of canals connecting Amsterdam and Rotterdam directly with the sea, free trade, and a reform of secondary education and the university system.

The Dutch Reformed Church was the official church, but it had little direct influence in politics. The liberal bourgeoisie belonged to the Dutch Reformed Church, but without actually paying much attention to its traditional Calvinist dogmas. The liberals were known as Modernists, a fitting name, as they were more committed to the nineteenth century than to the Protestant principles. One group of conservative Protestants, the Antirevolutionaries led by Guillaume Groen van Prinsterer (1801–1876), made themselves heard from time to time in religious as well as in political matters. The Roman Catholics, although a fairly large group, played a minor part in Dutch society, but in 1853 they succeeded in re-introducing the Catholic hierarchy, an event that marked the beginning of the re-entry of the Catholics into the national political scene.

To sum up, the major groups in the Netherlands were the Modernist liberals (the leading group, especially after the introduction of the Constitution of 1848), the Catholics, and the orthodox Protestants (Antirevolutionaries). None of these groups was yet organized into a political party. They tended to recognize a leader (Thorbecke among the liberals, Groen among the Antirevolutionaries), but the leader had little contact with his followers. To lay a program of action before the voters, for instance, was still frowned upon as an attempt to bribe the electorate.

The Holland of this period has aptly been called a "dominocracy," for intellectual life was to a great extent dominated, both locally and nationally, by the Protestant ministers (*Dominees*). Literature, for instance, was written by incumbent, functioning, or retired ministers. The other arts had practically fallen asleep.

In this rural, middle-class, church-going society a few freethinkers, rejecting religion, advocating the use of reason, and often professing a materialistic world view, caused quite a scandal. They offered, in fact, an alternative conception of the traditional mental and social world. Many readers of their journal later became socialists. The freethinkers did not form a group, properly speaking, but I will treat them as such because of the intellectual influence they exerted.

DARWIN AND THE UNIVERSITY WORLD

Before the Publication of the "Origin of Species"

In 1848, the year in which the Netherlands became a liberal constitutional monarchy, the new and still young professor of physiology Franciscus Cornelis Donders (1818–1889) gave his inaugural address to the assembled community of the faculty and students of the University of

Utrecht and the prominent members of the town's citizenry. As his topic he chose "The harmony of animal life: the manifestation of laws."[2] As Darwin himself later acknowledged, Donders closely anticipated the concept of natural selection in this address.[3]

Donders began his address by pointing to the order and harmony in nature. Everything is part of the great natural organism, a link in the immense chain that has neither beginning nor end and is basically indivisible. The position of the earth in the universe, its surrounding atmosphere, and the circulation of water from the sky back to earth and sea demonstrate the interdependence of all phenomena of nature and show how everything serves the development of life. Nowhere is this harmony more striking than in the interdependence of the animal world and vegetation. The animals give carbon dioxide to the plants, which in return give oxygen to the animals.

But the scientist, according to Donders, is not content with the description of the harmony of nature. He wants to discover the laws governing the development of this harmony. In astronomy, physics, and meteorology, striking advances have been made in establishing laws, but, Donders said, living nature seems too complicated and too rich to yield to laws. Plants and animals therefore tend to be considered products of God's wisdom, an approach that leads many people to study God's purpose in creating each species. But science does not occupy itself with the question of goals. Science deals with empirical knowledge, with facts. What does man know about God's goals? Goals, if they exist at all, can never be proven. Instead of studying the goals for which living nature is supposed to have been created, Donders proposed to demonstrate the laws according to which the harmony in living nature, especially in the animal world, is formed. He compared his enterprise with the work of Johannes Kepler, who discovered many laws governing the heavenly bodies, though only with Isaac Newton did these laws become founded

2 F. C. Donders, *De harmonie van het dierlijke leven: De openbaring van wetten; Inwijdingsrede bij de aanvaarding van het hoogleraarsambt aan de Utrechtse Hoogeschool* [The harmony of animal life: the manifestation of laws; inaugural address delivered at the occasion of the acceptance of a professorship at the University of Utrecht] (Utrecht: C. van der Post, 1848).

3 Jan and Annie Romein, *Erflaters van onze beschaving: Nederlandse gestalten uit zes eeuwen* [Testators of our civilization: Dutch personalities from six centuries], 4 vols. (Amsterdam: Em. Querido, 1946), IV, 44.

on the one fundamental principle that explained them all. Donders wanted to establish the laws that govern living nature, although he realized that these did not furnish the final unifying explanation.

Donders distinguished two aspects of the harmony reigning in living nature: the relation of the organism to outside conditions, such as its hunger for food, and the relation between the organism's vital needs and its inner organs, as for instance the length of its intestines as determined by its food intake. Donders suggested three laws to explain this harmony. The first law runs as follows: *Every animal organism is transformed by the permanent conditions it finds itself in. It is transformed in such a manner that the organism responds harmoniously to these influences.*[4] Commenting on this law, Donders concluded "that, given the gradual development of animals on the surface of our planet, the constitution of the various germs was determined by these influences, that is, by the circumstances; and that the step-by-step change in the circumstances caused permanent transformations, maybe even the division into the species we know today, in such a way that in each period the organization of the animals corresponded with the outside conditions." As examples, he mentioned the variety of human races, descending from the original couple; the adaptation of the eye to changes in light intensity, as seen in prisoners who have spent years in darkness and develop great optical sensitivity; the adaptation to air pressure in high and low regions; and the relation between vegetation used for food and the species living in its proximity.[5]

The harmony of nature, according to Donders, is not yet perfect. When the circumstances change, so will the organisms. The circumstances are always ahead of the organisms, which at a given moment are products of previous circumstances. If the new circumstances are too alien, they cause disturbances in the organism that disrupt the relation between the various parts of the body. In such cases death occurs. Far from being the punishment of an angry father, Donders explained, the imperfections of nature result from ongoing changing conditions.[6]

Donders's second law is the law of habit: *Every organ, every part of the body will be transformed as a result of the constant influence of the will*

---

[4] Donders, *De harmonie*, p. 21. All translations are mine unless otherwise indicated.
[5] Ibid., pp. 22–30.
[6] Ibid., pp. 31–34.

or other circumstances in such a manner that it corresponds to what the will or the circumstances demand of it.[7] All animal habits are determined by the needs of life. Strength, agility, precision of movement have to develop in harmony with the way of life of each animal. The senses and the instincts of the animals also develop as a result of habits dictated by the circumstances. As one organ changes, the others also have to adapt to the change.

These two laws presuppose, however, a third one, the law of heredity, phrased thus by Donders: *The condition of the ancestors is inherited by posterity. The condition of the parents is inborn in the children.*[8]

The results of habit are made permanent by heredity, which, Donders stated, is the foundation of the ascending perfection of creation. In explaining this law he pointed to the varieties in animal species, often brought about in historical times by changes in circumstances. Agriculturalists have profited from this law in creating artificial races. Donders suggested that in earlier times the species lasted longer because man did not intervene. Most of the variations—apart from the domesticated animals—are found in man himself, as he lives in so many different places on the globe and has consequently developed different ways of life and civilizations.

At this point, Donders stated that he had proven the harmony of nature as the necessary result of laws: the court of scientific law rejects goals; within the pages of nature's law code are written the words *use, habit,* and *heredity.*[9]

Donders's mechanistic explanation of the causes of the organization of nature, with its pertinent denial of purpose in nature, is quite striking. Let us note, however, that this is not yet Darwinism. We miss the perspective of geological time and a sense of history in which the species were shaped during a long process. Nevertheless, we may anticipate that Donders would readily accept Darwin's explanation of the origin of species. We can only speculate about the origin of Donders's concept of evolution. One influence may have been Robert Chambers's *Vestiges of the Natural History of Creation,* which was translated in 1849 by a

[7] Ibid., p. 36.
[8] Ibid., p. 46.
[9] Ibid., pp. 55–56.

former colleague of Donders at the Military Medical School, J. H. van den Broek.[10] The translation was prefaced by the noted organic chemist at Utrecht, G. J. Mulder, who did not commit himself openly to its ideas but did say that he found "a treasure of knowledge" in it. Pieter Harting, the biologist, later mentioned the *Vestiges* as one of the texts that paved the way for Darwin's ideas.[11]

In 1858, the Nestor of Dutch natural historians, Jan van der Hoeven, professor at Leiden University since 1835, published a small brochure on the succession and development of animal organisms during the various geological periods.[12] In it he dealt with the question of whether the varieties of animals and plants could be explained by a gradual development in which nature itself created the many forms of life from earlier and lower forms. If developmentalism were true, he argued, there could not be any missing link in the chain of developing perfection. One would have to assume that some day the links that are not yet known will be found in living nature or in the remains of now extinct forms. Van der Hoeven felt that at this point developmentalism left too much to the imagination: "If one ventures oneself in the shady maze of imagination," he cautioned, "leaving the bright field of research which is illuminated

[10] Robert Chambers, *Vestiges of the Natural History of Creation*, trans. by J. H. van den Broek as *Sporen van de natuurlijke geschiedenis der schepping, of schepping en voortgaande ontwikkeling van planten en dieren onder den invloed en het beheer der natuurwetten* [Vestiges of the natural history of creation, or creation and continuing development of plants and animals under the influence and control of the laws of nature] (Utrecht: J. G. Broese, 1849; 2d ed., 1850; 3d ed., 1854). Van den Broek also translated Chambers's *Explanations: A Sequel to the Vestiges*, as *Vervolg der sporen van de natuurlijke geschiedenis der schepping: Antwoord van de schrijver op de menigvuldige tegen zijn leer gerichte recensien; nadere verklaringen en toelichtingen omtrent de belangrijkste bewijsgronden zijner theorie der voortgaande ontwikkeling van planten en dieren* [Sequel to the vestiges of the natural history of the creation: answer of the author against the many criticisms made against his doctrine; more explanations and information concerning the most important proofs of his theory of continuing development of plants and animals] (Utrecht: J. G. Broese, 1849). A cheap edition of the *Vestiges*, trans. J. van Egmont, Jr., was published in 1866 (Rotterdam: D. Bolle).

[11] Pieter Harting, *Mijne herinneringen* [My memories] (Amsterdam: N. V. Noord-Hollandse Uitgevers Maatschappij, 1961), p. 75.

[12] Jan van der Hoeven, *Over de opvolging en ontwikkeling der dierlijke bewerktuiging op de oppervlakte van onze planeet in de verschillende tijdperken van haar bestaan* [On the sequence and development of animal organization on the surface of our planet during the various periods of its existence] (Leiden, 1858).

by the light of observation, pictures emerge before our minds which dissolve like clouds into each other, taking on any form."[13]

As an example of arbitrary representations, van der Hoeven mentioned Benoit de Maillet, who in the early eighteenth century described how all life started in the sea and how birds developed from flying fishes that were trapped on the shore. Lamarck, he continued, came up with the same theory but defended it with better arguments.

Van der Hoeven admitted that the remains of animals in the different geological layers of the earth might point to the truth of the developmental theory. He was familiar with the works of Georges Cuvier, Richard Owen, Louis Agassiz, and Adolphe Brongniart, who agreed that different species of plants and animals had existed in earlier times. Pondering over the question of how these older forms of life became extinct, van der Hoeven rejected the catastrophic theory as unsatisfactory, but he also rejected the developmental hypothesis because, as he said, nothing in our experience shows that such an emergence of new forms is possible. Who proves that the new forms that are found did not exist before? Geology may indeed teach us that the earth is older than sixty centuries, but it cannot show that the earth has existed since eternity. Once there must have been a creation. Although van der Hoeven realized that the traditional assumption of separate creations of the species was not really an explanation, he decided in the end to remain faithful to it, even in the face of the recent developments in geology. If we confess our ignorance, van der Hoeven stated with resignation, the principles of science are at least not violated. He expressed the hope that we might eventually discover the laws according to which the Creator made living nature.

It is interesting to realize that these serenely written pages were published so shortly before the storm of 1859. Van der Hoeven clearly felt that he was confronted with a great mystery, a real paradox. "Who holds," he exclaimed in utter confusion, "that the development of an animal with eyes out of an animal without eyes is a lesser mystery than the creation of an animal with its eyes specially made for it? Science does not close its books at this point. Concerning these problems science has not even opened any books yet."[14]

[13] Ibid., pp. 29–30.
[14] Ibid., pp. 47–48.

Another natural historian who joined the debate on the evolutionary hypothesis was Pieter Harting. He came to the University of Utrecht in 1841 and soon became a good friend of Donders. In 1855 and 1856, Harting read before the Physical Society of Utrecht two papers, which were published together the following year under the title "A comparison between the prehistoric and contemporary creations."[15] In the preface Harting first pointed out that God gave reason to man and that it is consequently the duty of a religious person to study nature. But this study should go deeper than a mere observation of facts. To state that the world is a beautiful work of God is not enough, Harting said, brushing aside van der Hoeven's scruples about a more speculative approach to science. What we observe in the world is change. Change takes time. We notice also that changes are the effects of earlier causes. We see a series of causes and effects, links in a chain, the end of which is in the hands of the Almighty. Unfortunately we cannot study all the links of the creation, given the weakness of our senses and the shortness of our lives, but we should go as far as we can in the study of the infinities of space and time.

Harting undertook in his paper to explain the results of science as they affect our understanding of historical time. The world, he explained, has a past of millions of years, during which many species that are now extinct lived. A surprised listener asked Harting why he did not mention the Genesis story, but he replied that he did not mention the cosmogonies of the Greeks and the Indians either: such accounts have to be ignored by scientists.[16] In the published version, Harting added footnotes "for the more scientific reader," and in the last one, covering about thirty pages, he discussed the evolutionary hypothesis. In 1858 he was already convinced that the future of science was in the direction of evolutionary theory, and he would have pronounced himself more openly in favor of it at this point had not his senior colleagues, van der Hoeven and Willem Vrolik (Amsterdam), advised that he tone down his conclusions.[17]

When we compare Harting's position in 1858 with that of Donders

---

[15] Pieter Harting, *De voorwereldlijke scheppingen, vergeleken met de tegenwoordige* (Tiel: H. C. A. Champagne, 1857; German trans. by E. A. Martin, Leipzig, 1858).

[16] Harting, *Mijne herinneringen*, p. 104.

[17] Ibid., p. 74.

in 1848, we notice that Harting was not concerned with explaining the mechanism that governs the evolutionary process, but that he had absorbed the new discoveries of geology: Harting was aware of the immense time perspective that the development of species involves. Together, Donders's and Harting's treatises provided a fertile soil for the reception of Darwin's theories.

In 1858, Sir Charles Lyell came to Utrecht to visit Harting. They discussed the evolutionary hypothesis at length. Lyell told Harting that his friend Darwin was working on a book that would certainly cause a sensation. Lyell added that he was still hesitant to accept Darwin's theories, but his last words were, Harting recalled, "Cette hypothèse est une de celles avec lesquelles il faudra désormais compter."[18]

On November 1, 1859, the new professor of botany at the Atheneum of Amsterdam, C. A. J. A. Oudemans (1825–1906), gave his inaugural address, covering the history of botany from Aristotle to the contemporary discoveries of geology.[19] Evaluating the explanations for the extinction of certain forms of life, through catastrophic destruction or through natural causes, he seemed to opt for the latter alternative. To the question of how we know that the extinct forms did not exist from the beginning, he replied that the earth itself tells us how the different forms followed each other in the course of time. With the remark that the study of fossil plants is today an integral part of the study of botany, Oudemans left his audience with the distinct impression that a gradual development of plant life was quite possible. Three days later the *Origin* was published.

After the Publication of the "Origin of Species"

The *Origin* got a favorable review in the *Wetenschappelijke Bladen* [Scientific pages]—a translation of an article in the *National Review*—

[18] Ibid., p. 75. "This hypothesis is one that must be reckoned with from now on."

[19] C. A. J. A. Oudemans, *Inwijdingsrede over de plantkunde beschouwd in hare trapsgewijze ontwikkeling van de vroegste tijden tot op heden: Uitgesproken ter aanvaarding van het hoogleeraarsambt in de genees- en kruidkunde aan het Atheneum Illustre te Amsterdam, 21 Nov. 1859* [Inaugural Address on botany considered in its gradual development from the earliest times till today: delivered at the occasion of the acceptance of a professorship in pharmacology at the Atheneum Illustre at Amsterdam] (Utrecht and Amsterdam: C. van der Post, Jr., and C. G. van der Post, 1860). The Atheneum became the University of Amsterdam in 1877.

although the journal did not mention the name of Darwin again for the next ten years.[20]

Of those who debated the issue of a natural development of species before 1859, van der Hoeven was one who never considered Darwin's explanation of the origin of species satisfactory. He managed to publish a translation of W. Hopkins's negative discussion of Darwin's book together with the first Dutch translation of the *Origin*, by T. C. Winkler.[21] In the preface of his translation of Hopkins's article, van der Hoeven acknowledged that Darwin accepted a first creation and could not be accused of materialism. He also admitted that Darwin was an excellent scholar. Nevertheless, he felt that Darwin's proofs for natural selection were not sufficiently supported by facts. He referred to his own earlier publication on this matter and to Hopkins's treatment of it. The solution of an evolution according to a plan apparently did not occur to him: according to van der Hoeven, one had either to accept the idea of separate creations or that of a godless natural development, and the latter was an opinion that gravely endangered the position of man and his soul. In 1863, van der Hoeven repeated his point of view.[22] Why devote so much attention to the problem of the origin of living nature, he wondered, and not to that of minerals? He felt that it was strange that some called the creation a "hypothesis," as if it stood on the same level as arbitrary ideas like spontaneous creation and natural development. Van der Hoeven died in the same year. His successor, Emilius Selenka, who gave his inaugural address in 1868, pronounced himself in favor of Darwin.[23]

The Amsterdam natural historian Willem Vrolik (1801–1863) wrote sarcastically in the preface to his book on the life and structure of animals

---

[20] "Darwin, Over het ontstaan der soorten" [Darwin, on the origin of species], *Wetenschappelijke Bladen* 2 (1860): 1–38.

[21] *Het onstaan der soorten van planten en dieren door middel van de natuurkeus, of het bewaardblijven der bevoorrechte rassen in de strijd des levens*, trans. T. C. Winkler, 2 vols. (Haarlem: A. C. Kruseman, 1860, 1869, 1883). A corrected edition, ed. H. Hartogh Heys van Zouteveen, appeared in 1890 as no. 1 in the series Biologische Meesterwerken [Biological masterpieces] (Arnhem and Nijmegen: Gebr. E. en M. Cohen, 1890, 1891).

[22] Jan van der Hoeven, Review of Pieter Harting's *Leerboek* [Textbook], *Nederlands Tijdschrift voor Geneeskunde* [Dutch journal of medicine] 7 (1863): 90–92. (See note 27 below for publication data of the *Leerboek*.)

[23] Emilius Selenka, *Oratio de animalium distributione geographica, quam habuit Emilius Selenka* (Leiden, 1868).

that it puzzled him that not all species preceding man had died out.[24] He also wondered whether the existing fishes had to be considered as mammals who did not make it. He granted that scientists had a hard time classifying species but said that that should not tempt anybody to fall for the Darwinian fantasies. But his colleague Oudemans, who in 1859 had already seemed in favor of a gradual development of plant forms, published a "Textbook of botany" (1880–1884) in which the common descent of plants was accepted.[25] The similarities and differences among plant species were explained as the result of different environments and of the process of natural selection. Oudemans wrote the handbook together with Hugo de Vries, who later became well known for his discovery of mutations in the evolution of plant species.

Pieter Harting (1812–1885) became Darwin's most convinced defender in the Dutch academic world. He studied medicine in Utrecht in order to make a living, but he was much more interested in the study of nature. As a high-school student he had already built his own microscope. One of his teachers at Utrecht was J. C. C. Schroeder van der Kolk (1795–1862), professor of anatomy and physiology, who introduced him to the anatomy of the human body, emphasizing the relationships between the function and the shape of human and animal organs. When Harting's medical practice proved to be a failure, he decided to publish some of his microscopic research in physiology and asked van der Hoeven for advice. In 1839 his first publication appeared, and with it he attracted enough attention to be named professor at the small atheneum at Franeker in Friesland, and later, after its closure, at the University of Utrecht. From this point on Harting was able to devote himself to his microscopic research, working closely together with the organic chemist G. J. Mulder (1802–1880), Donders, and van der Kolk. In 1848–1850 he published a major work on the microscope.[26]

---

24 Willem Vrolik, *Het leeven en maaksel der dieren* [The life and structure of animals], 3 vols. (Amsterdam: M. H. Binger en Zn, 1852–1860; Rotterdam: D. Bolle, 1852–1860).

25 C. A. J. A. Oudemans and Hugo de Vries, *Leerboek der Plantkunde* [Textbook of botany], 3 vols. (Zaltbommel: Joh. Noman en Zn, 1883–1884; Amsterdam: C. L. Brinkman, 1880–1884).

26 Pieter Harting, *Het microskoop, deszelfs gebruik, geschiedenis en tegenwoordigen toestand* [The microscope, its use, history and contemporary state] (Tiel: H. C. A. Champagne, 1848–1854; German trans., 1859; enlarged ed., 1866; Utrecht: van Dad-

In 1848, Harting gave a paper, "The Force of What Is But Small," on the contribution of small organic creatures to the formation of the earth's crust. This subject, new at the time, drew his attention for the first time to the fields of geology and zoology. Geological research on Dutch soil stimulated him to go ahead with these studies. Around 1853, Harting decided to leave the fields of plant physiology and geology to devote the rest of his life as a scholar to zoology and comparative anatomy; he realized that those were the fields in which most research had to be done as a result of the new developments in geology.

Harting was particularly interested in the "philosophical" aspects of zoology, that is to say in the general considerations based on the observation and comparison of facts. During the fifties Harting started to work on a textbook on the principles of zoology.[27] With the growing knowledge in the field of zoology, he decided that his students needed an introduction to zoology that would explain its underlying principles rather than describe the species. Harting had another important reason for writing the book: he wanted to familiarize his students with the evolutionary approach to zoology, which he defended in the 1859 papers mentioned earlier.

In the first volume of his textbook (1862) Harting introduced the evolutionary hypothesis. He realized, we read there, that many still felt that the idea of the gradual development of species is a heretical thought. But Harting stated that the hypothesis as such is not part of science proper. Science is not yet advanced enough to pronounce a well-founded judgment on such matters. Such ideas, he added, are mainly subjective, and he did not want to attach more importance to the subjective idea of

denburg en Co., 1848–1849); idem, *De nieuwste verbeteringen van het mikroscoop en zijn gebruik sedet 1850* [The latest corrections of the microscope since 1850] (Tiel: H. C. A. Champagne en Zn, 1857).

[27] Pieter Harting, *Leerboek der dierkunde in haren gehelen omvang* [Textbook of zoology in its totality] (Tiel: H. C. A. Champagne, 1862–1874): vol. 1, *Schets der algemene dierkunde: Eene inleiding tot hare wetenschappelijke beoefening* [Outline of general zoology: an introduction to its scientific study] (1862); vol. 2, *Natuurhistorisch overzicht der ruggemergdieren of gewervelde dieren* [Natural-historical survey of the Vertebrata] (1864); vol. 3, *Morphologie der ruggemergdieren of gewervelde dieren* [Morphology of the Vertebrata] (1867); vol. 4, *Natuurhistorisch overzicht der ongewervelde dieren* [Natural-historical survey of the Invertebrata] (1870); vol. 5, *Morphologie der ongewervelde dieren* [Morphology of the Invertebrata] (1874).

evolution than to other subjective ideas. But, as everyone is entitled to his
own opinion, he felt free to express his feelings on the subject. After all,
it is the purpose of a textbook to introduce students to all aspects of
their field, explaining its strong as well as its weak sides, and "to point
out the directions that have to be taken to bring it to a higher level."[28]
It is clear where Harting's sympathies were. According to him, it was
then generally accepted that every organism had had its developmental
stages. The period of Carolus Linnaeus, with his ideas of fixed species,
seemed to have come to its end. New insights were emerging, but, just as
in politics, in the world of science, too, conservatism and radicalism opposed each other. With his textbook Harting hoped to provide a bridge
between the two groups. He wanted to delineate exactly where the facts
ended and the hypothesis started, in order to give the hypothesis its due
consideration and to show that, if used with prudence, it could lead to a
better insight into the truth.

In the preface to the second volume of the textbook (1864), Harting
frankly called the species "temporary transient carriers of a certain more
or less determined form." "Species," he continued, "are for the science
of organic nature what words are for a language."[29] The descriptive
zoologist does not know what to do with forms that seem to belong to
two species at the same time; but for the more philosophical zoologist
these forms are very important because they show genetic relations between species. Harting now declared confidently that the genetic system
of zoology is the right one. He expected that in a few years nobody
would doubt this theory, although opinions might vary as to how the
transformation of species came about. Natural selection was not mentioned in the textbook.

In his autobiography Harting mentioned that he planned to add a
short history of zoology to the textbook, because, as he said, "the revolutionary history of zoology and its impact on many fields of life is one of
the most important chapters of the general evolution of human civilization." He was well aware of the importance of the new direction that his
science had taken.[30]

28 Ibid., I, xv.
29 Ibid., p. 6.
30 Harting, *Mijne herinneringen*, pp. 74–75. For the evaluation of evolutionary
thought by Harting, see also his essay, "Over de betekenis der zoologie voor de heden-

Harting made his last statement on Darwinism at the occasion of the publication of Herman Hartogh Heys van Zouteveen's Dutch translation of *The Descent of Man* in 1871.[31] He read a paper on J. E. Doornik, a medical doctor who had practiced medicine in Amsterdam at the turn of the century and had pronounced himself in favor of gradual formation of the earth and the species living on it.[32] Harting then summarized in twenty-one theses what, according to him, the generally accepted results of the natural sciences were concerning the evolutionary hypothesis. He mentioned among other things that the organic creation forms a unity from the beginning of time to the contemporary age; that the succeeding forms of life correspond with the conditions of the earth; that a common origin of these forms is possible. Harting also accepted the thesis that men and apes are descended from a common stock. The expressions "natural selection" and "struggle for life" do not, however, figure among his theses.

By his moderation in defending the theories of Darwin and his tolerance for other opinions, Harting contributed more than anyone else to the spreading of Darwin's ideas at the universities in the Netherlands.

---

daagsche beschaving" [Significance of zoology for contemporary civilization], *Album der Natuur* 1 (1880): 1–15.

[31] Pieter Harting, *Iets over J. E. Doornik en zijn aandeel in de ontwikkelingshypothese, gevolgd door eenige opmerkingen aangaande den tegenwoordigen staat der laatste: Voorgedragen in de gewoone vergadering, 25 Feb. 1871, bij gelegenheid van het aanbieden der Nederlandsche vertaling door Dr. H. Hartogh Heys van Zouteveen van het werk van Ch. Darwin, "De afstamming van den mensch"* [Remarks on J. E. Doornik and his share in the developmental hypothesis, succeeded by some remarks on the contemporary state of the last-mentioned hypothesis: delivered in the regular meeting, Feb. 25, 1871, at the occasion of the publication of the Dutch translation by Dr. H. Hartogh Heys van Zouteveen of Charles Darwin's book "The Descent of Man"], in *Verslagen en Mededelingen der Koninklijke Akademie van Wetenschappen, afd. Natuurkunde* [Reports and communications of the Royal Academy of Sciences, Department of Physics], 2d ser., vol. 5 (Amsterdam: C. G. van der Post, 1871), pp. 367–380.

[32] J. E. Doornik, *Wijsgerig-natuurkundige verhandelingen* [Philosophical-physical essays] (Utrecht: S. Alter, 1816). In the essay "Algemene beschouwingen van eene natuurlijke geschiedenis van het menselijk geslacht" [General considerations on natural history of the human species], Doornik opposed the catastrophic theory of geology. He viewed the totality of the organic creation as an interconnected whole, of which the lower divisions had emerged first. In this process, he noted a tendency toward ever-growing perfection. According to Harting, Doornik had nothing to do with German *Naturphilosophie*. Harting concluded that Doornik had anticipated Lamarck, Prevost, and Lyell.

His services to the cause of Darwinism and modern biology were acknowledged in the 1880's by his students and friends, who offered him a bust of Darwin.

Generally speaking, Darwin's theories were well received by the academic community. In 1859 the two scientists who rejected Darwin, van der Hoeven and Vrolik, were both at the end of their careers and about to be replaced by a younger generation of scientists. In the sixties most professors in natural history were Darwinists: in Utrecht, Harting (zoology) and Donders (physiology); in Leiden, Selenka (zoology); in Groningen, H. C. van Hall (botany and zoology), Derk Huizinga (physiology), and Matthijs Salverda (botany and zoology); in Amsterdam, T. C. Place (physiology) and C. A. J. A. Oudemans (botany); in the Institute of Technology at Delft, W. C. H. Staring, "the father of Dutch geology" (geology and mineralogy); and, at the Haarlem Teylers Institute,[33] the natural historian T. C. Winkler.

Conclusion

Before 1859 the Dutch natural scientists connected with the State Universities of Leiden, Utrecht, and Groningen and with the Atheneum Illustre of Amsterdam were clearly puzzled by the new paleontological and geological data concerning natural history, and many felt that these facts pointed somehow in the direction of a gradual development of the earth and of the plant and animal species. The *Origin of Species* crystallized the tentatively entertained notions of a gradual development of species. Especially among the younger scientists, the Darwinian theory of an evolution of the species quickly changed in status from a daring hypothesis to an undisputed fact.

The reorientation of the natural scientists from the traditional view of the world, with its teleologically structured living nature that was once and for all created by God in the exact way it was at the moment, to the Darwinian conception of a mechanically developing nature occurred, indeed, in a remarkably fast tempo, which cannot be explained by the state of scientific theory in the Netherlands around 1859 alone. In the next

---

[33] A foundation instituted by Pieter Teyler van der Hulst (1720–1778), consisting of two societies, one to promote the study of theology (Teylers Godgeleerd Genootschap), the other to promote the study of nature (Teylers Tweede Genootschap).

section of this study I will indicate the extent to which the change in scientific theory was facilitated and supported by profound sociological changes that took place in Dutch society at large.

DARWIN AND DUTCH SOCIETY

The Liberal Bourgeoisie

In the fifties and sixties the liberal elements of the Dutch middle class had taken to the philosophy of Modernism, a trend in Protestantism that sought a *rapprochement* between faith and modern culture, science in particular.[34] Significantly enough, Modernism's main spokesman was a colleague and close friend of the Darwinists Harting and Donders: Cornelis Willem Opzoomer (1821–1892), who since 1846 had been professor of philosophy at the University of Utrecht. His influence on society was great because he trained Protestant ministers, and ministers at this time were still an important intellectual force in the community. Opzoomer was a deist: he never denied the existence of God nor that of a spiritual world next to the natural world; but in his philosophy he was a positivist. He founded the "philosophy of experience," which accepted only empirical knowledge as certain and trustworthy. Opzoomer's scientific friends, Harting and Donders, doubtless stimulated him to work out his philosophy of experience.[35]

The goal of science as well as of philosophy is, according to Opzoomer, to discover laws of nature, of history, and of society. Everything in creation obeys the Great Law of Causality. Opzoomer and the Modernists felt that this law and the other laws were the work of the Supreme Being: they did not accept the possibility of a conflict between the truth offered by Christianity and that presented by science. Assuming a harmony between religion and science, they favored a monistic conception of the world, and they rejected all forms of supranaturalism, miracles in particular. Opzoomer believed strongly in progress, as guaranteed by

[34] See K. H. Roessingh, *De moderne theologie in Nederland: Hare voorbereiding en eerste periode* [The modern theology in the Netherlands: its preparation and first period] (Groningen: Erven B. van der Kamp, 1914).

[35] F. L. R. Sassen, *Geschiedenis van de wijsbegeerte in Nederland tot het einde van de negentiende eeuw* [History of philosophy in the Netherlands till the end of the nineteenth century] (Amsterdam and Brussels: Elsevier, 1959), p. 379.

the development of reason. He was optimistic about the world and history, since, as the world is guided by God by means of his laws, good will triumph everywhere. Darwin confirmed the Modernists' optimism and faith in progress, for now it had been proved by science that everything obeys the law of perpetual progress. After the publication of the *Origin*, Opzoomer accepted the law of evolution as the fundamental law of the universe.[36] The scientists following Darwin all adhered to the philosophy of Modernism.

One of the links between the university community and the general public was such a Modernist, the geologist T. C. Winkler (1822–1897), curator of the geological-paleontological collection of Teyler's Museum at Haarlem and a student of fossils. Winkler translated the *Origin* a few weeks after its publication in England. The translation was, as we have seen, accompanied by van der Hoeven's translation of an article critical of Darwin's ideas. Both were reviewed in *De Gids* [The guide], an important liberal magazine of general cultural issues, by A. Winkler Prins, a minister at Veendam in the province of Groningen.[37] As van der Hoeven had feared, Darwin's theories led immediately to speculations on a natural origin of man: Winkler Prins suggested in his review that Darwin's evolutionary theory might throw light on the origin of man. He felt that van der Hoeven and Hopkins took Darwin's scientific evidence too lightly. The least one can say, according to Winkler Prins, is that Darwin's conception deserves a place among other scientific theories. But there is no reason at this point, he tried to reassure readers worried about their animal descent, to assume that man comes from the apes.

In 1864, Winkler, the translator of the *Origin*, published a long article on life and evolution from a geological point of view in *De Gids*.[38] Since the publication of the *Origin*, he wrote, nobody doubts any more that the different species developed gradually, but the question is how

[36] Ibid., p. 321.

[37] A. Winkler Prins, Reviews of Darwin's *Het ontstaan der soorten* [The origin of species], trans. T. C. Winkler (1860), and W. Hopkins's *Over natuurkundige theorieen* [On physical theories], trans. Jan van der Hoeven (1860), in *De Gids* 1861: 718–740. (A. Winkler Prins was not related to T. C. Winkler.)

[38] T. C. Winkler, "Leven en ontwikkeling uit een geologisch oogpunt beschouwd" [Life and development considered from the point of view of geology], *De Gids* 1864: 218–245, 436–462. Winkler also translated works on geology by Lyell and D. Page.

this happened. Two methods of explanation are at our disposal: first, to investigate the transformations of the organism and to conclude from them the causes of which these transformations are an effect; second, to search for a law of nature. To anyone who might object to God's playing a part in nature, Winkler answered that science simply has nothing to say about God. Quite inconsistently, he stated further on that the law of evolution was used by the Creator in creating the world. God's greatness, he said, manifests itself not in miracles, but in supporting by the laws he made all that is and will be. Winkler pointed to exterior conditions, the use of the body, and Darwin's natural selection as natural causes for the development of species. Winkler saw natural selection almost as a person, actively selecting the best species to be preserved: "Every day and every hour," he wrote, "natural selection roams all over the world sniffing out changes in even the minutest of creatures. She [note the pronoun] rejects what is bad and preserves and collects what is good. Silently and without mercy she works everywhere and always for the amelioration of every organic being whenever the occasion presents itself."[39] Winkler considered natural selection as one of the most effective means of transformation of species, but not the only one. He was confident that the missing links in the evolutionary chain would be found. Embryology, he suggested, is also useful for the further study of the development of organic beings.

Winkler also touched on what constituted the heart of the matter for most people: the place of man on the evolutionary scale. Darwin did not discuss this point in the *Origin*, but according to Winkler man is part of nature and subject to its laws like everything else in nature. If we accept the evolution of species, we have to accept the descent of man from lower species. But Winkler set man reassuringly far apart from the other natural beings: man has, after all, much higher capacities of the mind than the mere animals, and these he cannot have inherited from them. Nor has man developed these capacities by the mechanism of evolution, as that is merely a method of functioning. Man received his higher capacities from the Legislator who decided upon the method of evolution. Winkler's conclusion was that there is certainly an evolution of life, but that life did not develop at random: there is a plan in the evolutionary

[39] Winkler, "Leven en ontwikkeling," p. 238.

process. Man may descend from the apes, but he is nevertheless very different because of his reason.

Winkler resumed his treatment of the subject of man in the context of Darwinism in *De Gids* in 1867.[40] He stated there that the evolutionary process of development has not yet reached its end: man will develop into an even higher being. Winkler repeated that what distinguishes man from the animals is not his physical structure but his capacity for mental development and his language. In the future more animal species will develop, but in the kingdom of man we will witness differentiations in mental development with only slight variations of the body. The evolution of mankind will be furthered by the fact that, where higher human races meet lower ones, the latter will become extinct.

Winkler felt that he had successfully saved the honor of man by vindicating his reasonable soul. But he could not answer the questions of how and at which point of the evolutionary process the manlike animal had received a soul, thus becoming a real man. But, for that matter, those who believe in the creation of man by God face the same problem in establishing the exact time when the human fetus receives its soul. If people would realize that their origin from the apes lies millions of years in the past, they would not be so upset by having apes in the family. Religious sensibilities quite mistakenly became mixed up with Darwinism: evolutionism in no way denies the existence of a creator; science simply keeps silent on God because it has no means of verifying religious statements. Far from leading to materialism, Winkler felt that the notion of the evolution of life, of instincts, of the intellect, of the reason of man in the successive geological periods "unfolds a picture before our eyes of the ever increasing role of the spirit over matter."[41]

Winkler provides an excellent example of the attitude of the Modernists, with their effortless compromise between religion and science. But, when the Modernists became acquainted with the interpretation of Darwin by more radical thinkers, they became concerned lest Darwinism lead too far toward the denial of God's intervention in the world. We see this, for instance, in the writings of Allard Pierson.

Pierson (1831–1896) studied theology under Opzoomer in Utrecht

40 T. C. Winkler, "De leer van Darwin" [The teaching of Darwin], *De Gids* 1867: 22–70.
41 Ibid., p. 70.

and Jan Hendrick Scholten, also a Modernist, in Leiden. After having been a minister for twenty years, he left the church, since he no longer felt at home in the dogmatic church atmosphere. But he felt equally unhappy about the Modernist compromise with the traditional church doctrines. During the following years he taught in Heidelberg as Privatdozent, and from 1877 to 1895 he was a professor of esthetics in Amsterdam. Pierson gladly admitted that man as we now know him had slowly developed into his present state, learning to walk upright, losing his body hair, and so on.[42] Darwin, he wrote approvingly, saw a progressive development of man, but not a teleological development tending toward the human form. Darwin thus "turned teleology upside down" in stating that man had to see in order to survive and that he therefore developed eyes. In the same way Darwin demonstrated that birds fly because they have succeeded in developing wings, not because they were destined to fly. So far, so good. But such developments are, according to Pierson, only of relative and limited importance. Man is very different from the animals in spite of his descent from them, for he is also an ethical being. According to Pierson, Darwin's theory is an effective means of combating naïve supernaturalism, but it should not be used to encourage materialism and atheism.

From 1848 on, the liberal bourgeoisie invaded many positions in the social-economic scene. They dominated the universities more and more. This explains why Darwinism was accepted so easily by the academic community. From a sociological point of view, Darwinism was received by and on behalf of the liberal bourgeoisie. Van der Hoeven, for instance, still belonged to the conservative generation of the Restoration period. He saw God's hand everywhere in nature. He could not really believe in a natural development, although as a scientist he had to admit that many factors pointed in that direction. Darwin's explanation of the development of species, removing God's influence so radically from natural processes, was repulsive to him. On the other hand, Harting and Donders (both from poor families, and Donders moreover from a

---

[42] Allard Pierson, "Een keerpunt in de wijsgerige ontwikkeling" [A turning point in the philosophical development], in *Uit de geschriften van Allard Pierson, 1865–1873, Tweede Reeks* [From the works of Allard Pierson, 1865–1873, second series] (The Hague: Martinus Nijhoff, 1904), I, 117–159. The essay was first published in *De Gids* 1871: 455–487.

Brabant Catholic family) were young and successful men in 1848 and identified readily with the liberal principles. Their Modernism is in line with their political and scientific views. The same was true of Winkler.

The cultural climate brought about by the dominance of the liberal bourgeoisie gave the scientists of the mid-nineteenth century the social position of torchbearers of progress. It allowed them to speak out with great assurance on social questions without being pinned down to a political party or social program. Donders, for instance, was concerned about the poor diet of the common people and urged the government to educate the population to use more nutritious foods under the guidance of the medical and physiological sciences. He considered the Royal Academy of Sciences (of which he was president for many years) an institution to guide the government in bringing about progress. Donders shared his paternalism with the liberal bourgeoisie. At the occasion of Moleschott's publication of *Die Physiologie des Stoffwechsels* (1851), Donders wrote to him that he did not think it appropriate to use physiological insights to preach materialism to the people: that only confuses a public that does not know the facts. "Better it is," he wrote, "to have faith than atheism on the basis of authority."[43]

Harting was equally aware of the responsibility of the scientist in society. In 1851 he founded a popular journal, *Album der Natuur: Een werk ter verspreiding van natuurkennis onder beschaafde lezers van allerlei stand* [Album of nature: a journal to spread the knowledge of nature among civilized readers of all ranks of society].[44] But he barred controversial subjects like Darwinism from its pages, because he judged the public not yet ripe for it. Harting wanted to further truth only among the elite. In contrast to the situation in Germany, the Dutch sci-

---

43 Quoted by Romein and Romein, *Erflaters van onze beschaving*, p. 74.

44 The journal was very successful. It helped to prepare the ground for the reception of a new type of school organized in 1863: the Hogere Burgerschool or H.B.S. This school was a creation of the liberals, founded to further the education of "the numerous citizens who want general knowledge, culture and preparation for the many business enterprises of the industrious society." Natural history occupied an important place in the curriculum, since "its educational value is today recognized by everybody," as Harting wrote optimistically in his autobiography. Inasmuch as the educational value of evolutionism was doubted by Catholics and Protestants, these groups protested against public schools where natural history might be taught from an evolutionary point of view.

entists had no radical inclinations whatsoever. A man like Jacob Moleschott (1822–1893), the famous Dutch physiologist, materialist, and
democrat, preferred to live in Germany and later in Italy.

As a Modernist, Harting did not have difficulty in combining science
and religion. "The plan of creation," he wrote, "is manifest in the creation itself. It [the creation] offers itself to our view as a unified and continuous whole and the plan realized in it is nothing else than a continuously progressing perfection." Man is the last link in the endless chain of
beings, and he is still in the process of developing. The device of nature
is "excelsior," and the scientist "cooperates with the Creator in reaching
God's great goal" by increasing the treasure of knowledge.[45]

A scientist who feels so much in harmony with God cannot but deny
that science leads to the decline of morality and religion. Harting defended science against the accusation that it leads to materialism and
atheism by pointing out that there are two kinds of scientists: those who
study facts and occupy themselves with matter only, and philosophical
scientists who look farther than the facts, "penetrating into the realm of
the immaterial, the realm of the spirit and of the eternally true and immutable"—for everything is governed by immutable laws.[46] To those
who are shocked by the idea of the rule of laws replacing the rule of God,
Harting explained that "the laws of nature are the thoughts of God." He
admitted that most defenders of materialism are scientists, but he ex-
cused their atheism as an understandable overreaction to the narrow-
minded theological conceptions of the church. In their anger they throw
out the baby with the bath water. Unfortunately, he sighed, there have
always been such hotheads, who compromise a good cause by their rash
actions. The trouble with these radical scientists is, according to Harting,
that they do not see that "most people are still children as far as their
reason is concerned."[47] True "apostles of science" respect the common

[45] Pieter Harting, "Fragment uit eene redevoering uitgesproken bij gelegenheid van
het vijftigjarig bestaan van het Natuurkundig Genootschap te Utrecht, de 6e Nov.
1863" [Fragment from an address delivered at the occasion of the fifty years of existence of the Physical Society at Utrecht, Nov. 6, 1863], *Album der Natuur*, n.s. 1864.

[46] Pieter Harting, "Iets over materialisme en materialisten, in verband met opvoeding
en onderwijs" [Some remarks on materialism and materialists, in relation to education
and teaching], *Album der Natuur* 1869: 129–150.

[47] Pieter Harting, *Ernstige woorden tot zijne landgenoten* [Serious words addressed
to his compatriots] (Amersfoort: A. M. Slothouwer, 1885), p. 38.

man and do not toss around ideas (Darwinism, for instance) for which the people are not yet ready.

Harting had a profound sense of time and of the long, arduous, but in the end successful, march of history. He saw a long road ahead for science, and he knew that each person is only allotted a short span of time. The scientists of the present, those "co-workers of the steady progress of humanity on the road that slowly but steadily leads to certain truth," collect the materials for the building posterity will erect.[48] Harting did not doubt that the development of history would be in the direction of ever-increasing perfection, for the progress of civilization depends on the progress of ideas, and, when these ideas are scientific insights about nature, they are the permanent possession of mankind. In this way the progress of science, in particular Darwinism, guarantees the progress of civilization.[49]

In spite of his scientific outlook, Harting was a romantic idealist: ideas set the world into motion and prompt men into action. His idealism went well with his romantic love for nature and his unshattered belief in a good Creator.

The atrocities of the Franco-German war of 1870 came as a cold shower on a generation that had learned to believe so confidently in progress. The war called forth in Harting some harsh thoughts about society. The many ideas of past golden ages and future millennia indicate that all human beings strive after happiness, Harting wrote in 1870 in an address entitled "The struggle for life."[50] But can these dreams ever come true? The answer depends on whether one proceeds according to wishful thinking or according to reason, accepting only what is proved by experience. As a scientist, Harting was obliged to listen only to his reason, but what he perceived was not encouraging. Life, he said, has always been a struggle, and our knowledge of life teaches us that as long as there is life there will be struggle, for unfortunately nature is not a good mother who takes care of her children. "On and under the earth, in the air, in the fresh waters and the sea, everywhere war reigns. The earth is an immense battlefield. The weak ones become a prey of the

48 Ibid., p. 39.
49 Harting, *Leerboek*; see also note 30 above.
50 Pieter Harting, *De strijd des levens: Rede gehouden 26 Sept. 1870* [The struggle for life: address delivered Sept. 26, 1870] (Utrecht: J. Greven, 1870).

stronger, the latter in their turn become the prey of even stronger ones. Everywhere we find that force triumphs over justice. Of the trillions of animals that are born daily, most die at an early age." The fossils teach us that this has been true since the earliest times. There is a continuous transformation of all the phenomena of nature, a steady metabolism that is the prerequisite of life, a continuous circle in which the death of the one is life for the other. "Struggle, a continuous struggle for life and for death, that is what nature shows us in the animal kingdom."[51]

In the human kingdom the situation is no better. In fact, no animal is as ferocious as man. In human society the strongest and healthiest will win the struggle for life. Man's history is the "history of the ever-recurrent struggle for life, and it teaches us that whole populations have disappeared from the stage of the world to make room for stronger ones." This happens, Harting continued naïvely, when Europeans settle in areas in which the climate and the soil allow their posterity to develop freely: the original population from a lower race will be pushed back till they have vanished completely. "We may feel sorrow about the fate of the many innocent and unfortunate; if the occasion presents itself we may even try to mitigate it. But the laws of nature have no compassion, and it is a law of nature that the weak succumb to the force of the strong." Harting saw the Franco-German war in the light of the struggle for life: "We observe in it [the war] a phase of the great struggle for life, in which the lower [the Celtic race] succumbs to be replaced by the higher [the German race]." The war, however cruel, will be beneficent in its consequences.[52]

Harting did not avoid the words "struggle for life" this time, although he had hardly used them before. It seems that the war woke Harting up from the dream of a happy and automatic progress and that all of a sudden this expression, so distasteful at first, made sense to him. In 1880, Harting had somewhat softened his stand on the struggle for life in society: at that time he admitted that, although experience teaches us that in many cases force triumphs over justice, the ideal of justice still exists in the human mind.[53]

Harting found more reasons to be disappointed after 1870. The har-

---

[51] Ibid., pp. 8, 9.

[52] Ibid., pp. 15, 21.

[53] Harting, "Over de betekenis der zoologie," p. 3.

mony between religion and science, in which he had believed so strongly, broke down at many points. In 1880, Dr. Abraham Kuyper founded the Free University of Amsterdam, organized upon fundamentalist Protestant principles. Harting could not understand how one could teach, not only theology and the humanities, but also physics, astronomy, and natural history on an orthodox basis. How little did he realize the close ties between his own scientific insights and his liberalism. His enemies were more perspicacious, as we shall see.

In his later years Harting had a foreboding of a social revolution, prepared by the alliance of religious groups and socialists, who had given too much influence to the uncivilized masses. Harting began to doubt even the progress of the human race, and he was particularly disappointed that the progress of the last hundred years was so limited: civilization more than ever before was threatened by "mysticism" and socialism.[54]

Harting's exasperation with the stubbornness of religion in the face of science and with the failure of science to guide the world to progress indicates the decline of the liberal regime in the Netherlands. It was caused by the emancipation of new social groups: labor, the Catholics, and the orthodox Protestants. Next, we will look briefly at them.

The Freethinkers

The liberal bourgeoisie saw in Darwin a fellow deist, an ally in the struggle for progress. The freethinkers, on the other hand, welcomed Darwin as the herald of materialism and atheism, as the scientist who provided an effective weapon against the opulent Modernist liberals with their easy belief in God and science. They were attracted to Darwinism because of its mechanical explanation of the development of nature. The freethinkers were more aware of what was going on in Germany than in England, and they tended to see Darwin through the glasses of the German materialists. Carl Vogt, for example, came to Rotterdam in 1868 to give a series of lectures on the descent of man.

The Dutch freethinkers were few in number, and they had no uni-

---

54 From a letter by Harting, quoted by A. A. W. Hubrecht, in "P. Harting herdacht" [In memoriam P. Harting], *Jaarboek van de Koninklijke Academie van Wetenschappen* [Yearbook of the Royal Academy of Sciences] 1888: 32–33.

versity professors as their spokesmen.[55] But they had an impact on society through their writings, particularly in their journal *De Dageraad* [Dawn], founded in 1855. During the first years its tone was deist and socially conservative. After 1860 the journal became the mouthpiece of materialists and atheists, who were in open rebellion against the Modernists and the liberal bourgeoisie. During the first period of publication, 1855–1867, *De Dageraad* spoke mainly for the small bourgeoisie, who resented the social and intellectual domination of the liberals. In its second period, 1870–1879, it was allied intimately with the new labor movements. In the Netherlands, socialism was introduced by the men of the *Dageraad*. Its influence in this period can be deduced from the large number of subscribers, many of them workers.

For the freethinkers, Darwin could not have published his *Origin* at a better moment. Some of the freethinkers who testified to the impulse given by Darwinism were Heys, Johannes van Vloten, and Multatuli.

Herman Hartogh Heys van Zouteveen (1841–1891) had been forced by his father to study law, but after his final exam he threw his law books into one of Leiden's canals and devoted himself to the study of nature: chemistry, geology, and zoology in particular. One of the theses of his dissertation (1866) was that Linnaeus's assumption of the fixity of species was wrong. Shortly after finishing his dissertation (on a chemical topic) Heys became a lector in zoology at the University of Leiden. In this position he did not hide his strong feelings for Darwin, which brought him into conflict with the theologians at the university and also with the church. In 1867 a new chair for zoology was proposed, and Heys's name was first on the list of nominations. But the conservative elements, spearheaded by an unidentified professor, prevailed on the government not to confirm the nomination, and Heys was passed over. Was it indeed his Darwinism that thwarted Heys's career, as rumors had it?[56] Or was it Heys's impetuous character that made people fear that he

---

55 See Oene Noordenbosch, *Het atheisme in Nederland in de negentiende eeuw: Een critisch overzicht* [Atheism in the Netherlands during the nineteenth century: a critical survey] (Rotterdam: W. L. en J. Brusse, 1931).

56 H. F. A. Peypers, "Levensbericht van Mr. Dr. Herman Hartogh Heys van Zouteveen" [Biographical note on Dr. Herman Hartogh Heys van Zouteveen], *Werken Koninklijke Academie van Wetenschappen* [Publications of the Royal Academy of Sciences] 1894: 351. Another obituary was published in *De Dageraad*, 1894.

would not fit into the community of detached scholars? As far as I know he never attempted to get a university position elsewhere, although that may also be explained by his poor health. Curiously enough, Selenka, who got the position in 1868, was a Darwinist.

After this defeat Heys lived as a private scholar. He traveled to Egypt to be present at the opening of the Suez Canal and, in 1872, to the United States to found a colony of Dutch farmers in the Santa Rosa Valley of California. Gout forced him to go back to Holland, where he retired to the countryside in Drenthe.

As a student, Heys was already a freethinker, but after his career had been thwarted by the orthodox Protestants he became a militant adversary of the church and of religion in general. In this role he got into the orbit of *De Dageraad*, in which he published frequently. In the freethinker circles he belonged to the older generation of antireligion writers; politically he was a conservative.

Heys published articles and books on many subjects, ranging from chemistry to banking and anthropology, and he was in correspondence with the leading scholars of his age, A. O. Voght, Moleschott, George John Romanes, and Charles and Francis Darwin. His greatest contribution lies, however, in his translations of Darwin.[57] He started with the *Descent.* Darwin sent him the pages of the book one by one as they came from the printer, and he wrote commentaries on these pages, which he then sent back to Darwin. Heys's comments are printed as footnotes in the Dutch translation. According to a letter from Darwin, Darwin appreciated these additions and used them in the next edition.[58]

[57] *De afstamming van den mensch* [The descent of man], 2 vols. (Delft: Joh. Ykema en van Gijn, 1871–1872; 2d cheap ed., Haarlem: J. J. van Brederode, 1882; trans. of the 2d English ed., Nijmegen and Arnhem: Gebr. E. en M. Cohen, 1871, 1884, 1885, 1890, 1891, 1893, 1895; cheap ed. of same, 1898); *Het uitdrukken der gemoeds-aandoeningen by den mensch en de dieren* [The expression of the emotions in man and animals] (The Hague: Joh. Ykema, 1873; Arnhem and Nijmegen: Gebr. E. en M. Cohen, 1891); *Het varieren van huisdieren en cultuurplanten* [Variation of plants and animals under domestication] (Arnhem and Nijmegen: Gebr. E. en M. Cohen, 1874, 1890, 1893, 1898). All Heys's translations are printed together in the series Darwin's Biologische Meesterwerken [Darwin's biological masterpieces], 5 vols., 5th ed. (Arnhem and Nijmegen: Gebr. E. en M. Cohen, 1893–1894).

[58] Letter from Darwin to Heys, quoted by Peypers, "Levensbericht," pp. 339–340: "My Dear Sir, I must have the pleasure of thanking you for a copy of your translation of my Expression book. It is by far the most beautiful edition which has been anywhere

In his preface to the translation of the *Descent*, Heys wrote that Darwin had collected so much evidence to prove his hypothesis that by now it was almost a certainty. To offset the negative feelings brought about by man's descent from the apes, Heys offered the readers a bright perspective for the future. Man's brain had developed tremendously since the Middle Ages, he claimed, and if he exercised his brain he would develop its quality and power and thus contribute to the perfecting of the human race. This would help his children in the struggle for life when their time came.

Heys's translation was introduced by Harting, who, as we have seen, read a paper on the occasion of its publication, but the relation between the two Darwinists was poor. Harting used to preach moderation to Heys, no doubt one of the "hotheads" to whom he referred. Once he rejected a paper by Heys for the *Album der Natuur* because the material was too controversial.[59] In 1871, Heys founded his own journal, *Isis: Tijdschrift voor Natuurwetenschappen* [Isis: journal for natural sciences], to spread the knowledge of Darwinism. It counted university professors (in particular from the University of Groningen) among its contributors. In *Isis*, Heys reported on all Darwin's works and on the major works on Darwinism published in other countries; in addition, he translated articles by T. H. Huxley and Voght, as well as others. He defended Darwin against Dr. P. J. F. Vermeulen, a Catholic biologist, in a series of articles on the development of the *Ascidia*.[60] On the occasion of Darwin's seventieth birthday, Heys organized a presentation to the master of an album containing portraits of about 250 of Darwin's Dutch followers.

Johannes van Vloten (1818–1883), also a freethinker, was a man of quite different interests.[61] He was not a scientist but a materialist philos-

---

published. I see that there are some notes by yourself at the end and these I will have translated by one of my sons who is as good a linguist as I am a bad one. You formerly sent me some corrections and additions of my *Descent of Man*, and I have found these extremely useful in preparing a new and corrected edition, etc., Ch. Darwin."

59 Peypers, "Levensbericht," p. 35.

60 *Isis*, nos. 10, 11 (1874).

61 See Mea Mees-Vemwey, *De betekenis van Johannes van Vloten* [The significance of Johannes van Vloten] (Santpoort: C. A. Mees, 1928); Gerard Knuvelder, *Handboek tot de geschiedenis der Nederlandse letterkunde* [Textbook for the history of Dutch literature] ('s Hertogenbosch: L. C. G. Malmberg, 1967), III, 428–430.

opher, a student of Dutch history, and a social radical. He studied theology and was the first in the Netherlands to defend David Strauss's book *Das Leben Jesu* in a review in *De Gids* (1842). He left the church in 1849. He became a professor at the small Atheneum of Deventer, was passed by for a professorship at Amsterdam, and retired to the countryside as a private scholar. There he published his own magazines, *De Levensbode* [The messenger of life], 1865–1880, and *De Humanist*, 1880–1883. In these journals van Vloten defended Darwinism warmly, mostly on the occasion of new publications on the subject in other countries.

Van Vloten rejected the concept of God as unscientific. The idea of God, he said with Ludwig Andreas Feuerbach, is merely a response to man's psychological needs.[62] Van Vloten attacked the Modernists for trying to preserve an honorable place for religion in this "century of Darwin." It is truly modern to give up supernaturalism and to fight, for instance, diseases with fresh air, light, a healthy diet, and better housing, instead of with praying and fasting. "Nature," he wrote, "is a blind force and it is not responsible for good and evil."[63] The insensitivity of nature is not a reason for despair, for man has the capacity to understand increasingly the relationship between all things, to control nature, and, aroused by love, to mitigate the evil to which all creatures are subjected. Van Vloten considered it Darwin's accomplishment to have shown that the efficiency of nature developed without the interference of God. He rejected the idea that the human race would become purified in the evolutionary process.

One of the contributors to van Vloten's *Levensbode* was Dr. H. J. Betz, a materialist à la Haeckel, but lacking van Vloten's social consciousness. Betz was a Social Darwinist. He considered society to be shaped by the blind forces of nature,[64] a view that did not encourage him to intervene to remedy existing abuses. He believed that nature is blind necessity, without a plan behind it. Betz agreed with van Vloten that

---

[62] Johannes van Vloten, "De modern-Kristelijke godsdienstleer en het Darwinisme, naar Duitse geloofsbespiegeling" [The modern Christian theology and Darwinism, according to German religious thinking], *De Levensbode* 7 (1874): 440.

[63] Ibid., p. 442.

[64] H. J. Betz, "Een nieuw boek tegen het Darwinisme" [A new book against Darwinism], *De Levensbode* 7 (1874): 286–428.

Darwin showed efficiency in nature to be caused by unconscious, material forces. Betz saw the struggle for life in society too. In the social struggle for life "the wealthy will win: their children are healthier, and people with healthy bodies are likely to have healthy ideas. Poverty and stupidity reinforce each other."[65]

Most freethinkers, however, did not draw the conclusion that the struggle for life operated in society on the same basis as in nature. The atheist Frederik Feringa (1840–1890), for instance, was in favor of Darwinism, but he wrote an angry reply to Harting's address "The struggle for life."[66]

The most influential of the freethinkers was Multatuli, pseudonym of Eduard Douwes Dekker (1820–1887).[67] In 1860 he published the book that made him famous, *Max Havelaar of de Koffieveilingen der Nederlandse Handelmaatschappy* [Max Havelaar or the coffee auctions of the Dutch Trading Company]. In this novel Multatuli depicted the cruelty of the Dutch colonial regime in the East Indies. He had been an administrator in the Indies himself, and he had been fired after an attempt to defend the Javanese against oppression. From 1860 on, Multatuli worked as a free-lance writer and journalist, mostly in Germany. In 1862 he started to publish his "Ideas," a collection of aphorisms on political and religious subjects. He was a sarcastic adversary of the liberal bourgeoisie, and he became the champion of lower-class interests. The *Dageraad* was the first channel through which he voiced his opinion. In 1861 he published there his famous "Prayer of an Ignorant Man" (many times reprinted on cheap leaflets). The prayer is an accusation of a God who does not manifest himself and abandons his children in their "grave struggle for humaneness and justice." It ends with the words: "The father is silent . . . O God, there is no God."[68] According to Multatuli, nature does not show traces of a loving God: nature is stupid; everything is as randomly put together as the goods of a department store.[69] He utterly despised the Modernists, who combined coal with the Bible.

[65] Ibid., p. 424.

[66] Frederik Feringa, *Democratie en Wetenschap: Vertoogen en Opmerkingen* [Democracy and science: considerations and remarks] (Groningen, 1871), pp. 270–275.

[67] See Noordenbosch, *Atheisme*, passim; Knuvelder, *Handboek*, pp. 365–386.

[68] Multatuli, *Volledige Werken* [Collected works], ed. Garmt Stuiveling (Amsterdam: G. A. van Oordschot, 1950), X, 59–61.

[69] Ibid., II, 383–384.

Although Multatuli had no first-hand knowledge of Darwin, he contributed much to making the results of Darwin's biological explorations known among the public. The hilarious story about Miss Laps in his novel *Woutertje Pieterse* is a good example. Miss Laps is a lower-middle-class lady, who almost faints when she is told that she is a mammal: "Others may be mammals, but she is a decent lady, for her father dealt in grains."[70] Multatuli had to admit that Darwin was not Darwinistic enough, for he still believed in a creator. "Darwin is childish," he wrote, and he called Darwin's books "patchwork."[71] He had his doubts about natural selection. In one of his masterful anecdotes Multatuli gave an evolutionary history of the jacket, from loincloth to dress coat. The two buttons at the back, he explained, "have the same meaning as the nipples of male mammals or the rudera of the gills in animals with lungs." "It is clear," he concluded, "that everything points to a gradual evolution, making the idea of creation completely superfluous. It should be noted, however, how little is solved by Darwin's natural selection: buttons do not copulate."[72]

Multatuli's writings mark the extent to which Darwinism had penetrated colloquial speech: he is one of the first to use the anglicism *evolutie* ("evolution") instead of *ontwikkeling* ("development"). He also speaks of a "free darwinian development" of a language[73] and of the "struggle for life" of the elements of a problem.[74]

With Multatuli, Darwin entered literature. The freethinker Carel Vosmaer,[75] a classical scholar, writer, materialist, friend, and admirer of Multatuli, also a friend of van Vloten and Betz, made propaganda for a radical Darwinism in his column "Vlugmaren" in the *Nederlandse Spectator*, of which he was one of the chief editors. In one of these he wrote: "See, here I give you for your enlightenment the Third Testament, in which are written the Acts of Baur, the Good Tidings of Strauss and the Book of Darwin."[76]

---

[70] Ibid., VI, 259.
[71] Ibid., VI, 115.
[72] Ibid., IV, 188–189.
[73] Ibid., VI, 6.
[74] Ibid., VII, 37.
[75] See Knuvelder, *Handboek*, pp. 433–438.
[76] Quoted in Noordenbosch, *Atheisme*, p. 72.

The Roman Catholics

The criticism from Roman Catholic circles during the nineteenth century has been characterized by L. J. Rogier, the historian of the Catholics in the Netherlands and a Catholic himself, as "an instinctive rejection by people who do not know the facts."[77] The priest B. H. Klönne (1834–1921) did not even accept the new geology, and he called paleontology mostly a creation of fantasy. In an 1869 publication, Klönne stated bluntly that if the descent of man from the apes is a necessary consequence of Darwin's theories, as Winkler had stated, the theses of Darwin are already condemned on that basis.[78] The philosopher Franciscus Becker, teacher at the seminary at Culemborg, wrote an article in 1873 on the limits of experience, rejecting the descent of man from the apes as not proved by science. To be on the safe side, he did not treat the specific biological aspects of Darwinism. The Dutch Catholics seem not to have known St. George Mivart's book *Genesis of Species* (1870), which announced the Catholic conciliation with the evolutionary hypothesis, even though it did not include natural selection.[79] Dr. P. J. F. Vermeulen, a natural scientist and conservative member of Parliament, who rejected Darwin on biological grounds, condemned Darwin mostly for the immoral consequences of his theories.[80]

The Orthodox Protestants

The attitude of the Protestants toward Darwin was, of course, in the context of Dutch society, of more importance than that of the Catholics, who had just begun to feel themselves part of that society. The Netherlands emerged, after all, as a Protestant nation after its revolt against

[77] Quoted by J. G. Hegeman, "Darwin en onze voorouders: Nederlandse reacties op de evolutieleer, 1860–1875" [Darwin and our ancestors: Dutch reactions to the doctrine of evolution, 1860–1875], *Bijdragen en Mededelingen betreffende de Geschiedenis der Nederlanden* [Contributions and communications concerning the history of the Netherlands] 85 (1970): 273.

[78] B. H. Klönne, *Onze voorouders volgens de theorie van Darwin en het Darwinisme van Winkler* [Our ancestors according to the theory of Darwin and the Darwinism of Winkler] ('s Hertogenbosch: Henri Bogaerts, 1869); see Hegeman, "Darwin en onze voorouders," p. 291.

[79] Hegeman, "Darwin en onze voorouders," p. 275.

[80] Ibid., p. 295.

Spain in the sixteenth century, and ever since its intellectual and moral climate has been permeated with Calvinism. The Modernists claimed to be the true heirs of Protestantism, defending human conscience against authority. In spite of their rejection of the "man-made" dogmas of Calvin and the Synod of Dordrecht (1618), they felt that they retained the kernel of the Protestant faith. Strauss's radicalism was too much for them. But, in reality, the loyalty of the Modernists was more to the nineteenth century than to Protestantism.

Now we have to inquire about the reaction of those Protestants who stuck to the creed of their fathers and refused to follow the spirit of the times. In these circles we find much resistance against all manifestations of modern science and of the modern spirit in general. The reception of Darwinism by these orthodox Protestants was, as is to be expected, as negative as that of the Catholics, but it showed more variety.

Although there were fundamentalists who took the Bible literally, Lambert Tinholt (1825–1886), minister at Koudum in Friesland and chief editor of the magazine *Stemmen voor Waarheid en Vrede: Evangelisch Tijdschrift voor de Protestantse Kerken* [Voices for truth and peace: evangelical magazine for the protestant churches], suggested a compromise. He was of the opinion that the Bible did not have to be taken completely literally. The exegesis of the Bible, he explained, has to investigate the extent to which the Bible is at odds with the proven facts of the natural sciences.[81] Then it has to explain to the faithful how the Bible must be understood so that contradictions with well-established facts of science are avoided. In this way the new insights of geology can be assimilated: one day in Genesis can, for instance, be explained as one geological period. The concordist point of view, which was not found among the Catholics, was also represented by Jan Lodewijk ten Kate (1819–1889), in his later years a minister in Amsterdam and an extremely popular poet in his time.[82] He wrote a long and tedious poem, *De Schepping* [The creation], in which he tried to work modern geology into the biblical story of creation (1867). The orthodox Protestants never gave up the idea of separate creation of species, and consequently natural selection and natural development of species could in no way

[81] Ibid., p. 273.
[82] Knuvelder, *Handboek*, pp. 303–304.

become incorporated into the orthodox system of thought till Kuyper showed a way of accepting progress and, with many restrictions, Darwinism.

Dr. Abraham Kuyper (1837–1920), a minister, theologian, and statesman of striking qualities, became the major spokesman for the fundamentalist Protestants in the eighties.[83] He studied theology in Leiden with Scholten, a Modernist, but he found his spiritual home in the strong and simple faith of the rural folk of his first parish. From 1872 to 1874, Kuyper was the chief editor of the antirevolutionary newspaper *De Standaard* [The standard]; from 1874 to 1877, he was a member of Parliament, and, after 1880, a professor at the Free University of Amsterdam, founded by himself. Later he became prime minister, returning after his term to Parliament.

With the freethinkers, Kuyper realized the urgency of the social question. Kuyper was a true democrat, in favor of extended suffrage, the organization of labor, the protection of labor by the government, and an active social policy in general. He opposed the rugged individualism and social conservatism of the liberal bourgeoisie, an intellectual and financial oligarchy whose power he wished to see ended.[84] Kuyper passionately wanted to bring the country back to obedience to God's laws, and at the same time he wanted to bring the orthodox Protestants back to the position of influence they had had before the rise of the liberal bourgeoisie. By conceiving the doctrine of "common grace" (*gemeene gratie*), he proved that the modern world was not necessarily bad. By common grace Kuyper meant that, even after the Fall, man still had the capacity to build a world that was acceptable in God's eyes. On the basis of common grace, Kuyper could accept modern science and the progress of civilization. Kuyper founded the Free University of Amsterdam to provide a place where modern science could be studied by Christians within the framework of orthodox Protestant principles.

Among religious responses to Darwinism, Kuyper's was doubtless the most creative one. In 1899 he delivered the address to the Free University quoted at the beginning of this essay.[85] In it, he explained his orthodox Christian position in regard to evolutionism. Before the rise of evo-

---

[83] Romein and Romein, *Erflaters van onze beschaving*, pp. 145–178.
[84] Ibid., p. 158.
[85] Kuyper, *Evolutie*.

lutionism, we read there, science occupied itself only with what is empirically observable, leaving the mysterious aspects of life to religion. But recently evolutionism has conquered all areas of human knowledge, leaving nothing to religion. Thanks to Herbert Spencer and Ernst Haeckel, even the Anti-Christians now have an all-embracing religious system. But this new faith is haunted by the "ghost of Decadence," for the Darwinian struggle for life leads to the usurpation of power, to Bismarckianism, to imperialism and militarism. As a result the world is coming to a situation comparable to that of the dying Roman Empire: our fall will not be long. Is there no way to stop this downward trend? Yes, Kuyper exclaimed, by ceasing to "whore with Evolutionism." We must look at scientific theories, such as evolution, from a Christian point of view; that is to say, we have to consider such theories as hypotheses, not as doctrines that pretend to tell the truth about reality. The "sin" of taking scientific hypotheses as factual statements was especially prevalent in Germany, from which Kuyper took most of his examples. He pointed hopefully to the dissent existing in the evolutionary camp itself, looking forward to the speedy disintegration of evolutionism altogether.

Kuyper rejected evolutionism as a world view because it was incompatible with the teachings of the Bible, but he ingeniously accepted it as a scientific hypothesis tentatively made up by limited human beings. It is amazing how close Kuyper came to the position of a liberal like Harting. In his book on common grace he wrote: "In spite of the changes brought about by the fall, the great plan of God is continuing. His mankind comes; that mankind gets its history; in this history it [mankind] is involved in a process; that process develops what God's command has hidden in its very beginnings; and common grace is the holy instrument by which God brings this process to an end, in spite of sin."[86] Kuyper compared the system of *election* (by God) with that of *selection* (by the blind forces of nature). The difference between the two systems he stated is, in more than one way, only that of one letter.[87]

[86] Abraham Kuyper, *De gemeene gratie* [Common grace] (Amsterdam and Pretoria: vrhn Höveker en Wormser, 1902–1903), II, 24.

[87] Abraham Kuyper, *Het Calvinisme: Zes Stone-lezingen in Oct. 1898 te Princeton (N.J.) gehouden* [Calvinism: six Stone Lectures delivered during Oct. 1898 at Princeton, N.J.] (Amsterdam and Pretoria: Boekhandel vrhn Höveker en Wormser, 1898), pp. 194–195.

Conclusion

Darwin's progress in the Netherlands was easy. It came at a moment in which the country was preparing itself, under the leadership of the liberal bourgeoisie, for its take-off into the age of industrial capitalism. Darwin served in this situation as a catalyst for the liberals as well as for the freethinkers, helping them to outline their ideas and stance on society. Opposition came only from religious quarters.

The liberal bourgeoisie assimilated Darwin's ideas of evolution in nature easily as additional support for their already existing faith in progress, reason, and science. In their eyes, Darwin explained the mechanism of progress in nature (and in civilization) that was planned and guided by the Creator. As a result, Darwin corroborated strongly the establishment's faith in itself.

As for Darwinist biology, the research of Harting and others was certainly disinterested and not designed to justify any social purpose. Nevertheless, one cannot but notice that the professional opinions of the scientists coincided to a remarkable degree with their social position and their attitudes toward society. And Darwinist biology would not have been so generously supported in the universities if it had worked against the liberal establishment.

Why had a biological theory like Darwin's elicited so much response in society? The new physics of Albert Einstein, for example, seems to have had little or no effect on public life. One reason is that the opinion-making public was ready to accept the idea of an evolutionary progress. On the basis of the situation in the Netherlands, where the ground was so well prepared by the new geology and paleontology and by the philosophy of Modernism for a favorable reception of Darwin, I am inclined to think that Darwin was very much in touch with the general disposition of his age. Even if we cannot find the missing links that connect him with the other thinkers on evolution, we must assume that he was influenced by the general trend of genetic thought in his time. Another reason for the repercussions of Darwinism in society was that it touched upon the question of man's origin and destination. An appeal to the "democratic" principle of getting better than one's parents often served to sweeten the medicine of a merely natural origin.

To the freethinkers, Darwin served as a bomb to explode the existing social and intellectual order. For them, the nineteenth century was truly "the age of Darwin." In their eyes, Darwin furthered the cause of materialism and atheism. In this circle Darwin paved the way for socialism.

Although one might presume that the liberals were prone to Social Darwinism (Kuyper suggested this, too, in his address on evolution), and the freethinkers were not, the lines are not that easy to draw. Pierson was against Social Darwinism; Betz and Heys were in favor of it.

In the 1880's, Darwinism entered its second phase in the Netherlands. Darwin's theses were viewed less emotionally. Scientists like A. A. W. Hubrecht (1853–1915), a student of Harting and his successor in Utrecht, and Hugo de Vries, professor of botany at Amsterdam, started to take a fresh look at Darwin's theories. De Vries, while keeping the idea of an evolution in nature, challenged Darwin's concept of natural selection with his discovery of mutations. The Antirevolutionaries managed, thanks to Kuyper, to accept evolutionism as long as it was stated as a working hypothesis. Among the Catholics, Pater Wassman accepted the idea of evolution, although he made an exception for man. The socialist Ferdinand Domela Nieuwenhuis mentioned Darwin's law of natural selection as one of the greatest laws of the universe, but he put Peter Kropotkin's law of mutual help on a par in analyzing human society. He preferred to use De Vries's mutationism rather than Darwinian evolution as theoretical support for revolution.

By 1900, evolutionism had acquired its citizenship, but the hypnotic spell it had cast on an earlier generation had begun to wane.

SPAIN

THOMAS F. GLICK

That the reception of Darwinism in Spain was linked to a social and political event of the first order—the Revolution of 1868—was a fact that was consciously and explicitly appreciated by the first generation of Spanish evolutionists and their opponents. Indeed, commentators on the left and right alike agreed that the ideology of the "intellectual mesocracy" who substantially altered the intellectual physiognomy of the nation in the 1870's and 1880's was dominated by the overweening presence of Darwin, Ernst Haeckel, and Herbert Spencer. It would seem therefore that the Spanish case might offer a valid test of Karl Mannheim's hypothesis that the structure and objectives of social groups underlie the idea-sets that they elaborate and assimilate.

Scientific life in prerevolutionary Spain was characterized by stagnation and marginality. Such scientific activity as existed was relegated to the sidelines of intellectual and academic endeavor, confined by an immobile and authoritarian orthodox educational structure. José M. López Piñero describes the tortured existence of an "intermediate generation" of scientists of mediocre accomplishments, who nevertheless kept the torch alive and created the institutions in which the first generation of Spanish evolutionists were trained.[1]

References to Darwin between 1859 and 1868 are almost nonexistent. The earliest allusion to evolution of species in a Spanish journal seems to be a satirical "Scale of Transformations" appearing in *El Museo Uni-*

[1] José M. López Piñero, "La literatura científica en la España contemporánea," *Historia general de las literaturas hispánicas* (Barcelona: Editorial Vergara, 1968), VI, 679–684.

versal in 1863. This was a series of four line drawings illustrating the "origin of certain species of animals," specifically (the first) the evolution of a pig into a bull and then into a man and (the last) that of a goose into a donkey into a "jackanapes" (*mequetrefe*). The jackanapes resembles Herbert Spencer, an identification borne out by the description of "a certain individual" who spoke about things he did not understand—philosophy, religion, politics, and elections—talking such nonsense that his ears began to grow long. There is no direct mention here of Darwin or his works (to be sure, the process described is a parody of Lamarckism).[2]

In his novel *Fortunata y Jacinta*, Benito Pérez Galdós sums up the academic milieu in Madrid during the student days of his character Juanito Santa Cruz in the years 1864–1869 by observing that the students discussed transformism and the ideas of Darwin and Haeckel, even though it was not yet fashionable to do so.[3] The first serious discussion of evolution is found in a discourse delivered by José de Letamendi (1828–1897) in the Ateneo Catalán on April 13 and 15, 1867, in which the Catalonian physician criticized the mutability of species from a Thomistic viewpoint, lumping together in a general criticism of monism sweeping condemnations of Comtian materialism, Lamarck, and Darwin.[4] These scattered references do not significantly alter the conclusion that Darwinism made few inroads before 1868.

That the revolution opened the floodgates for a host of new ideas, among which evolutionism was the most salient, was a leitmotiv of the

[2] *El Museo Universal* (later, *La Ilustración Española y Americana*) 1863: 160, 176, 192, 208. This kind of joke was widely disseminated in British journals of the 1860's; see Alvar Ellegård, *Darwin and the General Reader: The Reception of Darwin's Theory of Evolution in the British Periodical Press, 1859–1872* (Göteborg: Elanders Boktryckeri Aktiebolag, 1958), pp. 238–241. I believe these sketches are not originally Spanish, although they seem not to be of English provenance either.

[3] Benito Pérez Galdós, *Obras completas*, 6 vols. (Madrid: Aguilar, 1950), V, 14. See also the discussion in Leo J. Hoar, Jr., *Benito Pérez Galdós y la Revista del Movimiento Intelectual de Europa, Madrid, 1865–1867* (Madrid: Insula, 1968), pp. 67–71; Hoar's conclusions of Darwinian influence in a series by Galdós on creation seem unwarranted.

[4] José de Letamendi, *Discurso sobre la naturaleza y origen del hombre* (Barcelona: Ramírez, 1867). See discussion in Tomás Carreras y Artau, *Médicos-filósofos españoles del siglo XIX* (Barcelona: Consejo Superior de Investigaciones Científicas, 1952), pp. 200–204.

scientific literature of the day.[5] As Baltasar Champsaur Sicilia, a Canarian socialist who took part in an early Darwinian polemic, reminisced: "The so-called Revolution of September had opened many windows, and we breathed an air of liberty and of cheering life which makes one think of something like a people newly master of itself. . . . There were discussions everywhere. . . . It was like a true awakening after a long somnolescence."[6] Ten years later, the anthropologist Francisco Tubino reflected that around 1867, when he had first begun to lecture at the Economic Society of Madrid about the new anthropological science, he feared for the poor reception of these ideas in Spain, a country where "emotion carried the day in everything" to the detriment of reason, and he failed to realize at the time that so much interest in the subject would soon be aroused.[7] The chemist José R. Carracido recalled the intellectual excitement at the University of Santiago de Compostela when Augusto González de Linares first lectured on evolution in 1872. The "murmurs of protest and applause of counter-protest," the palpable signs of the deep schism in the audience would have been inconceivable, in Carracido's view, had not the general intellectual milieu of Santiago been first prepared, "owing to the mental stimulation consequent to the political revolution." With the same vigor that intellectuals began to discuss such issues as national sovereignty and the separation of church and state (Carracido's narrative continues), they also began to talk about mutability of species and the origin of man. You could hear groups of students walking the ancient streets around the university discussing the struggle for life, natural selection, and adaptation, invoking the names of Darwin and Haeckel.[8]

But, by 1872, not only the intellectual climate, but also the very struc-

---

[5] Thomas F. Glick, "Science and the Revolution of 1868: Notes on the Reception of Darwinism in Spain," in *La Revolución de 1868: Historia, pensamiento, literatura*, ed. Clara E. Lida and Iris M. Zavala (New York: Las Américas, 1970), pp. 267–272.

[6] Baltasar Champsaur Sicilia, *Transformismo* (Las Palmas: Imprenta Miranda, 1928), p. v. All translations are mine except where otherwise indicated.

[7] Francisco Tubino, "La ciencia del hombre según las más recientes é importantes publicaciones," *Revista Contemporánea* 11 (1877): 407.

[8] José R. Carracido, *Estudios histórico-críticos de la ciencia española*, 2d ed. (Madrid: Imprenta de "Alrededor del Mundo," 1917), pp. 276–277. See also Benito Madariaga, *Augusto González de Linares y el estudio del mar* (Santander: Institución Cultural de Cantabria, 1972).

ture of education itself, had changed. As a result of the Free Education Act, proclaimed in the first month of the revolution (October, 1868), censorship was abolished, modern science courses were introduced into the university curriculum, and new departments were founded (for example, those of physiology and histology at the University of Madrid). Moreover, the act empowered local governmental agencies to join with private institutions to develop new scientific programs independent of the control of the central government. This act, as Temma Kaplan notes, is considered to have supported the development of empirical science in Spain and, especially, to have underwritten the beginning of modern biology.[9]

Although the restoration (1874) re-established "official science," reintroduced religion into the university curriculum, and brought back censorship, and in spite of the fact that most of the Darwinists who had gained university science chairs in the wake of the revolution lost them during the Second University Crisis of 1875, the permeation of evolutionary ideas was so pervasive that Catholic revanchism was unable to roll back the tide.[10] In the sections that follow, I present an overview of the permeation of Darwinian ideas in various intellectual endeavors and in various areas of the country, in the hopes of gaining a broad perspective on the diffusion of evolutionary ideas in Spain and of suggesting lines of future research.

Not unsurprisingly, the *Descent of Man* was translated into Spanish (at least by 1876) *before* the *Origin of Species* (1877). Moreover, a survey of the literature reveals that, although Darwinian evolution was discussed between 1868 and 1871, it was not until after the French version of the *Descent of Man* reached Spain that the discussion became a full-fledged polemic. In 1872 there was a flurry of attacks on Darwinism on all fronts: an intelligent critique of evolution literature in English, German, and French by Emilio Huelin in the *Revista de España*; a speech at the Ateneo de Madrid by the conservative political leader Antonio Cánovas del Castillo, attacking Darwinian ethics; and, as the year drew to a close, Gaspar Núñez de Arce's famous polemical poem, "A Darwin," in-

[9] Temma Kaplan, "Positivism and Liberalism," in *La Revolución de 1868*, ed. Lida and Zavala, p. 258.

[10] Ibid., p. 261; Glick, "Science and the Revolution of 1868," p. 268.

cluded in his orthodox call to arms, *Gritos del combate*.[11] The pro-Darwin counterattack did not really build until about five years later, or at least until 1876, when the *Revista Contemporánea*, the major mouthpiece of evolutionism in Spain, was founded by the Cuban positivist José de Perojo. It was Perojo who published the first Spanish translation of the *Origin of Species* in 1877 and who was one of the translators of the second Spanish version of the *Descent of Man*.[12]

The publication of a Spanish edition of the *Origin* was hailed by Manuel de la Revilla in the *Revista Contemporánea* as "a most happy symbol of our progress," an event that proved that "those times had passed in which the transcendental doctrine [of evolution] was received with puerile fears by the common folk and with empty denunciations or insipid jokes by the learned." The theory, he continued, was now heard with respect, despite its incitement of conservatives' pride, "fed by anthropocentric error."[13] It will be noted, however, that, whether translations were available or not, most Spanish participants in the evolution debate preferred to read the *Origin* and *Descent* in French.

The works of Ernst Haeckel were the prime conduit for the diffusion of evolutionary ideas in Spain. On the pro-evolution side, positivists read his works enthusiastically. The positivists were sensitive to hard arguments based in the biological sciences; but they seem to have preferred Haeckel to Darwin because the former spoke more directly to

[11] Emilio Huelin, "Los brutos, supuestos engendradores del hombre," *Revista de España* 25 (1872): 5–29; Antonio Cánovas del Castillo, "Discurso pronunciado el día 26 de noviembre de 1872," in *Problemas contemporáneos*, by idem, 3 vols. (Madrid, 1884–1890), I, 167–171 (see Enrique Gil y Robles's critique of this speech, "Sobre un discurso de Cánovas," *Ciencia Cristiana* 19 [1881]: 10–12, 410–413); Gaspar Núñez de Arce, "A Darwin," in *Gritos del combate*, 11th ed. (Madrid: Fernando Fé, 1904), pp. 135–145. The poem is dated December 24, 1872. See also Núñez de Arce's comments on the theory of evolution in *Gritos del combate*, pp. 382–383, and Juan Valera's comment in his *Obras completas*, 3 vols. (Madrid: Aguilar, 1944–1958), II, 453–455.

[12] Charles Darwin, *Origen de las especies por medio de la selección natural o conservación de las razas en su lucha por la existencia*, trans. from the 6th English ed. by Enrique Godinez (Madrid: José de Perojo, 1877; reprint ed., Madrid: Lucuix, 1880); idem, *Descendencia del hombre y la selección en relación al sexo*, trans. José de Perojo and Enrique Camps (Madrid, 1885). An earlier translation, *El origen del hombre: La selección natural y sexual*, had been published in Barcelona (Llop, 1876; Trilla y Serra, 1880; other editions in 1892 and 1897).

[13] Manuel de la Revilla, "Revista crítica," *Revista Contemporánea* 10 (1877): 117–119.

the extension of Darwinian models to the social sciences, the prime area of concern to the Spanish positivists. A series of articles by or about Haeckel appeared in the pages of the *Revista Contemporánea* and *Revista Europea* in the later 1870's. The message was more or less the same in all: Haeckel had created a total system that explained the evolutionary development of the cosmos.[14] Anti-Darwinists did not fail to get the message and uniformly branded Haeckel as the chief promoter of anti-Christian materialism. Haeckel loomed so large that a pro-evolution Catholic like Juan González de Arintero could arrive at a view according to which Darwin had made no statements at odds with the Bible in the *Origin*, only to suffer a "conversion" at the hands of Haeckel, under whose influence the unacceptable positions in the *Descent* were articulated![15]

Haeckel's more properly scientific views were also widely disseminated. Augusto González de Linares both wrote and lectured on Haeckel's notions of morphology in the 1870's,[16] and several of Haeckel's books were translated into Spanish.[17]

[14] The best statements are by the leading evolutionary social theorist Pedro Estasen, in "La creación, según Haeckel," *Revista Contemporánea* 17 (1878): 148–166, esp. pp. 164–165; and by the anthropologist Francisco Tubino, in "Darwin y Hackel [*sic*]," *Revista de Antropología* 1 (1874): 238–256, and "La ciencia del hombre según las más recientes e importantes publicaciones," *Revista Contemporánea* 11 (1877): 408–413. Other articles of significance include José de Perojo, "Haeckel juzgado por Hartmann," *Revista Contemporánea* 1 (1875–1876): 358–369, in which the leading Spanish positivist casts doubt on Eduard Hartmann's prediction that a reconciliation can be effected between mechanistically oriented natural science and theology; Jules Soury, "La antropogenia de Haeckel," *Revista Contemporánea* 9 (1877): 178–187; Eduard Hartmann, "Haeckel," *Revista Europea* 7 (1876): 7–15, 65–73; Ernst Haeckel, "Los adversarios del transformismo," ibid. 8 (1876): 737–750; and more than a dozen articles by Haeckel in the *Revista Europea*, e.g., "Sentido y significación del sistema genealógico, o teoria de la descendencia," ibid. 12 (1878): 1–9. See also note 50 below.

[15] Juan González de Arintero, *La evolución y la filosofía cristiana: Introducción general* (Madrid: G. del Amo, 1898); cited by Rafael Sanús Abad, *Positivismo y ciencia positiva en el siglo XIX español* (Burjasot: Colegio Mayor de San Juan de Ribera, 1963), p. 24.

[16] Augusto González de Linares, "La morfología de Haeckel," *Revista Europea* 11 (1878): 32, 62–63. In 1877 a course on Haeckel's morphology was offered by Linares at the Institución Libre de Enseñanza in Madrid; see *Revista Contemporánea* 9 (1877): 117. The notes to the course were published the same year in the *Boletín de la Institución*.

[17] Ernst Haeckel, *Historia de la creación natural o doctrina científica de la evolución*, trans. Claudio Cuveiro González, 2 vols. (Madrid, 1878–1879); idem, *Ensayos de*

The most representative Spanish disciple of Haeckel was the Valencian anatomist Peregrín Casanova Ciurana (1849–1919). Casanova was a man of the revolution, receiving his *licencia* in medicine at Valencia in 1871 and his doctorate at Madrid in 1875. In 1873, he began to correspond (in imperfect French) with Haeckel and may have spent some time in study at Jena, perhaps in 1876.[18] Casanova wrote repeatedly to Haeckel concerning the best way to orient his courses at the medical school in accord with the latest evolutionary theory, seeking to combine classical human morphology with the evolutionary method. From Haeckel he received news of Carl Gegenbaur's latest researches in advance of their publication, and it was owing to Casanova that Gegenbaur's works became the standard anatomical texts at the University of Valencia.[19] From Haeckel, too, Casanova received copies of the *Jenaische Zeitschrift* and *Kosmos*, in which journals he was able to keep abreast of the master's latest researches. In 1877, Casanova wrote a general biology text, which was saturated with Haeckelian concepts.[20] The book bore a salutatory introductory letter from Haeckel, and Casanova, in his introduction, presented an evolutionary view of the history of science in which the strongest ideas win out over the weak. The past is strewn

---

psicología celular, trans. Oswaldo Codina (Valencia: Pascual Aguilar, 1882; reprint ed., Madrid, 1889). With reference to Haeckel's writings on the cell, see also Enrique Serrano Fatigato, "Estudios sobre la celula," *Revista Europea* 4 (1877): 166–173, 513–520, 568–571; Haeckel, *Morfología general de los organismos*, trans. Salvador Sanpere y Miguel (Barcelona: La Academia, 1885; rev. ed., 1887); idem, *La evolución y el transformismo* (Madrid, 1886).

[18] Nine letters from Casanova to Haeckel, dated 1873–1883, survive in the collection of the Ernst Haeckel House, Jena; copies were supplied to me by Professor Georg Uschmann. The letters give no clue concerning Casanova's stay in Germany, the sole reference for which is Manuel de Espinosa Ventura's *Memoria apologética del Dr. Peregrín Casanova Ciurana* (Valencia: Tipográfica Moderna, 1921), p. 11.

[19] On the works of Gegenbaur that formed part of the library of the Valencia medical school, see Pilar Faus Sevilla and José M. López Piñero, *Catálogo de la Biblioteca Histórico-Médica de la Facultad de Medicina de Valencia*, vol. 1, *Anatomía* (Valencia: Catedra e Instituto de Historia de la Medicina, 1962), pp. 41–42, nos. 166–174. When I visited the library in 1965, I found that all the works of Gegenbaur were still on reserve for students of anatomy. The role of evolutionary anatomy and histology texts by Gegenbaur, Mathias Duval (*Catálogo*, p. 33, nos. 121, 122), and Jean-Léon Testut (p. 91, no. 438) as conduits of Darwinian ideas cannot be underestimated.

[20] Peregrín Casanova Ciurana, *Estudios biológicos*, vol. 1, *La biología general* (Valencia: Ferrer de Orga, 1877).

with outmoded cosmological and scientific ideas, and modern science must still struggle against surviving medieval notions. Biology[21] must participate in these advances, and its morphological section ought not to be merely descriptive, but dynamic, and not only ontogenic, but phylogenic—investigating the history, not only of individuals, but of the species. In sum, his study aims at nothing less than the "reduction of all phenomena to natural, mechanical efficient causes: the monistic conception of the universe." The text itself is a popularization of Haeckel. A section on the dynamic character of biology begins with a citation on the biochemical basis of life from the *Natürliche Schöpfungsgeschichte*. Here Casanova presents Haeckel's ideas on the formation of life from nonlife in the form of granulations or plastids. The chapter on the results of morphological development discusses integration and differentiation considered in a phylogenetic series and includes descriptions of the processes of natural selection, heredity, and adaptation. An extensive chapter on the theory of evolution closes with the lament that in Spain, a country that leaves much to be desired in moral and scientific culture, the doctrine of evolution has few devotees.[22]

In 1881, Casanova penned an introduction to a Spanish version of two of Haeckel's essays, "The Perigenesis of Plastids" and "The Soul of Cells." Here he introduced Haeckel's notions of heredity (opposed to those of Darwin), stating that the attempt to reduce the etiology of heredity and reproduction to physicochemical causes was bound to meet with entrenched hostility. Casanova, a freethinker, especially admired the second essay, for it was posited on the notion that forms of life lower than man were also capable of "intellectual activities." "Who can confine will and thought to the narrow circle of the human brain?" he asked rhetorically.[23] Casanova's library, preserved at the University of

---

21 It was common in the 1870's to define the term *biology*. A typical definition (copied from St. George Mivart, I believe) is that of Huelin: "Biología es la ciencia de la vida. Comprende la zoología o ciencia de los animales y la botánica o de las plantas." (Huelin, "Los brutos," p. 5, n. 1. "Biology is the science of life. It includes zoology, or the science of animals, and botany, or that of plants.") Casanova's definition is striking both for its departure from this simple model and for its Haeckelian construction: biology is "the science of matter in continuous and colossal molecular motion" (*Biología general*, p. xii).

22 Casanova, *Biología general*, pp. vii–xv, 34–41, 239–245, 388–405.

23 Haeckel, *Ensayos de psicología celular*, pp. v–xvii. Casanova's essay is dated November 8, 1881. Casanova also translated a speech by Haeckel, "La teoría de la

Valencia Medical School, contains a representative selection of evolutionary works.

The effects of evolutionary theory on Spanish science of course transcended the bounds of mere popularization. Of most significance, the work of the great Spanish school of neurohistologists was evolutionary to the core, and the founding fathers of the school first read Darwin in the immediate postrevolutionary period. Spanish neurohistology was, properly speaking, composed of two interrelated schools. The first, associated with Luis Simarro (1851–1921), was primarily concerned with the pathological anatomy of nervous disorders, while the second, led by Santiago Ramón y Cajal (1852–1934), cultivated the normal histology of the nervous system.

Simarro first read Darwin in a French translation just prior to the revolution in 1868, while still a medical student in Valencia. When discovered with the book in his quarters, he was dismissed from his job as a science teacher by the monks who ran the school.[24] In Simarro's early articles on comparative anatomy and morphology, dependence on Haeckel and his disciples was great; all of his works were laced with references from Gegenbaur's *Morphologische Jahrbücher*. From 1880 to 1885, Simarro worked in Paris with various French scientists, including the evolutionary anatomist Mathias Duval, on whose anthropology course (organized along evolutionary lines) Simarro reported to the Spanish public in an 1880 article.[25] Returning to Spain, he lectured frequently on the theory of evolution and incorporated notions of evolutionary changes

---

evolución en sus relaciones con la ciencia general," *La Crónica Médica* 1 (1877–1878): 329–335, 364–368, 393–396.

[24] Kaplan, "Positivism and Liberalism," p. 256.

[25] Luis Simarro, "La enseñanza superior en París: La Escuela de Antropología, curso de Mr. Matías Duval," *Boletín de la Institución Libre de Enseñanza* 4 (December 3, 1880): 91, 173. Duval (1844–1907) succeeded Paul Broca as Professor of Zoological Anatomy in 1880. His work on the microscopic structure of the central nervous system was strongly influenced by Darwin, and he believed that embryological studies had to be organized along transformist lines. On Duval, see Charles Coury, "Mathias Marie Duval," *Dictionary of Scientific Biography* 4 (1971): 266–267. Duval's physiology text was translated in Spanish as *Curso de fisiología*, trans. D. A. Espina y Capo (Madrid, 1884). On Simarro's histological work, see Temma Kaplan, "Luis Simarro, Spanish Histologist," in *Actas del Tercer Congreso Nacional de Historia de la Medicina*, 3 vols. (Valencia: Sociedad Española de Historia de la Medicina, 1971), II, 523–533.

and adaptive developments into his work on the pathology of the nervous system. Because of Simarro's rigorous evolutionism, the influence of Darwin and Haeckel is also palpable in the work of his most important disciple, Nicolás Achúcarro (1880–1918). In Simarro's laboratory around the turn of the century, Achúcarro was introduced to evolutionary histology, beginning his studies with the structure of the nervous systems of fish and other lower animals before progressing to the more complicated structure of the human nervous system. His work on the structure of neuroglia was also conceived in phylogenetic series.[26]

Ramón y Cajal recalled that he first began reading Lamarck, Spencer, and Darwin around 1874–1875, after he had received his *licencia* in medicine. At the same time he was also able "to savour the juicy and elegant, although frequently unacceptable or exaggerated, biogenic hypotheses of Haeckel, the spirited professor of Jena."[27] Reflecting on his scientific orientation at the beginning of his doctoral studies, Cajal lamented having felt "disdain toward interpretive norms drawn from comparative anatomy, ontogeny and phylogeny." He recalled how his own provincial candor and scientific ingenuousness had contrasted with the propensity of some of his young colleagues to expostulate on the philosophical implications of evolution and vitalism, which seemed to Cajal far from the business of learning the details of anatomy. But by 1879 he had learned German, begun the study of comparative anatomy, and now in seriousness immersed himself "in the modern theories of evolution, of whom the standardbearers at the time were Darwin, Haeckel and Huxley; I increased my embryological notes substantially."[28] According to Pedro Laín Entralgo, Cajal's early reticence with regard to evolution (1876–1880) initially impeded him from following Gegen-

26 Nicolás Achúcarro, "De l'évolution de la névroglie, et spécialement de ses relations avec l'appareil vasculaire," *Trabajos del Laboratorio de Investigaciones Biológicas* 13 (1915): 169–212. See also Pedro Laín Entralgo, "Vida y significación de Nicolás Achúcarro," in *Nicolás Achúcarro, su vida y obra*, ed. Gonzalo Moya (Madrid: Taurus, 1968), pp. 58, 63.

27 Santiago Ramón y Cajal, *Obras literarias completas* (Madrid: Aguilar, 1961), pp. 173–174. In the same passage, Cajal remarks that the first refutation of the *Origin of Species* to reach his hands was the Ateneo discourse of Cánovas del Castillo, "as eloquently written as [it was] lightly documented." Since the speech in question was delivered in November, 1872, Cajal may have been introduced to evolutionary ideas earlier than he later recalled.

28 Cajal, *Obras literarias completas*, pp. 265–267.

baur and resolving the problem of anatomical research through the study of comparative anatomy. By 1883, however, when he published some popularizing articles in journals in Valencia and Zaragoza, his thinking had become manifestly impregnated with the mentality and rhetoric of the evolutionism that was to guide his future career as a histologist. The crux of his methodology was, simply, to apply to the study of the histology of the nervous system the idea that what is morphologically less differentiated is ontogenetically prior to the more differentiated. Or, in his own colorful characterization: "Since the adult jungle—the jungle of the cerebral cortex—is impenetrable and indefinable, why not recur to the study of the young grove?" His famous investigations on the genesis of neurons were decidedly evolutionary, as he traced the development of the nervous system from embryological to adult forms in phylogenetic series. This work he described in another expressive rhetorical question: "In this evolutive [embryological] trajectory won't there be revealed, perhaps, something like an echo and recapitulation of the dramatic history lived by the neuron in its millenarian travels through the animal series?" His belief in the individuality of the neuron and in Haeckel's biogenetic law led him to the systematic examination of embryonic medulae of different ages. His evolutionism, however, was strictly limited. Where others (in Laín Entralgo's analysis) saw evolutionary continuity—purely quantitative differences between species—Cajal was more interested in qualitative morphological differences, that is, in innovations along the phylogenetic trajectory. He thought it improbable, in spite of his Haeckelian methodology, that quantitative differences alone could explain the differences between the brains of mammals and that of man.[29]

Evolutionary themes pervade Cajal's monumental *fin-de-siècle* comparative study of the nervous system of men and vertebrates, which system Cajal believed to represent "the last terminus of evolution of living matter," the most complex production of nature.[30] Throughout this study Cajal recurs insistently to evolutionary concepts, especially when synthesizing comparative material. He reduces the processes of historical

[29] Pedro Laín Entralgo, *España como problema* (Madrid: Aguilar, 1957), pp. 298–304.

[30] Santiago Ramón y Cajal, *Textura del sistema nervioso del hombre y de los vertebrados*, 3 vols. (Madrid: Nicolás Moya, 1897–1904), I, 1.

development of the nervous system to three evolutionary laws: (a) the multiplication of neurons, toward the end of multiplying the connections among diverse organs and tissues; (b) morphological and structural differentiation of the neurons to adapt them better to their role as transmitters; (c) unification or concentration of the nerve masses, or law of conservation of protoplasm and of the time of conduction. Cajal duly notes the similarity of this third principle to Spencer's "law of longitudinal and transversal integration" of the nervous system, which he gleaned from Miguel de Unamuno's 1896 translation of Spencer's *Progress: Its Law and Cause*. In summary, Cajal notes that "In good evolutionary doctrine, and with even greater cause admitting the principle of natural selection as the efficient cause of morphological and functional adaptation, it is necessary to justify all structural phenomena appearing in the phylogenetic or ontogenetic series, by reason of the real utility which the organism might gain from it; for if it should turn out to be useless, the same selection would soon end up discarding it. The utilitarian aim pursued by nature in this case is simply the conservation of protoplasm combined with economy of time."[31]

At the turn of the century, Cajal's views of the development of the nervous system were both Darwinian and teleological. If, he posited, multicellular animals had not been able to create a nervous system at some point along the path of their evolution, they would not have progressed in complexity much beyond what we observe in the vegetable kingdom, "because the excessive division of labor requires the supreme direction of the nerve cells in order to conserve the harmony and solidarity of the diverse associated parts." Moreover, in the struggle for life, extensive and energetic nervous responses will be more efficacious than weak ones: hence, the evolutionary basis of differentiation of nervous functions in the animal series. In subsequent years, Cajal was to become less Darwinian and more qualifiedly teleological.[32]

[31] See the evolutionary discussions of the morphology of nerve cells and of vertebrate spinal medulae, ibid., 75, 505–507. The three laws are discussed in ibid., I, 10–12. The quotation is from p. 12.

[32] Ibid., I, 2. In the definitive French version of the same work, Cajal qualifies the teleological position, stating that Spencer and others have proved, "par voie téléologique, il est vrai, que chez l'être multicellulaire, pour qu'il y ait progrès, un système nerveux doit se différencier, qui, le mettant en relations constantes avec le monde extérieur, assure sa conservation propre et sa défense à travers les âges." (*Histologie du système*

The influence of evolutionism on other branches of Spanish science is perhaps more to be sought in works of popularization than in scientific works of a highly original nature. One such example is Balbino Quesada's general physiology text, written in 1880. Quesada's exposition is based on a thoroughly evolutionary perspective, particularly as he repeatedly stresses the role of adaptation throughout the entire physiological process, from cell to organism. His discussion of reproduction is based largely on Spencer, especially the eugenic conclusion that cultivated men are obliged to contribute to the betterment of the species. The volume ends with a general discussion of transformism and its detractors, with Quesada concluding that evolution is the only theory that is capable of explaining the facts of science without contradiction.[33]

There is no doubt that men like Quesada and Cajal were rigorous evolutionists. One finds in their works, nevertheless, strong hints of vitalism and finalism. Cajal's propensity for anthropomorphizing neurons and other cells he described is well known; and part of his early reticence in accepting evolution was related to his feeling that a strict Haeckelian interpretation of the relationship between man and the lower animals was diminishing to human dignity. Quesada, more a finalist than a vitalist, was a believer, not in straight-line evolution, but nevertheless in an evolution leading to progressively higher forms "which lead to the realization of a universal end."[34] In an evolutionary philosopher, such as Miguel de Unamuno, vitalist-finalist trends became much more central and explicit than these hints, suggesting that such harmony as was

---

nerveux de l'homme et des vertébrés, trans. Léon Azoulay [1909; reprint ed., 2 vols., Madrid: Consejo Superior de Investigaciones Científicas, 1952], I, 3. ". . . in teleological fashion, it is true, that, for progress to occur in a multicellular organism, a nervous system must be developed, which, by putting the organism into constant communication with the outside world, assures its individual survival and its defense through the ages.") In later works, too, he took a less Darwinian view of adaptive processes in cellular differentiation: "However, it is important not to exaggerate, as do certain embryologists, the extent and the importance of cellular competition to the point of likening it to the Darwinian struggle which has been rigorously proved for certain organisms" (*Studies on Vertebrate Neurogenesis*, trans. Lloyd Guth [Springfield, Ill.: Thomas, 1960], p. 400; first published in French, 1929).

[33] Balbino Quesada, *Tratado elemental de fisiología general* (Madrid: Eduardo Cuesta, 1880), pp. 36–57, 244, 328.

[34] Ibid., p. 2.

achieved between evolution and traditional Spanish values was effected on a philosophical, rather than a theological, level of explanation.[35]

The role of scientific societies founded after the Revolution of 1868 seems to have been crucial in the reception and promulgation of evolutionary ideas in Spain, although at the present stage of research we have only general notions as to the role played by the various institutions. We know from the careers of the early neurohistologists, in particular, that the Academies of Anatomy and Histology, both in Madrid and both founded after 1868, were centers of evolutionary research. (The latter organization, founded in 1874, bore the significant appellation of *Free Society of Histology*, a symbol of its independence from traditional academe, just as the Institución *Libre* de Enseñanza, founded in 1875, stressed its independence after the Second University Crisis.) The Histology Society served as a place for discussion of European scientific developments as well as a springboard for evolution-oriented research.[36] The Ateneo de Madrid was also a center of evolutionary debate. An older institution, it was nevertheless dominated by the men of 1868 (especially after 1876, when the conservatives deserted en masse). In spite of the fact that the Krausist leadership tended to be antievolution, the Ateneo provided a forum for extensive, and often acerbic, discussion of Darwinian theory.[37] The Spanish Anthropological Society, founded in 1865, was also a center of evolutionary discussion, and some of the early discussions of the evolutionary impact on anthropology were published in its journal. Old-line humanist societies such as the Spanish Academy

[35] On Unamuno's evolutionism, discussion of which is beyond the scope of this paper, see Peter G. Earle, "El evolucionismo en el pensamiento de Unamuno," *Cuadernos de la Catedra Miguel de Unamuno* 14–15 (1964–1965): 19–28; Carlos París, *Unamuno: Estructura de su mundo intelectual* (Madrid: Ediciones Peninsula, 1968), pp. 133–167; Thomas F. Glick, "The Valencian Homage to Darwin in the Centennial Date of his Birth (1909)," in *Actas del Tercer Congreso*, II, 577–601.

[36] Kaplan, "Luis Simarro," pp. 529–530.

[37] Krausists were anti-Catholic but religious, and they objected to the positivists' emphasis on the total secularization of all ideas and institutions, as well as to the positivist stress on empiricism at the expense of preordained moral tenets; see Kaplan, "Positivism and Liberalism," p. 260. On the antievolutionism of Gumersindo de Azcárate, a leading Krausist figure in the Ateneo, see *Revista Contemporánea* 3 (1876): 127. On the antievolutionism of José Moreno Nieto, see ibid. 6 (1876): 368. On debates on Social Darwinism in the Ateneo in 1882–1883, see Carmelo Lisón Tolosana, *Antropología social en España* (Madrid: Siglo XXI, 1971), p. 125 n.

and the Academy of History seem to have been antievolution bastions, in spite of allegations that the admission to the latter of the geologist Juan Vilanova y Piera, a Spanish follower of Armand de Quatrefages, was a sign of its lack of hostility to Darwin.[38]

It was among positivist social scientists that Darwinian evolution made its greatest impact. The first writers on anthropological subjects in nineteenth-century Spain were all influenced to some degree by Darwin and Haeckel and by evolutionary anthropologists, such as John Lubbock and Edward B. Tylor.[39] Among prehistorians, one could cite Manuel Sales y Ferré (1843–1910), a Krausist by training who nevertheless was strongly influenced by positivism and evolutionism. He was the founder of another new academic institution, the Instituto de Sociología, in Madrid. Underlying his view of the development of early civilizations was the notion of the struggle for existence between primitive tribes and societies.[40]

The most significant evolutionary sociologist in late nineteenth-century Spain was Pedro Estasen y Cortada (1855–1913), who wrote a series of articles (in 1876–1878) designed to demonstrate the total evolutionary nature of social development.[41] "The positivist, transformist theory," wrote Estasen, "proceeds from observation, interprets the law of evolution which it finds in social, just as in natural, organisms, and applies the laws of selection, of heredity, of adaptation to the study of human development." The nature of this evolution is, moreover, finalistic, in

[38] The allegation is that of Carracido, in *Estudios*, p. 290. Vilanova is better seen not as a reluctant evolutionist but as the establishment's answer to evolutionism.

[39] Lisón Tolosana, *Antropología social*, p. 125, n. 126.

[40] See Manuel Sales y Ferré, *El hombre primitivo y las tradiciones orientales* (Seville: El Mercantil Sevillano, 1881), pp. 220–221. Sales wrote many books on prehistory and the origins of civilization. He also published a book of commentaries on de Quatrefages's *Natural History of Man* and translated a volume by Eduard Hartmann, under the title *La verdad y el error en el darwinismo*.

[41] Pedro Estasen y Cortada, "La teoría de la evolución aplicada a la historia," *Revista Contemporánea* 4 (1876): 447–464; 5 (1876): 218–234; idem, "El positivismo y la teoría de la evolución," ibid. 11 (1877): 420–448; idem, "Contribución al estudio de la ciencia social: Filosofía de la aristocracia," ibid. 13 (1878): 418–434; idem, "Noción del derecho según la filosofía positiva," ibid. 7 (1877): 505–520; 10 (1877): 322–347; idem, "Contribución al estudio de la evolución de las instituciones religiosas o materiales para llegar a la síntesis transformista de las instituciones humanas," ibid. 13 (1878): 62–74.

that the unchangeable laws of nature strive toward incessant perfection in everything, including social laws. Natural and organic transformations have proceeded to the point where humanity is tending each day toward the ideal of a true felicity. The mechanism attending these transformations is selection—"that law which rejects what is irregular and inharmonic; that force which sustains good and makes it persevere."[42] The role played by heredity in natural evolution corresponds, in social evolution, to law, those precepts by which societies organize themselves according to some moral coordinates. Moreover, the nature of *derecho* ("law") in the sense of *right* relates directly to the struggle for existence, the right to live—a right inhering in the best adapted, which in social terms means the society with the best system of juridical sanctions, or, what is the same in this view, morality. Such sanctions are passed along from generation to generation, and this process is the law of social heredity: ". . . the sum of elements and institutions which one generation inherits from another." By this process, societies transform themselves "by virtue of the adaptation and variability of social species."[43]

Variation (and natural selection) in Darwinian evolutionary theory have their social counterparts in the division of labor, analyzed by Estasen in a study of the function of aristocracies. This study denotes in all societies a constant phenomenon of diversity of classes and categories, just as in nature. All beings are unequal, and this gives rise to an inequality of functions, as well as to the adaptation of organisms, through continuous changes of form, to the division of labor that nature requires. In the general history of organisms, those that appear later, with greater morphological complexity and thus with greater faculties with which to adapt to the varied conditions of the environment, occupy the principal station—this is as true of plant and animal species as of human societies. Before making his main point about the way in which elites are selected, Estasen makes a brief and fascinating digression on the biogenetic law: just as ontogeny recapitulates phylogeny within the womb, so does the extrauterine development of the human being recapitulate the social history of mankind.[44] The socialization process of a child is a "sociological

---

[42] Estasen, "Teoría de la evolución aplicada a la historia," pp. 452, 463–464.

[43] Estasen, "Noción del derecho," esp. p. 344.

[44] For a later example of Social Haeckelism in Spain, see Pio Baroja's novel, *El árbol de la ciencia* (Madrid: Espasa Calpe, 1932), p. 134: ". . . adopting the principle of

résumé of the life of humanity." The perfection of motor control in the individual is therefore seen as analogous to the development of social control. Aristocrats are nothing less than the select, who have proven themselves fit to occupy the highest role in social life. Estasen makes it clear that the only aristocrats worthy of the title are aristocrats by virtue of intelligence. The money aristocracy of his day represents not a fulfillment of the natural order but an unjust distortion of it. To explain why his evolutionary view may not be understood as a justification of tyranny, he recurs again to the notion of law, by means of which society guarantees equal opportunity to all and through which an intellectual elite rises to assume control.[45]

There was much discussion in Spain of the 1870's over the implications of the chapter on "Moral Faculties" in the *Descent of Man*. It was a topic discussed by almost every social philosopher of note, the most prominent being the future prime minister, Antonio Cánovas del Castillo. Cánovas was rather pleased that Darwin had found a place in his theory for religious sensibilities in the evolutionary history of mankind, although he criticized him for not indicating that such sensibilities were "indispensable."[46] A few years later the same problem was subjected to a lengthier critique by another conservative politician, Joaquín Sánchez de Toca, who thought that the Darwinists' discussion of the evolution of moral faculties was historically inaccurate, since it was based almost exclusively on observations of primitive and ancient peoples (and thus omitted advances in the Christian era). He criticized as well Darwinian views of marriage and sexual selection (as applied socially) from a Christian point of view, attacking the moral bases of Darwinian eugenics.[47] Estasen, in his article on the evolution of religious institutions

---

Fritz Muller that the embryology of an animal reproduces its genealogy, or as Haeckel says, that ontogeny is a recapitulation of phylogeny, one could say that human psychology is no more than a synthesis of animal psychology."

[45] Estasen, "Filosofía de la aristocracia," passim; idem, "Noción del derecho," pp. 344–345. Spanish Social Darwinists were continually defending their theories from the charge that they lent theoretical support to tyranny; cf. report of a speech at the Ateneo de Madrid by Manuel de Revilla, *Revista Contemporánea* 2 (1876): 123.

[46] Cánovas, "Discurso," p. 171.

[47] Joaquín Sánchez de Toca, "La doctrina de la evolución de las modernas escuelas científicas," *Revista Contemporánea* 21 (1879): 273–288. See also his interesting discussion of the debate over linguistic evolution, ibid., pp. 69–84.

(interestingly subtitled "Materials to Arrive at a Transformist Synthesis of Human Institutions"), also developed the theme of emotions in animals as possible antecedents of primitive religious practices of humans, a discussion based not only on Darwin, but on Lubbock and Walter Bagehot as well.

Most of the intellectual activity thus far described took place in Madrid, the political, intellectual, and scientific capital. But the diffusion of evolutionary ideas in the provinces was, both for depth and for rapidity of permeation, remarkable. In Seville, the physician Antonio Machado y Núñez contributed to the propagation of Darwinian ideas with a series of popularizing, pro-evolution articles in the positivist *Revista Mensual de Filosofía, Literatura y Ciencias* (1869–1874). Machado y Núñez particularly stressed geological themes supporting evolutionary theory, citing ongoing geological processes (e.g., formation of dunes and polders) conducive to the deduction that the earth is millions of years old.[48] In other short articles he compared Darwin to Newton and insisted on Darwin's openness and lack of dogmatism.[49] In 1874, Machado y Núñez launched a campaign to popularize the works of Haeckel with a long article on ontogeny and phylogeny, coordinated with the publication of a translated article by Haeckel himself.[50] Seville was also the site of the pro-evolutionary activities of Sales y Ferré. In Granada, a polemic was engendered when, in the autumn of 1872, a teacher of natural history in the provincial *instituto* (secondary school), Rafael García y Alvarez, initiated the semester with an explicit and comprehensive presentation of the theory of evolution, which he supported as the best explanation of natural, as well as social, development. García's address was con-

[48] Antonio Machado y Núñez, "Darwinismo," *Revista Mensual de Filosofía, Literatura y Ciencias* 4 (1873): 523–528.

[49] Antonio Machado y Núñez, "Apuntes sobre la teoría de Darwin," *Revista Mensual de Filosofía, Literatura y Ciencias* 3 (1872): 461–470. See also idem, "Teoría de Darwin: Combate por la existencia," ibid. 4 (1872): 3–8; idem, "Teoría de Darwin," ibid., pp. 129–133.

[50] Antonio Machado y Núñez, "Leyes del desenvolvimiento de los grupos orgánicos y de los individuos: Filogenia y ontogenia," *Revista Mensual de Filosofía, Literatura y Ciencias* 6 (1874): 145–153, 193–208, 241–249, 289–297, 337–342; Ernst Haeckel, "Origen y genealogía de la raza humana," ibid., pp. 26–32.

demned by the bishop and placed on the index.[51] In 1880 another academic joust was initiated in the same city when Miguel Rabanillo Robles, a professor of pharmaceutical organic chemistry, inaugurated the academic year at the Literary University of Granada with a lecture in which he indicated that human emotions were qualitatively similar to those of animals, bringing down upon himself the condemnation of the orthodox.[52]

Valencia was the scene of a large-scale debate on evolution, held under the auspices of the Ateneo Científico in 1878. At the session of February 4, Dr. Joaquín Serrano Cañete spoke of the transcendence of the question under debate, lamenting that the polemic had fallen into the hands of two groups of extremists, "the party of the impatient" and that of the retrogrades, each having aggrandized Darwinian theory for its own ends. "The experimental sciences," he admonished, "do not admit such mentors." He pointed out that the idea of evolution was an ancient one but that Darwin's glory was to have put it on the solid base of natural selection and the struggle for life. He concluded his remarks with a critique of the exaggerated materialistic ideas of Ludwig Büchner and Carl Vogt.[53]

The next three sessions were devoted to a tripartite lecture on Darwinism by Amalio Gimeno, the physiologist, who initiated his exposition by claiming that, in spite of the great interest in the subject, few volunteers came forward to lecture openly, owing to the "apparent con-

[51] Rafael García y Alvarez, *Discurso en la solemne apertura del curso académico de 1872 a 73 en el Instituto de 2ª Enseñanza de la Provincia de Granada* (Granada: Indalecio Ventura, 1872). See also his *Estudio sobre el transformismo* (Granada: Ventura Sabatel, 1883), based on a series of articles in the *Revista de Andalucía*, vols. 6 and 7. On the basis of these latter, García y Alvarez was criticized in a Catholic journal as a blind enthusiast of Darwin, "more radical than his master" in his acceptance of pantheism, atheism, and materialism. See J. M. Orti y Lara, "El catecismo de los textos vivos: Sobre el origen del hombre," *Ciencia Cristiana* 12 (1879): 209–213. The discourse of 1872 was condemned in a Pastoral of November 1, 1872 (*Arzobispado de Granada: Boletín eclesiástico* 28 [1872]: 394). On its inclusion in the Spanish index, see León Carbonero y Sol, *Indice de los libros prohibidos por el santo oficio de la Inquisición Española desde su primer decreto hasta el último, que espidió en 29 mayo de 1819, y por los Rdos. obispos españoles desde esta fecha hasta fin de diciembre de 1872* (Madrid: A. Pérez Dubrull, 1873), p. 210.

[52] See Arturo Perales, "Observaciones a un discurso universitario," *El Sentido Católico en las Ciencias Médicas* 3 (1881): 17–23.

[53] *El Mercantil Valenciano*, February 14, 1878.

flict between religious ideas and the theory of transformism." He ventured that, by the time science was able to prove the truth of Darwinism, there would no longer be any conflict between evolution and religion, as such evolution-leaning Catholic clerics as Angelo Secchi and Hyacinthe de Valroger had already stated. The religious opposition, in his view, was mainly owing to the repugnance that the idea of the development of life from nonlife inspired. The first lecture ended with a discussion of recent divisions in the Darwinian camp among materialists, such as T. H. Huxley, Haeckel, Büchner, and Vogt, on the one hand, and theistic evolutionists, such as Clémence Royer, Jean Albert Gaudry, and Asa Gray, on the other. The second lecture began on this note, with a discussion of the views of Gray (who said that evolutionists should eschew the problem of cause), Secchi, and Valroger. There followed an exchange between Gimeno and Rafael Rodríguez de Cepeda, who claimed that Darwinism was incompatible with Catholicism, inasmuch as it denied the human spirit, making all humanity emanate from *matter*. Gimeno's final talk dealt with arguments and proofs from embryology, anatomy, physiology, and psychology. He mentioned Haeckel's gastraea theory and asserted that the recapitulation argument alone was a solid enough support for Darwin's theory, even though some dismissed it by simply calling this process proof of the consubstantiality of matter. Gimeno concluded on an admonitory note, saying that the theory of evolution was of greater importance than many Spaniards believed and that evolutionists should not be condemned without a hearing. On the other hand, the younger generation appeared to have enslaved itself to a new scientific messiah, forgetting that the final scientific judgment belonged to the future.[54]

The session of April 25, chaired by Celso Arévalo, began with a thinly veiled reference to the political jeopardy encountered by professed evolutionists, who had been removed from office in the wake of the Second University Crisis. "The Spanish professoriate in general, and the Valencian in particular," Arévalo declared, "has sufficient independence and firm enough convictions not to be imposed upon by anyone." Arévalo

[54] *El Mercantil Valenciano*, March 28, 1878; April 11, 1878; April 25, 1878. See also Gimeno's satirical dialogue, "¿Darwinismo cristiano?" in *Antología de trabajos científicos, literarios, políticos y sociales del Profesor Amalio Gimeno y Cabañas* (Madrid: J. Cosano, 1935), pp. 747–757.

devoted most of his speech to the concept of species, which was one of the greatest obstacles to acceptance of the theory of evolution by dogmatic Catholics. He deduced that species were variable within as yet undetermined limits and that in past epochs there probably were greater links between natural groups than exist presently. In conclusion, he recurred again to the theme of the need for free scientific inquiry, be it contrary or not to ideas emitted by theologians.[55]

The final speech of the cycle, delivered by José María Escuder at the session of May 9, began with an evocation of the spirit of the Revolution of 1868, which had swept away all preconceptions, both social and religious. Escuder did not mince words. "There exists," he declared, "a complete opposition between revelation and science." Revelation cannot suffer any contradiction whatsoever, whereas science subjects itself to one dogma alone, *the fact*, which it investigates through observation. There followed a long discussion of whether matter was eternal or created, Escuder quoting Father Secchi on the absurdity of postulating a creation *ex nihilo*. As for arguments that Darwinism was contrary to Scripture, he retorted that a book whose authorship is in doubt cannot be advanced as a criterion for truth. Escuder was answered by two gentlemen who protested "in the name of their wounded Catholic sentiments," and the session was hurriedly adjourned by Gimeno, who occupied the president's chair.[56]

Valencia, as indicated above in the discussion of the role of Peregrín Casanova, remained in the forefront of Spanish evolutionism, and was in 1909 the site of a centennial homage to Darwin, in which Casanova and his leading disciples participated.[57]

The intensity of the debate over Darwinism in the Canary Islands in the 1870's is indicative of the rapid diffusion of Darwinism in Spain in the years after 1871, a diffusion so pervasive that it penetrated even the remotest provinces of the nation in less than a decade.

The intellectual debate over evolution was centered in the city of Las

[55] *El Mercantil Valenciano*, May 2, 1878. The next session (ibid., May 9) was devoted to a vague discourse by Vicente Calabuig y Carra concerning the relation between Darwinism and positivism.

[56] *El Mercantil Valenciano*, June 15, 1878.

[57] See Glick, "The Valencian Homage to Darwin."

Palmas on the island of Gran Canaria, a city of some twenty thousand inhabitants and the intellectual center of the archipelago. The polemic began early in 1876 with the appearance of the first fascicle of a natural history of the islands written by Gregorio Chil y Naranjo.[58] In the introduction to his treatise, Chil drew a picture of the Quaternary epoch, during which the simian mammalian form was modified until it developed into man, being distinguished from other animals owing to its faculty of abstract thought. He pointed out that prehistorical studies were in their infancy, having been given a tremendous impetus from the work of Darwin, "who opened the gates," and Haeckel, who "proved the unity of organic and inorganic nature, the identity of the basic elements of one and the other, . . . leading the genealogical [i.e., historical] science to a vision of the origin of all creation."[59]

By the time the tenth fascicle of the work, which contained no other allusions to evolution, had appeared in May, a special synod appointed by the Archbishop José María de Urquinaona y Bidot had met to examine the contents of Chil's opus.[60] The synod filed its negative report on June 12, 1876, and on June 21 the archbishop issued a pastoral letter prohibiting the reading of the book.[61]

In his letter, Urquinaona first lamented the fact that Chil, the scion of a distinguished family, who had been educated in church schools, should have strayed along the path of materialism. Most of the rest of the letter was devoted to the report of the synod, which declared that Chil was a freethinker of the stripe of Hobbes, Diderot, and La Mettrie. Most of the refutations of Chil's evolutionary arguments were based on Scripture, although Buffon was cited on the superiority of man over animals, spirit

[58] Gregorio Chil y Naranjo, *Estudios históricos, climatológicos y patológicos de las Islas Canarias*, 3 vols. (Las Palmas: La Atlántida, 1876–1891).

[59] Ibid., I, 13–15. The introduction ends with an apostrophe to the effect that true knowledge is found in the eternal and absolute God (p. 17).

[60] *La Prensa* (Las Palmas), May 26, 1876. *La Prensa* was a liberal daily that was anticlerical and undogmatic, proscience but opposed to evolution. It condemned the appointment of a synod on the grounds that any action taken would be counterproductive because it would incite curiosity about Darwinism.

[61] *Carta pastoral que el ilmo. y rmo. Sr. D. José María de Urquinaona, obispo de Canarias y administrador apostólico de Tenerife dirige al clero y fieles de ambas diócesis, con motivo de la obra, que ha empezado a publicarse en esta ciudad, con el título de "Estudios históricos, climatológicos y patológicos de las Islas Canarias;" prohibiendo su lectura* (Las Palmas: Víctor Doreste, 1876).

over matter.[62] The synod accused Chil of ignorance of theology and warned that if his ideas were accepted they would spell the doom of the doctrine of original sin ("la caída del primer hombre") and, along with it, the promise of the Messiah and the establishment of the church. Those who favor freedom of thought, the report continued, pretend to love humanity but would deny to the poor the hope of eternal glory; the poor man, denuded of his religious beliefs, would curse his poverty and ready his knife for the rich. Such are the fruits of materialism, which, in the view of the synod, is but "a sad symptom of social disorganization."[63] In sum, the work was adjudged "false, impious, scandalous, and heretical."

As a result of the report, the archbishop ordered the faithful to abstain from reading the cited book, and the fascicles that had already been purchased were to be remitted either directly to the archbishop or to the parish priests. No faithful Christian was to read the book, unless licensed by the Holy See. In closing, Urquinaona stressed the hope that Chil would learn his error and issue a public retraction.[64]

The censure of Chil became a *cause célèbre*. Apparently, the church harassed the naturalist personally, since it was reported in early July that he had had to travel to Madeira to be married because of the opposition of certain authorities of the local diocese.[65] Subsequent events illustrate the execution of anti-Darwinian strictures on the parish level. *La Prensa* reported in August that the priest of the village of Santa Brígida had ordered the faithful of his parish to hand over "some papers" of the work of Chil, immediately. The priest further ordered all Catholics to appear at his house to sign some sort of affidavit—and this (the paper editorialized) in the last third of the nineteenth century![66] Several days later the priest of San Telmo in Las Palmas read the ban from his pulpit, advising parishioners to turn in their copies of the proscribed work. No sooner had he pronounced the name of Chil than a dog, which had been sitting quietly up to that point, bounded up to the altar and began bark-

[62] Ibid., p. 17.

[63] Ibid., p. 18. "Social disorganization" in conservative parlance meant demands by the lower classes for social justice.

[64] Ibid., pp. 21–22.

[65] *Las Palmas*, no. 66, July 5, 1876. *Las Palmas* was a reformist Republican newspaper, the Canarian organ of the party of Manuel Ruiz Zorrilla.

[66] *La Prensa*, August 14, 1876.

ing at the celebrant. The beast, *La Prensa* observed, "seemed to be combating openly the ecclesiastical veto, as the priest, harassed by the dog, could scarcely finish reading the famous document, producing hilarity among the audience. This episode," the paper continued sardonically, "has been interpreted in two ways, each more miraculous than the other: one group believes that Satan, through the mouth of the dog, has declared himself against the episcopal order; others hold that common opinion was expressed through the canine's conduct."[67]

The Chil affair, which induced a general discussion of evolution in Las Palmas, gained a certain notoriety abroad. In June, 1877, it was reported that Professor René Verneau of the Museum of Natural History had arrived from Paris bearing letters to Dr. Chil from de Quatrefages and Paul Broca. Several days later it was learned that Broca had read aloud in a session of the Anthropological Society of Paris a passage from Chil's book in which Chil expressed the need for more work on the prehistory of the Canaries.[68] The French Anthropological Society's interest in Chil went beyond this. Chil's *Estudios*, though bearing a Las Palmas imprint, had actually been printed in Paris by Ernest Leroux. The work was then reviewed in the organ of the Anthropological Society by Ludovic Martinet,[69] and the same author followed up with an ample reportage on the Chil affair, discussing the text of the episcopal censure and asking rhetorically: "A quand notre tour?"[70]

The evolution polemic set off by Urquinaona's censure was a three-cornered battle among Baltasar Champsaur Sicilia, an anticlerical, pro-Darwin socialist writing in *Las Palmas*; the lawyer Rafael Lorenzo y García, an anti-Darwin, anticlerical freethinker, writing in *La Prensa*; and the clerical anti-Darwinians, led by a young priest named José Roca y Ponsa, writing in the clerical weekly *El Gólgota* and supported editorially in right-wing papers such as *La Lealtad* and *Gran Canaria*. The initial phase of the polemic was an exchange of articles during the winter of

---

[67] *La Prensa*, August 17, 1876.

[68] *La Prensa*, June 2, 1877; June 6, 1877.

[69] *Revue d'Anthropologie*, 2d ser. 5 (1876): 696–698.

[70] Ludovic Martinet, "Le cléricalisme aux Canaries," ibid. 7 (1878): 181–184. "When will it be our turn?" Martinet's report was picked up by *The Academy* for April 6, 1878 (13: 306), and thence by Andrew D. White, *A History of the Warfare of Science with Theology in Christendom*, 2 vols. (New York: D. Appleton, 1896), I, 85.

1876–1877 between Champsaur and Lorenzo, motivated by the publication of the latter's book, *Estudios filosóficos sobre la especificación de los seres*, a critique of Darwinian concepts of speciation.[71] Lorenzo's argument, adumbrated in this work and embellished in succeeding articles and books, was that species were invariable and couldn't be transformed into others (although they were modifiable to a limited extent), a position that he supported by citing such anti-Darwinian scientists as Louis Agassiz, Pierre Flourens, and others.[72]

Lorenzo was also supported editorially by the newspaper's editor, Pablo Romero. The position of both of these men is difficult to understand. Both continually combined their attacks on evolution with familiar platitudes about science conquering all, and the newspaper had been an outspoken supporter of Chil.[73] Throughout the controversy it was apparent that the real enemy was the church; both Lorenzo and Romero strove to maintain an intellectual discussion with Champsaur, on the one hand, while flailing the church's ignorance and obstinance, on the other. Thus, Lorenzo took pains to introduce his own anti-Darwinian treatise by alluding to John William Draper on the conflict of religion and science and by asserting that the theory of evolution was not antispiritual. Nevertheless, he said, theologians persecute as false, impious, scandalous, and heretical anyone promoting what is taken as an antispiritualist idea, and this without knowing the language of the natural sciences—a thinly veiled attack on Urquinaona's board of inquiry.[74] Lorenzo appears to have been a finalist agnostic, characterizing species as "divine types" and

---

[71] Rafael Lorenzo y García, *Estudios filosóficos sobre la especificación de los seres* (Las Palmas: Viuda de Romero, 1876).

[72] Further works by Lorenzo: "Contestación al artículo 'Rápida ojeada por el opúsculo Especificación de los seres,'" *La Prensa*, January 10, 1877; "Ulteriores argumentos contra el Darwinismo del oculto articulista de *Las Palmas*," *La Prensa*, February 11, 1877; "Continuación de los argumentos contra el darwinista de *Las Palmas*," *La Prensa*, March 20, 1877; "Las contradicciones del darwinista de *Las Palmas* al descubierto," *La Prensa*, April 20, 1877; "Más sobre axolotes," *La Prensa*, May 30, 1877 (in this installment Lorenzo says he has read Haeckel, the compendium of Darwin's works by Giovanni Omboni, and various treatises of Büchner); *Estudios filosóficos sobre el origen y formación de los seres vivientes* (Las Palmas: La Verdad, 1877); and *Triunfo de las ciencias* (Las Palmas: El Independiente, 1879).

[73] On August 22, 1876, *La Prensa* attacked the rightist *La Lealtad* with tongue in cheek for not turning in its copies of Chil's work.

[74] Lorenzo, *Especificación de los seres*, p. 8.

alluding to a "superior idea" informing the origin of man. Romero, for his part, pointed out that Lorenzo, although opposed to Darwin on the question of the origin of human species, also opposed theologians who interpret *their book* in a literal, routine way. According to Romero, Lorenzo proved that Darwinism, although wrong on scientific grounds, did *not* lower man in dignity, was not antispiritual, and did not discredit the idea of divinity. In a review of Lorenzo's book of 1877, Romero mentioned that Lorenzo had two detractors, a Darwinist and an anti-Darwinist, but reserved his invective for the latter, whom he accused of undermining the solid principles of science with "grotesque forms and pretentious observations."[75]

Baltasar Champsaur Sicilia attacked Lorenzo in a series of articles in *Las Palmas*.[76] He hit hard on the anti-Darwinian discussion of speciation, pointing out that species are merely artificial classifications and that arguments supporting fixity are specious. A socialist, Champsaur also stressed the social linkages of anti-Darwinism, pointing out, for example, that it is not difficult to understand how a person who proclaims his shame at being descended from a monkey might also be ashamed of relationship with a black or a savage. Champsaur was typical of a class of nonscientist radical intellectuals who were almost universally self-proclaimed Darwinians. His library, which has been preserved in the Museo Canario of Las Palmas, is representative of the range of books owned by a lay evolution enthusiast.

The clerical attack was directed more against Lorenzo than Champsaur. (It has been wisely said that attempts at compromising the positions of science and religion horrified conservatives more than atheism itself.[77] Lorenzo consistently proclaimed that his aim was to harmonize scientific and theistic positions.) Roca's book, based on a series of articles in *El Gólgota*, is a run-of-the-mill anti-Darwinian tirade: Darwinism is antiscientific, anti-Christian, and contrary to the dignity of man.[78] In the

[75] *La Prensa*, December 6, 1876; November 20, 1877; November 30, 1877.

[76] *Las Palmas*, December 17, 1876; January 17, 1877; January 22, 1877; February 2, 1877; March 7, 1877; April 12, 1877; May 12, 1877; February 7, 1878. These articles, originally published anonymously, were reprinted as *Transformismo* (see note 6 above).

[77] Willard B. Gatewood, *Controversy in the Twenties* (Nashville, Tenn.: Vanderbilt University Press, 1969), p. 33.

[78] José Roca y Ponsa, *El Sr. Licdo. Lorenzo y García ante la fé y la razón* (Las Palmas: Víctor Doreste, 1878). See also the anonymous book, probably also by Roca,

book, Roca leveled a blast at Canarian journalists of the left who availed themselves of the Chil polemic to ridicule the faith.[79]

The Canarian polemics highlight the diversity of opinion expressed on the subject of evolution in provincial Spain in the late 1870's. The discussion was characterized by amateurism, in that the participants were not scientists or even physicians. Judging by front-page coverage given the polemic, one can assume a substantial interest in the subject on the part of the reading public, indicating broad popularization of the issue. By 1881, when Darwinism was discussed from the podium of the Sociedad de Amigos de País on the island of La Palma, we can reckon the primary regional diffusion of evolutionary ideas amongst literate Spaniards as completed.[80]

Manuel de la Revilla, a leading positivist spokesman, pointed out with good reason that, by the mid-1870's, for multiple and complex reasons, the Darwin controversy had generated a great deal of philosophical confusion, to the point where such movements as positivism, materialism, and Darwinism had achieved virtual identity in the eyes of the casual observer. Since most scientists and most members of the positivist and materialist schools of philosophy accepted Darwinism, it appeared that these schools had engendered it. And, indeed, there were in both the positivist and the Darwinian camps thinkers who constructed a complete materialist metaphysics, "which they present with notorious flightiness as the inevitable result of scientific observation and as an unavoidable consequence of the theory of evolution." But evolution, Revilla continued, need not necessarily lead to a monist philosophy or exclude any high religious conception.[81] Here Revilla was doubtless speaking more to Krausist anti-Darwinists than to Catholic ultras, but he touched on a point that was also at the heart of the Catholic position. It was a standard assumption of Spanish anti-Darwinism that evolution must lead to

Cuatro palabras sobre un reciente folleto del Licdo. D. Rafael Lorenzo y García titulado Estudios filosóficos sobre la especicación [sic] *de los seres* (Las Palmas: Víctor Doreste, 1877).

[79] Roca, *Lorenzo y García ante la fé*, p. 18.

[80] Antonio Rodríguez López, *Consideraciones sobre el Darwinismo* (Santa Cruz de la Palma: El Time, 1881).

[81] *Revista Contemporánea* 10 (1877): 119–121.

materialism and atheism and that positivists were to blame for all three. Sánchez de Toca remarked that the left applauded Darwinism because it was antireligious: thus, to discredit Darwin was to discredit at the same time all the other schools linked to it.[82]

By the early 1880's the Darwin polemic in Spain had reached immense proportions—more than fifty *extensive* articles or books, according to one account,[83] and several hundred perhaps, if shorter articles were included. The Catholic response may be conveniently divided into three groups: (*a*) exegetical anti-Darwinists, hard-line ecclesiastics who did not bother with the scientific details of the question; (*b*) learned hardliners, generally well-informed intellectuals, both lay and clerical, who sought to support their antievolutionary arguments with the best sources available; and (*c*) a small group of Catholic scientists and ecclesiastics who sought to harmonize the two positions and who moved cautiously toward a theistic evolutionary position. It is indicative of the dominant tone of Catholic antievolutionism that both groups (*b*) and (*c*) drew their main arguments from the works of de Quatrefages,[84] although each group made somewhat different use of his ideas. The intellectual hardliners used de Quatrefages as a scientific witness, using his objections to Darwinism to reject the entire concept of evolution. The moderates used him as a support for a limited theory of evolution, a line more in keeping with de Quatrefages's true stance.

Exegetical anti-Darwinism is best represented in official church publications, such as the synodal censures of Chil y Naranjo and García y Alvarez. In the former, no contemporary scientist was cited. In the latter, no scientists were mentioned save Buffon and de Quatrefages; the other forty citations were from Scripture and theological writings of diverse epochs.[85]

Nor could the informed intransigents claim any great originality of argument. Here we survey only a few of the more influential pieces. Emilio Huelin, writing in 1871 and 1872, played a somewhat ambivalent role. He understood the theory of evolution well and lamented both

[82] Sánchez de Toca, "Doctrina de la evolución," p. 301.

[83] Sanús Abad, "Apologética española," p. 12, n. 5.

[84] De Quatrefages was generally read in French, with the exception of his *Historia natural del hombre*, trans. A. García Montero (Madrid, 1874).

[85] *Arzobispado de Granada: Boletín eclesiástico* 28 (1872): 394.

the fact that no Spanish translation of the *Origin* had yet been made and the fact that those intellectuals who were discussing evolution in the Ateneo de Madrid "and other scientific circles of Spain" were ignorant of the writings of most of the German and English scholars who had written on the subject.[86] This lack he redressed in a review article, uniting for Spanish readers a great mass of Darwinian bibliography, heavily weighted on the anti-Darwin side. In particular he cited the works of the older generation of German anti-Darwinians, some of whom (e.g., Andreas and Rudolf Wagner) were already dead.[87] In introducing his survey of the literature, Huelin articulated a theme that underlay much of the subsequent Catholic literature: Darwinism dominated the entire educated world, had caused radical changes in biological sciences, and had made an impact on linguistics, politics, philosophy, and theology as well. If true, the Darwinian theory would cause nothing less than the total disruption of traditional values and result in moral anarchy. (In the works of less intelligent critics, the notes of fear and defensiveness would be more apparent, if below the surface of articulation.) After citing a large number of anti-Darwinian scientists, in addition to such liberal Catholics as St. George Mivart, Franz Heinrich Reusch, and Giovanni Battista Pianciani, and after stating that many Catholics could accept the idea that Adam's body, but not his soul, might have been the result of an evolutionary process, Huelin then condemned the whole theory as a "chimerical, arbitrary and gratuitous fiction," alleging that Darwin's observations, showing similarities between varieties, had produced only superficial likenesses. His conclusions were ambivalent: the debate over evolution had produced notable scientific progress, but the theory itself lacked any applicability to human beings. Nor did he endorse evolution of the lower orders.

The 1876 work of Emilio Reus y Bahamonde, *Filosofía de la creación*, I know only from a critique by Revilla.[88] Reus gave both theological and transformist arguments favoring evolution and then refuted both. Much of the discussion centered on extensive comparisons between the human and simian anatomy, a line that Revilla criticized on the grounds that no

[86] Emilio Huelin, "Revista científica," *La Ilustración Española y Americana*, March 15, 1871, p. 138.
[87] Huelin, "Los brutos," p. 22 n.
[88] *Revista Contemporánea* 2 (1876): 508–511.

serious evolutionist believed that man had descended from living anthropoid apes. Indeed, Reus was unable to controvert Huxley's thesis that there was less difference between men and apes than between apes and lower monkeys. Revilla then criticized Reus's discussion of the species question on the grounds that (*a*) no one agreed as to the definition of the term and (*b*) no serious scholar could sustain the absolute invariability of species, a *sine qua non* of the Catholic position. Nevertheless, Revilla praised the work as the first Spanish book of any importance to combat transformism with scientific reasoning.

The aristocratic novelist Emilia Pardo Bazán produced an influential critique of evolutionism in 1877 for the influential Neo-Catholic journal, *Ciencia Cristiana*.[89] This treatise is notable for its defensiveness, cloaked in the snide smugness of orthodox intellectuality. Pardo stated that Catholics appeared to have given up hope for their cause, terrorized by Darwin. *Everyone*, whether for or against the theory, accepted it (at a gut level, one assumes) as true; therein lay the root of Catholic panic. Pardo counseled calm. Evolutionism was not a law, like gravitation, which had a limited scope of action, but a hypothesis that sought to create an entire natural philosophy. She claimed that the *Origin of Species* had been roundly rejected by eminent scholars like Flourens and de Quatrefages. It was only when Haeckel emerged as its paladin and attempted to remake creation at his caprice (just as Alexander and Napoleon had tried to remake the world) by inventing evidence where there was none that transformism gained momentum. As a result of Haeckel's leadership, evolutionism in Germany had assumed a "transcendental, bellicose and utopian" direction not found in England. Pardo's essay may be seen as the attempt of an orthodox Catholic intellectual to come to terms with the evolutionary cataclysm. In the last analysis she weakened, admitting with the moderates that God was behind creation no matter what method he adopted. Later, she used the struggle for existence as a theme in her fiction, but her unwillingness to come to grips with the problems posed by the *Descent of Man* excluded her from the moderate camp.

[89] Emilia Pardo Bazán,, "Reflexiones científicas sobre el darwinismo," *Ciencia Cristiana* 4 (1877): 289–298, 481–493; 5 (1877): 218–233, 393–410, 481–495. See discussion by Harry L. Kirby, Jr., "Pardo Bazán, Darwinism and *La Madre Naturaleza*," *Hispania* 47 (1964): 733–737.

Two provincial secondary-school teachers contributed books to the polemic around 1880. The first, by Manuel Polo y Peyrolón of Valencia, is a fairly well-conceived run-down of the scientific evidence, point by point, introduced again by expressions of defensiveness and fear. Polo feared that the spate of pro-Darwin literature, including the discourse of García y Alvarez and another at the University of Madrid by Rafael Martínez Molina, was indicative of a pro-Darwin wave in Spain. He was also worried by institutional support for the theory. The Institución Libre de Enseñanza, a private school where many Darwinists who had lost their chairs had taken refuge, was supposed to have named Darwin an honorary professor. Worse still was evidence of state support: the natural-history collection of the School of Agriculture at Madrid was to include species, knowledge of which was necessary for information about "organic evolutions." This decree, included in the *Gaceta Oficial* for January 23, 1878, Polo took as proof that "official Spanish science is *evolutionist*, that is, enemy of independent creation and therefore of the doctrine of Genesis."[90] In this deduction he was probably close to the truth. Polo was further disturbed by the fact that his countrymen were "not content with appropriating the scientific impieties and heresies of our neighbors" but even exaggerated them, a measure of the implacable evolutionary stance of positivists. He devoted an entire chapter to proving that the diffusion of Darwinism was owing not to the scientific competence of Darwin but to support from the forces of materialism. His scientific case rested, again, on the question of species, which he believed to be immutable.[91]

The second book, by Luis Pérez Mínguez of Valladolid, was openly defensive in posture. There was throughout Europe "so dense an atmosphere of Darwinism that one can scarcely inhale anything else." The result was that it was difficult for Catholics to get hold of books that presented the scientific objections to the theory. Part of the problem, Pérez pointed out, was that Darwin's argumentation was so cautious and circumspect that it was difficult to counter. In support of this contention, he singled out a passage in the *Origin*, Chapter 7, where Darwin states

[90] Manuel Polo y Peyrolón, *Supuesto parentesco entre el hombre y el mono*, 2d ed. (Valencia: Manuel Alufre, 1881), p. x.
[91] Ibid., pp. 35–46.

that it is almost certain that an ordinary ungulate could be converted into a giraffe. Pérez noted this passage—one with obvious Lamarckian bias—as a statement wholly uncharacteristic of Darwin's usual cautious methodology.[92]

The most interesting anti-Darwin literature was that produced by moderate Catholics who attempted to accommodate the theory. Foremost among the Catholic scientists was the resolute follower of de Quatrefages, the paleontologist and geologist Juan Vilanova y Piera (1821–1893). A member of the "intermediate generation" of nineteenth-century Spanish scientists, Vilanova was an articulate spokesman for de Quatrefages's viewpoint of an evolutionary process limited to the lower orders. His views on the antiquity of man followed haltingly in the steps of his leader. Vilanova wrote an early critique of the *Origin* (1869), which stressed the incompleteness of the paleontological record. The presence of both simple and complex forms of life in the same strata led to the probable conclusion of the independent origin of species.[93] In the following decade Vilanova engaged in a polemic with Revilla, who characterized as "an absurd undertaking" Vilanova's campaign in the Ateneo de Madrid to harmonize science and religion ("a Utopia which we will never see realized") by popularizing geological and prehistorical studies but opposing Darwinism. Revilla alleged that Vilanova's protests hid a "profound fear" and that his intellect had been "petrified into a true fossil of an outdated science and a moribund faith."[94] For his part, Vilanova reacted defensively, expressing outrage at the liberals' habit of branding those opposed to Darwinism as reactionary and "persons of limited understanding." He reiterated his conviction that paleontological evidence was contrary to Darwin and thus took solace in the thought that "my conversion to Darwinism will still be much delayed." Moreover, he asked, how could Revilla say he was not fair-minded

[92] Luis Pérez Mínguez, *Refutación a los principios fundamentales del libro intitulado Origen de las Especies de Carlos Darwin* (Valladolid: Hijos de Rodríguez, 1880), p. 12. Pérez states that he owned the sixth edition of the *Origin* and that he had also read previous editions.

[93] Juan Vilanova y Piera, "Origen y antigüedad del hombre," *Boletín-Revista de la Universidad de Madrid* 1 (1869): 233–247, 449–462, 641–663. See discussion in Glick, "Science and the Revolution of 1868," p. 269.

[94] *Revista Contemporánea* 1 (1875–1876): 128; 3 (1876): 383–384.

when he had included a Darwinian essay by Tubino in a volume under his editorship?[95]

The campaign to harmonize science and religion in Spain gained impetus from the publication in 1876 of a Spanish translation of Draper's *History of the Conflict between Religion and Science*.[96] There was a great deal of discussion of Draper's book, especially in the Ateneo de Madrid, and finally, in 1878, the Real Academia de Ciencias Morales y Políticas organized a contest to reply to Draper.[97] Of the several books generated by the contest, the most influential was that by the Jesuit Miguel Mir. Although Mir, too, rejected evolution, he displayed enough confusion on the subject to indicate that he was caught in a trap between the fear of unorthodoxy and a genuine proclivity toward theological modernism. Nevertheless, Mir struck an important note when he proclaimed unequivocally that "In reality, we cannot see in which words of the Holy Scripture or in what dogmatic doctrine or authority the condemnation of the Darwinian hypothesis could be supported."[98] The Bible indicates a progressive creation, from simple to more complex, and even Genesis 1 does not indicate that all species were created immediately and by divine action. Mir's retreat from literalism must be seen as an important break with the intransigent Catholic defense. He stated that evolutionism was a question independent of Christian dogma and drew attention to the work of Christian evolutionists. He admitted that there was no doubt that animal and plant life had been modified through the influence of the environment and that some species had disappeared, while new ones had arisen. But that this had happened through Darwin-

---

[95] Juan Vilanova, "La doctrina de Darwin," *Revista Europea* 7 (1876): 356–358; idem, "La catedra de prehistoria en el Ateneo y su censor Revilla," ibid. 8 (1876): 219–223.

[96] John William Draper, *Los conflictos entre la ciencia y la religión*, trans. Augusto T. Arcimis (Madrid, 1876; reprint ed., 1885, 1886, 1888). This was the influential translation, from the English (there was another, from French), which boasted a prologue by the Krausist politician Nicolás Salmerón, who stressed the contradiction prevailing in Spain between science and Catholicism. Arcimis taught astronomy at the Institución Libre.

[97] María Dolores Gómez Molleda, *Los reformadores de la España contemporánea* (Madrid: Consejo Superior de Investigaciones Científicas, 1966), p. 69; Sanús Abad, "Apologética española," pp. 13–18.

[98] Miguel Mir, *Harmonía entre la ciencia y la fé*, rev. ed. (Madrid: Los Huérfanos, 1885), p. 332 (1st ed., Madrid: Manuel Tello, 1881).

ian mechanisms had not been demonstrated. His attribution to Darwin of the notion that development in nature is always progressive seems a misconstruction of the theory. He furthermore admitted a uniformitarian view that the same processes are at work in nature now as ever before, but one of these processes, he asserted, is, "as is known, the fixity of species"! Citing de Quatrefages, Mir held the line firmly on the direct creation of man by God.

The moderate school reached its apex in the works of Zeferino Cardinal González (1831–1894) and Juan González de Arintero (1860–1928). Both belonged, broadly speaking, to a group of Catholic theologians and scholars who tried to achieve a synthesis of orthodox theology with evolutionism, a group that included the Englishman St. George Mivart, the French cleric M. D. Leroy, and Father John Zahm of the University of Notre Dame. The history of this school begins, properly speaking, in 1876, when Pius X conferred a doctorate upon Mivart, thereby conferring implicit papal sanction on his ideas, and ends with the encyclical of scriptural infallibility (*Providentissimus Deus*) of 1893, subsequent to which Mivart's theory of man's corporeal evolution was condemned and both Leroy and Zahm were obliged to recant. Mivart, as a recent biographer states, "stood for the scientific opposition to the supposed materialism of the new biology and for the maintenance of Christian orthodoxy."[99] As such, he was a natural source of ideas and inspiration for the Spanish conciliationists.

Mivart, as Peter J. Vorzimmer has shown, was not simply the Catholic fanatic Darwin came to see him as. He raised valid objections to the role played by natural selection in the evolutionary process. To natural selection he counterposed an alternative: that the directing hand of God acts as the main agent in evolution, assuring the progressive direction of change.[100] As for the problem of the origin of man, Mivart interpreted the Biblical verse, "God made man from the dust of the earth, and breathed into his nostrils the breath of life," as meaning that man's body was not created in the primary and absolute sense but was instead evolved from pre-existing material (symbolized by "dust of the earth")

[99] Jacob W. Gruber, *A Conscience in Conflict: The Life of St. George Mivart* (New York: Columbia University Press, 1960), p. 115.

[100] Peter J. Vorzimmer, *Charles Darwin: The Years of Controversy* (Philadelphia: Temple University Press, 1970), pp. 230–232.

and thereby created through the operation of secondary laws. But man's soul originated in a different way, through the direct action of God.[101]

Zeferino González wrote about evolution in most of his major philosophical works, beginning in the 1870's; and his ideas successively were modified as he accepted more and more of the evidence for evolution. In his *History of Philosophy* (originally written in 1878 and revised extensively in 1886) he stressed the affinity of Darwinism, atheism, and materialism—the familiar triad. He introduced the Thomist concept of species and criticized Darwin for converting the logical and ideal relations of the classical concept of species into genealogical ones. He admitted that both animal species and man are subject to more or less important modifications, which result in varieties or races, but rejected qualitatively significant transformation of species. He distinguished between the Darwinism of the *Origin* and that of the *Descent*, declaring that "if we limit ourselves to the evolution or transformation of plant and animal species, which is what constitutes the fundamental and truly characteristic idea of the *Darwinism of Darwin*, it is licit to speak thusly; if from this Darwinism [i.e., natural selection] is excluded, in addition to its application to man, an application which science does not in any way justify, and if necessary reservations are made about the creation of the world and the rational soul, it can fit and does fit within Catholic dogmas."[102] Although at this time he rejected evolution on philosophical grounds, he agreed that a Catholic could accept the doctrine as a scientific theory so long as (*a*) the lower orders were disengaged from man and (*b*) Darwin was disengaged from Haeckel. He was particularly adamant when it came to the "Darwinian left," a grouping that included Haeckel, Büchner, and Matthias Schleiden; but he nevertheless presented an accurate and dispassionate analysis of Haeckelian monism.

In May, 1889, both Vilanova and González participated in the Catholic Congress, held in Madrid. It was during this period that a general debate over Tertiary man was taking place in Europe, with de Quatrefages and others moving toward acceptance. Here González hedged and was duly attacked in *El Imparcial* by José R. Carracido. In his reply to

---

[101] St. George Mivart, *On the Genesis of Species* (New York: D. Appleton, 1871), p. 300.

[102] Zeferino González, *Historia de la filosofía*, 2d ed., 4 vols. (Madrid: Agustín Jubera, 1886), IV, 278–296.

the latter, González admitted that man was older than had once been thought and then launched into a discussion of his views on the culture of primitive man, on which issue he remained skeptical. Anyway, these questions may be decided by science in the future, he added; they are in no way dependent on Genesis.[103]

By 1891, in his work *La Biblia y la ciencia*, González moved closer to a less ambiguous evolutionary position.[104] Here he stated that, with regard to the origin of the body of Adam, one had either to reject Mivart or admit that man was descended from animals. González believed that this latter alternative was not incompatible with dogma, but he rejected it anyway. Instead he suggested a modification of Mivart. Rather than viewing the body of Adam as nothing more than a developed ape into whom God infused a rational soul, González suggested that Adam's body was partly the product of evolution and partly the direct work of God himself. God took an imperfectly developed body, an unfinished product of evolutionary process, and added the finishing touches himself. This theory was discussed at length by Zahm, who dismissed it as an unnecessary modification of Mivart.[105]

Arintero must rank as one of the most committed Catholic evolutionists of late nineteenth-century Europe. As a youth he vigorously combated evolutionism on the basis of a priori theological arguments he had been taught in seminary. As his knowledge deepened, he reached a mental standstill, followed by an open conversion to Darwinism, which he regarded as inoffensive as a philosophical system, except when carried to extreme lengths. Between 1892 and 1898 he conceived and executed an eight-volume work covering all aspects of evolution, especially as it related to the origin of man, theology, and philosophy. The first volume, *La evolución y la filosofía*, published in 1898, was the only volume to

103 Carracido, *Estudios*, pp. 313–335. Carracido took González to task for allegedly opposing "anthropological Darwinism" and for condemning the exaggeration of the cultivators of prehistory. Carracido's article is on pp. 313–324; González's reply follows on pp. 325–335.

104 Zeferino González, *La Biblia y la ciencia*, 2 vols. (Madrid, 1891; 2d ed., Seville: Izquierdo, 1892).

105 J. A. Zahm, *Evolution and Dogma* (Chicago: D. H. McBride, 1896), pp. 359–362. Zahm had also read Mir. *Evolution and Dogma* was translated into Spanish as *La evolución y el dogma*, trans. Miguel Asúa (Madrid: Sociedad Editorial Española, 1905).

reach print. In it, he admitted that he had originally been opposed to evolution because he was unable to overcome the scholastic thesis of the metaphysical immutability of species. This reticence he later resolved by equating the Thomist concept of *species* with the modern biological *class*. To reject evolution totally, he stressed, is out of the question. What he admitted, however, was evolution only *within* classes, not between them. Moreover, although a self-proclaimed "evolutionist," Arintero was more Lamarckian than Darwinian. In this, as in his call for a new natural philosophy of living beings—he urged doing to Darwin what Saint Thomas had done to Aristotle—he was a precursor of the Catholic evolutionism of Pierre Teilhard de Chardin, aspects of whose thought were anticipated in the writings of more than one Spanish writer of the epoch.[106]

Looking back on the Darwin polemic in nineteenth-century Spain one must admit, first, the utter banality of the debate. Scarcely an original idea was uttered on either side, a reflection of the low level of scientific creativity (the neurohistologists were an outstanding exception), on the one hand, and the inane posturings of the orthodox, on the other. There are, too, elements of pathos in the portraits of typical figures of Spanish Darwinism: Peregrín Casanova's halting letters to Haeckel; the fact that a vulgar curate was able to show that long sections of García y Alvarez's controversial discourse had in fact been plagiarized from Clémence Royer's introduction to the French translation of the *Origin*.[107] It was as easy for anti-Darwinists to attack the mediocrity of the evolutionists as it was for Darwinists to attack the bigotry and obtuseness of the Catholics. On the other hand, in contrast to the banality, there was a certain vigor in the debate, which came from awakening intellectual excitement. To be a Darwinist in Spain was to live on the thin edge and to place oneself in certain jeopardy. The antievolutionists, for their part, gained a certain sense of *esprit* in picturing themselves as the last bulwark against

---

106 For a discussion of Arintero's evolutionism, see Sanús Abad, "Apologética española," pp. 23–32, and Alvaro Huerga, "La evolución: Clave y riesgo de la aventura intelectual arinteriana," *Studium* 7 (1967): 127–153.

107 Francisco de Asís Aguilar, *El hombre, es hijo del mono? Observaciones sobre la mutabilidad de las especies orgánicas y el darwinismo* (Madrid: Antonio Pérez Dubrull, 1873), p. 6, n. 1.

a German onslaught, the protectors of Spain from a Europe gone mad under the stimulus of a band of German materialist scientists, led by the nefarious Haeckel. Darwinism was part of the baggage of the positivist intellectual, just as anti-Darwinism was part of the baggage of the Catholic intellectual. The role of each was to support positions whose inspiration was drawn from sources and directed toward goals at some remove from the mainsprings of science. In the debate between the "two Spains," liberal and conservative, modern and traditional, Darwinism was a touchstone.

Elsewhere, I have stressed the tripartite processes of the polarization, normalization, and popularization of evolutionary thought in Spain, all three of which were substantially completed by the anniversary year 1909.[108] The polarization process, whereby Darwinism quickly became a creed of the left, in the wake of the Revolution of 1868, and anti-Darwinism a pillar of the orthodox right, has been described in this paper. By normalization, I mean the depolemicization of evolution among *scientists*—the men who gained university and secondary-school chairs in the last quarter of the century. This process needs more documentation. It appears that, in spite of the consistent loss of chairs by leading Darwinians, the replacements proved to be evolutionists also. Popularization was a process limited to the bourgeois political left, who by 1909 had recognized Darwin as the "great revolutionary of science."

In social terms, Darwinism was associated with the rising middle class, a group that was awakening intellectually during the span of the polemic. This is the class symbolized in Galdós's novel *Doña Perfecta* by Jacinto, who asks the town priest about Darwinism. The youth says that he hears that natural selection has many partisans in Germany, to which the priest replies that it is no wonder in view of the fact—he hints— that the figure of Bismarck lends credence to the theory of man's descent from monkeys.[109]

It is noteworthy too that so much of the discussion took place among secondary-school teachers. Pro-Darwin polemicists, such as García y Al-

---

108 Glick, "Valencian Homage," p. 601.

109 Benito Pérez Galdós, *Doña Perfecta* (Madrid: Hernando, 1961), pp. 70–71 (first published, 1876); in English, trans. Harriet de Onís (Woodbury, N.Y.: Barron's Educational Series, 1960), pp. 56–57. See discussion by Jorge Arturo Ojeda, "El siglo XIX en *Doña Perfecta*," *Comunidad* 3 (1968): 689.

varez and Máximo Fuertes Acevedo, of Badajoz,[110] were professors—
the latter a director—of *institutos*, as were such antievolutionists as Polo
y Peyrolón and Pérez Mínguez. An adversary of Fuertes hazarded the
guess that:

> El amibo o amiba,
> Que del agua nació con alma viva,
> Cuando le dió la gana
> En pez se trasformó, si no fue en rana;
> Ensanchando más tarde sus pellejos
> Formó . . . varios bichejos.
> De estas trasformaciones como fruto
> Resultó el Director de un Instituto.
> Si este sigue la norma
> Veremos en qué bicho se trasforma.[111]

There is significance in the fact that the discussion took place in part,
and typically, at this level. It demonstrates, first, the extent of the secularization of secondary education and, second, the furious Catholic reaction to the invasion of a formerly sacred bastion. Moreover, the fact
that these institutes were not in Madrid but in the provinces is a mark of
the outward, as well as downward, percolation of Darwinian ideas in
Spain.

[110] Fuertes was the author of *El darvinismo* [*sic*], *sus adversarios y sus defensores*
(Badajoz, 1883).

[111] Clara de Sintemores, *El darwinismo en Solfa* (Madrid: Aguado, 1887), p. 41.
"The amoeba / who was born from water with a living soul, / when he felt like it /
transformed himself into a fish, if not a frog; / expanding later his skin / he formed
various critters. / From these transformations as fruit / there resulted the Director of an
Institute. / If he follows the norm / we'll see into what critter he is transformed."
Catholic harmonizers, such as Arintero, found such blunt satire demeaning; see Huerga,
"Evolución: Clave y riesgo," p. 132 n. 14. Juan Valera made the same point about
Núñez de Arce's satirical poem "A Darwin" (see note 11, above); he praised its ironic
style but said that ridiculing apes did not do justice to the metaphysical points in
question (*Obras completas*, II, 453–455). For a further, more elegant elaboration of the
monkey joke by a Spanish poet, see Antonio Vinajeras, *El Congreso de Guinea* (Matan-
zas, Cuba: Imprenta El Ferro-Carril, 1879).

MEXICO

ROBERTO MORENO

When *The Origin of Species* appeared in 1859, Mexico was involved in the final phase of a long struggle for the republican federal system of government. The triumph of the liberals was delayed by the French intervention and the empire of Maximilian (1862–1867), whose execution marked not only the moment of liberal triumph but also the end of a series of foreign attempts to control the destinies of the nation. The subsequent reconstruction, which included everything from the political system to public education as well as all aspects of culture and science, was undertaken with enthusiasm by the most outstanding of the men associated with Benito Juárez up to 1872. The regime of Sebastián Lerdo de Tejada maintained the stability necessary for such a work of reconstruction. In 1878 the Revolution of Tuxtepec brought to power Porfirio Díaz, who initiated his prolonged dictatorship with a type of stability not totally appropriate to the cultivation of independent scientific research. The dictatorship lasted until 1910.

THE BIOLOGICAL SCIENCES IN MEXICO DURING THE NINETEENTH CENTURY

Alfonso L. Herrera divides the development of modern biology in Mexico into two stages: a prerevolutionary period (1821–1909) and the revolutionary period (1910–1921). According to Herrera, the first of these periods was characterized by "the accumulation of materials and a lack of coordination in research," in contrast to the activities of the

NOTE. José Ruiz de Esparza of the Universidad Nacional Autónoma de México collaborated in this research. The article was translated from Spanish by David Littlewood and Robert Lugo Adams.

revolutionary period, which really began in 1915 with the founding of the Dirección de Estudios Biológicos. Despite the absence of over-all coordination in the earlier period, Herrera nevertheless justly praises the distinguished Mexican naturalists Julián Cervantes, Joaquín and Juan Dondé, Pedro de la Llave, Gumesindo Mendoza, Melchor Ocampo, Leonardo Oliva, Leopoldo Río de la Loza, and José M. Velasco. In the second half of the century, biological research tended to coalesce in such institutions as the Sociedad Mexicana de Historia Natural (founded September 16, 1868), whose journal, *La Naturaleza*, was published until 1912. A cognate institution, the Instituto Médico Nacional, was founded in 1890 and published the results of biological research in its various organs (*El Estudio, Anales, Materia Médica*). Still, in Herrera's view, conditions for research were poor.[1]

But Herrera's judgment, colored by his revolutionary fervor, was not completely fair. Other modern scholars have treated the theme of Mexican science in the late nineteenth century and have concluded that conditions, although not entirely favorable, did allow many researchers to do good work.[2] The Mexican nineteenth century was filled with political struggles and grave problems, internal and external, that prevented the consolidation of a structured and uniform scientific effort. Even when political stability was achieved toward the end of the century, the dictatorship of Díaz, supported by the positivist "scientific party" (*científicos*), refused to grant any broad freedom to scientific dissidents. Nevertheless, in addition to the institutions mentioned by Herrera, there existed, from 1833, the Sociedad de Geografía y Estadística, which published a bulletin; the Sociedad Científica Antonio Alzate, which published its *Memorias*; and many other organizations of a more ephemeral nature. In the various states of the republic there were such institutions as the Colegio del Estado de Guanajuato and the Sociedad de Ingenieros de Jalisco, where the scientific spirit had an opportunity to express itself in a more or less organized fashion, however tentative.

[1] Alfonso L. Herrera, *La biología en México durante un siglo* (Mexico City: El Democrata, 1921), pp. 2–7.

[2] Eli de Gortari, *La ciencia en la Reforma* (Mexico City: Centro de Estudios Filosóficos, 1967); *La ciencia en México* (Mexico City: Fondo de Cultura Económica, 1963); Enrique Beltrán, "El panorama de la biología mexicana," *Revista de la Sociedad Mexicana de Historia Natural* 12 (1951): 69–99.

The Arrival of Darwinian Ideas

Mexico was in no sense unaffected by the scientific revolution instituted by Darwin and his disciples. The controversies that the new theory stirred up were reflected in the scientific and general intellectual thought of the nation. Most works on Darwinism that found their way to Mexico were written by Frenchmen. Mexico's cultural dependence on France was perhaps the cause of a minor delay in the introduction of Darwinism and of the polemical debates that introduction occasioned, and, at the same time, it sheds light on the positions that Mexicans took in the controversy, in accord with divergent currents of French opinion. The French model explains, too, why excessively violent reactions were not produced.

Only two studies have been published that examine the Mexican reaction to Darwinism. The conclusion of Santiago Genovés is that, for reasons of historical instability in Mexico, Darwin's work "had a weak and delayed effect."[3] Manuel Maldonado-Koerdell, for his part, studies the situation in greater depth and concludes that, "in spite of notorious propaganda (favorable or unfavorable) arising in response to Darwin's ideas in many countries, there appeared scarcely more than a couple of footnotes to his works in our country between 1870 and 1900." This he attributes to the slow acceptance of Darwinism in France.[4]

[3] Santiago Genovés, "Darwin y la antropología," *Revista de la Sociedad Mexicana de Historia Natural* 20, nos. 1–4 (December 1959): 31–41. Genovés claims to have culled the *Anales del Museo Nacional de Antropología* (henceforth referred to as *Anales del Museo*), the *Memorias de los Congresos Internacionales de Americanistas*, *La Naturaleza*, the *Gaceta de la Academia Nacional de Medicina*, *El Estudio*, *Anales del Instituto Médico Nacional*, and various bibliographies of Nicolás León. His earliest reference to evolutionary ideas is Alfredo Dugès's article on the classification of man (see note 6 below) in 1882–1884. There are two mentions of Alfonso L. Herrera: the catalogue of the anthropological collection of the Museo de Antropología and "Estudios de Antropología mexicana," by Herrera and Ricardo E. Cícero, in *Anales del Museo* 1: 127–128. Finally, Genovés cites Dr. José Ramírez, who at the Eleventh Congress of Americanists (1895) proposed that man was native to America, a hypothesis based on evolutionary ideas. It is peculiar that, having had access to the journals that he mentions, Genovés did not find the data that appear there.

[4] Manuel Maldonado-Koerdell, "Linneus, Darwin y Wallace en la bibliografía mexicana de ciencias naturales, I: Primeras referencias a sus ideas en México," *Revista de la Sociedad Mexicana de Historia Natural* 20, nos. 1–4 (December 1959): 63–78, presents magnificent data but curiously does not see the importance of his findings. In spite of his conclusion and somewhat bold affirmation that certain authors (Dugès and

The problem is that the study of Darwinism in Mexico necessitates a laborious search of all, or nearly all, the press—scientific and general—of the second half of the century, wherein one encounters only sporadic data. One cannot hope to find pro- and anti-Darwinian books, but still this does not mean that Darwinism was not an important issue in Mexico, as will be demonstrated.

On the basis of research done for this article, it appears that Darwinism made its entrance into Mexico during the 1870's. Although we cannot be absolutely sure that there are no earlier references, the absence of mention is probably attributable to the state of war, which lasted until 1867, and also to the fact that *The Descent of Man*, the work that most affected Catholics, did not appear until 1871.

The contours of the Darwinian theory received by Mexicans from principally French sources can be reconstructed, thanks to a stream of references encountered throughout the final part of the century. One illustration of Darwinian sources that were studied at the time is a list used by Santiago Sierra in the debate of 1878, analyzed below. Sierra reported that "we have in our possession and have read all of Darwin's work and we have even translated one, *La filiación del hombre* [*The Descent of Man*], that we will soon try to publish and popularize among us." He also mentioned works by Oskar Schmidt, Carl Vogt, and Charles Martins; Ernst Haeckel's *Natural History of Creation* and *Anthropogeny*; and T. H. Huxley's *Physiology, The Place of Man in Nature, Lay Sermons,* and *Anatomy of Vertebrates*. Sierra also mentioned Alfred Russel Wallace and listed among those he styled "detractors" Emile Blanchard, Karl von Baer, and Edmond Perrier, whose works he recommended to Catholics.[5]

In 1882, Alfredo Dugès mentioned Huxley, Haeckel, Paul Gervais, Armand de Quatrefages, and others in reference to the classification of

---

Jesús Sánchez, among others) kept "completely silent about evolutionary ideas," the data that he unearthed served as a point of departure for the present research. Maldonado-Koerdell presents information concerning the positivist polemic, the translation from French of Darwin's work on the formation of vegetable mold, and the translation of T. H. Huxley's article, "The Coming of Age of *The Origin of Species*" (1880), in the *Boletín de la Sociedad de Ingenieros de Jalisco*; he studied the work of evolutionist Alfonso L. Herrera with some attention and mentioned the anti-Darwinian treatise of Agustín Aragón. This is no small contribution, in spite of his unjustifiable conclusion.

[5] Santiago Sierra, "Confesión paladina," *La Libertad*, January 27, 1878.

man among the primates.[6] In 1884, Gen. Vicente Riva Palacio cited Darwin's *Descent of Man* and *Variation of Animals and Plants under Domestication*, Haeckel's *General Morphology*, and Vogt's *Lectures on Man* (all in French translation).[7] In 1895 Agustín Aragón, an anti-Darwinian positivist, revealed acquaintance with *The Origin of Species* and *Variation of Animals and Plants*, Wallace's *Darwinism* (in French translation), and Herbert Spencer's *The Factors of Organic Evolution*, besides mentioning Haeckel.[8]

In 1897 Alfonso L. Herrera, in the first explicitly Darwinian book published in Mexico, cited the French translations of Darwin's *Variation of Animals and Plants* (Paris, 1880), *Origin of Species* (Paris, 1882), and *The Descent of Man* (Paris, 1872). He cites as well Haeckel's *Natürliche Schöpfungsgeschichte* (translated as "Histoire naturelle générale des règnes organiques," Paris, 1869) and, again, Wallace's *Darwinism* in the French translation of 1891.[9]

Lists such as these could be multiplied. We shall mention, however, only the relevant books included in the collection of the National Library of Mexico, as revealed in the catalogues published by José María Vigil. Through them we know that in 1890 the following editions were at the disposal of the public in Mexico City: Darwin, *De la variation des animaux et des plantes sous l'action de la domestication* (2 vols., Paris, 1868) and *La Descendance de l'homme* (2d ed., 2 vols., Paris, 1873–1874); Léon A. Dumont, *Haeckel et la théorie de l'évolution en Allemagne* (Paris, 1873); Ernest Faivre, *La Variabilité des espèces et ses limites* (Paris, 1868); J. L. de Lanessan, *La lucha por la existencia y la asociación para la lucha: Estudio sobre la teoría de Darwin* (Madrid, 1884); Armand de Quatrefages, *Charles Darwin et ses précurseurs français: Etude sur le transformisme* (Paris, 1870); and Oskar Schmidt, *Descendance et Darwinisme* (Paris, 1880).[10]

6 Alfredo Dugès, "Consideraciones sobre la clasificación natural del hombre y de los monos," *La Naturaleza* 6 (1882–1884): 280–283.

7 Vicente Riva Palacio, "El virreinato," in *México a través de los siglos*, ed. idem, 5 vols. (Mexico City: Ballesca, 1884–1889), II, 472–476.

8 Agustín Aragón, "Apreciación positiva de la lucha por la existencia," *Memorias de la Sociedad Científica Antonio Alzate* 9 (1895): 141–161.

9 Alfonso L. Herrera, *Recueil des lois de la biologie générale* (Mexico City, 1897), pp. 145–146.

10 José María Vigil, *Catálogos de la Biblioteca Nacional de México, quinta división:*

In his listing of 1895, Vigil recorded the following editions: Darwin, *Les Récifs de corail* (Paris, 1878); Huxley, *Physiographie* (Paris, 1882); Wallace, *La Sélection naturelle* (Corbeil, 1872); Pelegrín Casabó, *Páginas de la creación* (Barcelona, 1888); Darwin, *Ueber die Entstehung der Arten* (Stuttgart, 1867) and *The Descent of Man* (London, 1871); Patrick Geddes, *L'Evolution du sexe* (Chateauroux, 1892); Haeckel, *Anthropogénie* (Paris, 1877); Robert Hartmann, *Les Singes anthropoïdes et leur organisation comparée à celle de l'homme* (Tours, 1886); Paul Topinard, *L'Homme dans la nature* (Paris, 1891); Angelo Vaccaro, *La Lutte pour l'existence et ses effets dans l'humanité* (Paris, 1892); Vogt, *Lecciones sobre el hombre* (Madrid, 1881); and August Weismann, *Essais sur l'hérédité et la sélection naturelle* (Paris, 1897).[11] And in 1897 the evening section of the library had at its disposal the following: Darwin, *The Descent of Man* (New York, 1873), *On the Origin of Species* (New York, 1873), and *Journal of Researches* (New York, 1873); Haeckel, *Natürliche Schöpfungsgeschichte*; Spencer, *The Principles of Biology* (New York, 1873).[12] So we see that neither Darwinian nor anti-Darwinian books were scarce in Mexico between 1870 and 1900.

In regard to Spanish translations of evolutionist works or even those that mention the evolution controversy, until now we have encountered only a few. We know that Santiago Sierra had translated *The Descent of Man*, but for reasons unknown to us the promised publication did not materialize, perhaps because Sierra died in a duel in 1880. At any rate, in 1878 he was a collaborator of the new weekly magazine of the daily *La Libertad*, in whose first years were published the following translated articles in which Darwinism was discussed: Spencer, "Los fundamentos de la sociología"; Emile du Bois-Raymond, "La historia de la civilización y la ciencia natural"; M. C. von Naegeli, "Los límites de la ciencia"; Edmond Domet de Vorges, "El reino humano"; and Haeckel, "Sentido y significación del sistema genealógico."[13]

Ciencias matemáticas, físicas y naturales (Mexico City: Secretaría de Fomento, 1890), pp. 78–90.

[11] Ibid., *Primeros suplementos* (1895), pp. 21–26.

[12] Ibid., *Biblioteca nocturna* (1897), pp. 89–96.

[13] *El Mundo Científico y Literario: Suplemento de La Libertad* 1 (1878): 1–4, 6–7; 3–5; 4–6; 7; 15.

In *La Naturaleza*, journal of the Sociedad Mexicana de Historia Natural, volume 6 (1882–1884), was published an article by Darwin, "La formación de la tierra vegetal por la acción de los gusanos," which in turn had been translated from *La Revue Scientifique* (no. 3, 1882). From the same French publication (no. 14, 1882) was drawn an anti-Darwinist article by Rudolf Virchow, published under the title, "Darwin y la antropología."[14]

Similarly, in 1882 the *Boletín de la Sociedad de Ingenieros de Jalisco* printed a translation of Theodore Gill's moderately anti-Darwinian treatise, "El hombre fósil de Mentone." Several months later an article by Huxley, "La mayor edad del 'Origin de las especies,'" appeared in the same journal.[15]

More translations could probably be unearthed in provincial newspapers. In any case, whether because of political, religious, or other sorts of pressure or because of the wide dissemination of French works among the Mexican middle class, there exist no Mexican editions of the principal evolutionary works in the nineteenth century. But Darwinism had indeed arrived in Mexico and was widely diffused among the educated. In no way can the notion that evolutionary ideas were weakly or poorly assimilated be sustained.

The Diffusion of Darwinism and Evolutionary Controversies

Although earlier mention may exist, the oldest unequivocal citation of Darwinism dates from 1875. It comes from the pen of Justo Sierra, a person of great influence in Mexican education, professor in the newly founded secondary school in the capital, a future force in the later reopening of the university, and a Spencerian positivist. In an article commenting on a round-table discussion of spiritualism and science, he defends observation and experimentation and points out that, the greater the precision of the observations and the independence of the observer, the more likely the deductions are to be true in any theory, "even those seemingly consecrated by science itself."

14 *La Naturaleza* 6 (1882–1884): 89–110; 183–190.

15 Theodore Gill, "El hombre fósil de Mentone," trans. Manuel Pérez Gómez, *Boletín de la Sociedad de Ingenieros de Jalisco* 2, no. 3 (March 15, 1882): 90–96; T. H. Huxley, "La mayor edad del 'Origen de las especies,'" trans. José M. Castaños, ibid., nos. 9–10 (September 15, 1882; October 15, 1882).

Therefore, for example, there exists in the world of scientific discourse a debate admirably suited to the revival of scholarly and lay interest. We refer to the theory of evolution of Darwin and Wallace. Up to now this theory may have counted in its favor many facts superior perhaps to those adduced by the proponents of the theory of fixity of species. One scholar, notably inclined towards Darwinism, at present is conducting explorations of the sea on a much grander scale than ever before. In one of his explorations he extracted from the ocean's depths a small mollusk identical to those of earlier periods. An orthodox Darwinian, shall we say, might have thrown the mollusk into the sea. The captain of the Challenger expedition, as a philosopher worthy of appellation, recorded the fact with the greatest scruples, without worrying about the consequences that might ensue.[16]

As this citation indicates, Justo Sierra was thoroughly familiar with Darwin's theory. Some months later during the same year of 1875, Sierra wrote on the teaching of history and noted the need to modify the traditional structure of the field, inasmuch as "Science has destroyed the supposed unity of the human family and has caused the retracing of our origin back beyond the animal kingdom, to the plant world, and back to the first manifestations of the vital power of the planet, and has formulated with Darwin and Wallace the great law of evolution."[17]

A year later we encounter the first Darwinian scientist in Mexico, Francisco Patiño, who showed himself partial to the new theory in a study of carnivorous plants. Patiño wrote that scholars had elaborated a new way of thinking, impossible to conceptualize before, which "already seems a dogma in the present time." He referred to the "chain of being" that is confirmed by the carnivorous plants. Thanks to Darwin, he asserted, "the notion that plants neither feel nor move is at the point of being proved incorrect." Each animal and plant is a link in the great chain of being.[18] Actually, Patiño seems to have been more of a well-intentioned materialist, taking what he needed from Darwin, than a thoroughgoing evolutionist.

The most significant controversies over Darwinism took place in 1877

[16] Justo Sierra, "El espiritismo y el Liceo Hidalgo," *El Federalista*, April 2, 1875. On Spencerian sociological thought in Mexico, see Moisés González Navarro, *Sociología e historia en México* (Mexico City: El Colegio de México, 1970), passim.

[17] *El Federalista*, November 10, 1875.

[18] Francisco Patiño, "Las plantas carnívoras," *Gaceta Médica de México* 11, no. 24 (December 15, 1876): 474–479.

and 1878. They allow us to pinpoint the resistance to the new natural system offered by the forces of religion and philosophy (in this case, Comtian positivism). The scientific controversy erupted later and was less raucous.

The first debate took place in the Asociación Metodófila Gabino Barreda, named for the donor who founded the organization, a disciple of Auguste Comte in France who returned to Mexico to found the Escuela Preparatoria. Barreda was a convinced positivist all his life and educated in the same philosophy an important group of Mexican public figures, discussed below. Darwinism was debated in the association's sessions of February 25 and March 4, 11, and 18, 1877, the published minutes of which filled up ninety pages of print.[19]

In less than four pages, Pedro Noriega outlined Darwin's theory. He mentioned the three laws of inheritance, adaptation to the environment, and the struggle for existence; and he explained how natural selection operated. He refuted the objection that all the intermediate forms of animals should be described and deemed the theory to be the only scientific one. This outline shows that Noriega had assimilated the theory well. At the beginning of the debate, a Señor Flores said he agreed with the work. Then, Luis E. Ruiz began to complicate the matter, asking that the "logical bases" of Darwinism be further detailed. Member Muñoz insisted that good theories explain all the facts and that it ought to be proved that only one theory does, in fact, explain them. Noriega replied that the proof is indirect: since it cannot be proven that species may have been formed independently, they must have evolved as Darwin said. At this juncture Gabino Barreda took the floor, holding it for a long interlude.

Barreda began by stating that what interested the membership was *method*, and it was from this point of view that he wanted to discuss Darwinian theory. To summarize his lengthy exposition, Barreda was

[19] *Anales de la Sociedad Metodófila Gabino Barreda* 1: 97–186. This polemic has been studied by Leopoldo Zea, *El positivismo en México: Nacimiento, apogeo y decadencia* (Mexico City: Fondo de Cultura Económica, 1968), pp. 162–165. Maldonado-Koerdell mentions it, but without conceding it much importance. He was not acquainted with it in complete form, since the copy that he consulted only reached page 120. Concerning Barreda, see Moisés González Navarro, "Los positivistas mexicanos en Francia," *Historia Mexicana* 9 (1959): 119.

disturbed that Darwinism was so widely accepted as a symbol of progress, an attractive thing for the young, and a weapon against theological cosmogonies. (Here he demurred, declaring that to be against Darwin was not to be in favor of the Biblical doctrine.) Moreover, he objected to Darwin's method in that the generalizations of the theory fell short when amplified to show details. He concluded that the theory was "arbitrary and irrational. . . . Darwin's theory, called the evolution of the species, genus, etc., has not to this day satisfied the conditions demanded by scientific method and, for the same reason, whatever be the cause that might inspire us, or the repugnance we might feel for its opponents, we should not accept it as a proven fact, but rather as a hypothesis whose proof has not yet been made and which has before it great obstacles, among them the fundamental law that governs the propagation of organized beings."[20] It should be emphasized that the continual admonitions made by Barreda to his disciples that they not accept Darwinism merely because of sympathy demonstrate that in those years evolutionism was a fad among Mexican intellectuals.

In the association's meeting held on March 4, Dr. Porfirio Parra, a curious positivist with Darwinian tendencies, and Gabino Barreda spoke. Parra stated openly that Barreda hadn't completely understood some of the ideas of Darwin, who used an "eminently metaphorical" language. He then defined the notions of the survival of the fittest and of natural selection, refuting many of the traditional objections against Darwinism and demonstrating a good understanding of the material. Barreda rebutted with complicated anti-Darwinian explanations, repeating examples and defining his fundamental objection, "which consists of the lack of positive proofs in favor of the theory" and of Darwin's "insistence on presenting as an objective fact, a purely subjective concept, satisfactory for describing and coordinating the data which observation shows us, albeit in incomplete form, relative to the fundamental similarity of certain types and the character of descent which can be described by types placed in a certain subjective order." Later, Barreda quoted Darwin, taking advantage of those instances in which the English naturalist confessed his lack of knowledge in order to refute him. He concluded by asserting that Lamarck's theory was philosophically superi-

[20] *Anales de la Sociedad Metodófila Gabino Barreda* 1: 102–109.

or in spite of his having ruined it later with metaphysical explanations. Darwinian evolutionary theory resembled the theories of the alchemists, which were based on facts that did not logically support the conclusions reached.[21]

The meeting of March 11 began with a speech by Señor Flores. He was a faithful Darwinist, familiar with Comtian philosophy, who succeeded in understanding the nature of Barreda's objections. Thus, he began by outlining the conditions that a hypothesis should meet in order that it might be considered scientific and legitimate. This outline was followed by a summary of Darwin's theory (citing Ludwig Büchner and Haeckel). In his long, well-reasoned disquisition, Flores demonstrated a thorough knowledge of the theory and its biological bases and concluded that "Darwin's theory is about hierarchies of causes. The fundamental laws of which it is made up are true causes; considered analytically they are capable of producing the effects that are attributed to them, and synthetically they manifestly tend to produce them. Darwin's theory is therefore an inductive one, or else there is no theory that can be so considered." Then, as usual, Gabino Barreda spoke again. There is no need to gloss his speech, similar in content to all of his previous ones. But it is worth noting that Barreda apologized for his perseveration by stating that he had not previously realized that the other members were followers of Darwin or that his theory "had taken such complete possession of their spirit and, one could almost say, made them into fanatics."[22] This shows the diffusion of Darwinism even among the positivist disciples of the indefatigable Barreda.

The session of March 18 was completely occupied with a speech by Barreda. His conclusion, however, is unexpected: "Does this mean that Darwin doesn't deserve the thanks of science? By no means. Darwin in his formulation of the theory was faithful to scientific method and called attention to some very important facts and has therefore given a powerful impetus to science, inspiring research of all types and bringing under discussion new facts of great interest. If some day a satisfactory theory is formulated concerning the origin of species Darwin will undoubtedly be considered as one of its most illustrious precursors."[23]

21 Ibid., pp. 111–123.
22 Ibid., pp. 125–156.
23 Ibid., pp. 157–186.

This debate among the positivists is of great importance, not only because it shows that there were several convinced Darwinists in 1877, but also because it was, as Maldonado-Koerdell has pointed out, the first extensive public commentary on Darwin's theory. It shows the resistance that the theory aroused, under the cloak of philosophical debate, among some of the most illustrious personalities of the era. On the other hand, it confirms our suspicion that the education received in the Preparatory School was far from being behind the times and that, whether favorably or unfavorably, evolutionary ideas were frequently discussed. Further proof of this will appear below.

The second debate considered here was of a distinctly different nature. It turned on an accusation made by the Catholic Society of Mexico in its newspaper that Justo Sierra's textbook on ancient history written for the Preparatory School contained Darwinian errors. The Catholics' objections concerned Sierra himself, his brother Santiago (whose belief in Darwinism we have already noted), and perhaps one other collaborator of the progressive newspaper *La Libertad.* The discussion, which was not too prolonged, also touched on other points currently under debate (e.g., liberalism) and soon degenerated into pure satire.

Justo Sierra had begun to publish, in serial form, a summary of ancient history, *Compendio de historia de la antigüedad.*[24] In it he felt obliged to outline—with neither praise nor censure—the basic nature of Darwin's theory. On January 5, 1878, the first chapter, containing the evolutionary material, was noted as a book received by the Catholic newspaper, *La Voz de México.*[25] On the following day *La Libertad* published a humorous article (who knows with what intention) in which apes demanded world power because mankind had at last recognized its origin.[26] But the *casus belli* was the criticism that *La Voz* levelled against

---

[24] Justo Sierra, *Compendio de historia de la antigüedad* (Mexico City: José María Sandoval, 1879). Edmundo O'Gorman has studied this work and concludes that it is the one Sierra began to publish in installments in 1878; see O'Gorman's edition (Mexico City: Universidad Autónoma de México, 1948), pp. 5–7. See also J. I. Mantecón, I. Contreras, and I. Osorio, *Bibliografía general de don Justo Sierra* (Mexico City: Universidad Nacional Autónoma de México, 1969), pp. 54–55.

[25] *La Voz de México,* January 5, 1873.

[26] "El porvenir de los gorilas," *La Libertad,* January 6, 1878.

Sierra's textbook.[27] The extensive review began by accusing the author of being anti-Catholic and then proceeded to the content of the book: "Choosing to expound the most extravagant theory of the utopian Darwinist school, positivist and incredulous in its system, it can be seen that the few affirmations of the work itself are erroneous and at times contradictory and that in the final analysis they explain nothing, no matter how much the scholars there cited be presented as oracles of modern science." *La Voz* asserted that this type of theory attacked Catholicism and offered Georges Cuvier as an example of the true scientist, while accusing Haeckel of uttering "conjectural absurdities."[28] On the same day a note appeared in *La Libertad* denouncing *La Voz* and *El Centinela Católico* as reactionary and ignorant: they attacked Darwin without reading him and attributed unjustly to Darwin the statement that man is a descendant of the monkey. The author was consoled in the knowledge that Darwin himself would never learn of the existence of such newspapers and attacks.[29]

The following day Justo Sierra anonymously answered *La Voz*'s criticism of his book. His extensive article presents a well-thought-out argument. Sierra thought it essential that, given the separation of church and state, the professor should teach the sciences even though they might conflict with theology. On the other hand, history had to provide an understanding of the origin of man, and in this instance the professor had the alternative choices of the religious doctrine of spontaneous generation and that of "the indefinite and perpetual evolution of all beings through all ages, all environments, and all forms." The choice was clear, and for this reason Cuvier was of no value, "for the simple reason that he did not know the prehistoric sciences as do Huxley, who is not a Darwinist [*sic*], Spencer, Tyndall, Haeckel, Broca, Giard, Schmidt, Vogt, Martins, etc., who do not recognize the authority of the eminent. French naturalist in questions which he had not mastered or had treated

27 Before *La Voz*'s criticism of Sierra, it had published (on January 20, 1878) an article entitled "El Darwinismo," in which the editors solaced themselves by referring to an anecdote, which they later reprinted, according to which someone had let down a monkey when Darwin received his doctorate.

28 *La Voz de México*, January 25, 1878.

29 "Contrastes," *La Libertad*, January 25, 1878. I have been unable to locate copies of *El Centinela Católico* for this year.

lightly." Sierra questioned the seriousness of the Catholic response to all problems and concluded: "It must be understood that we defend in this only the rights of science and in no way the pretensions of any philosophical school to teach a moral; but when we see men of knowledge and virtue like Tyndall and Darwin censured as materialists, for the sole crime of teaching science and not the Bible, he who so writes, who believes in the immortality of the intelligent and responsible principle, can do no less than have compassion for those who, with such great mastery, pass judgment on spirit and matter. Do they perchance know what is spirit and what is matter?"[30]

For its part *La Voz de México* took up the challenge of searching out the passage wherein Darwin had said that man descended from the monkey and accused the editors of *La Libertad* of not having read Darwin.[31]

On January 27, articles appeared in each of the contending dailies. In *La Libertad* Santiago Sierra took up the challenge offered by *La Voz*. He stated that his colleague from *La Voz* had surely thought a bit and had searched for the citation from Darwin concerning man's descent from monkeys and that what he found was that they descended from a common trunk, which is quite different. As for the accusation that he had not read Darwin, Sierra presented a list of the works he owned by or about Darwin (which we have examined above), and he offered to make these available so that his opponents could inform themselves. For the rest, he said that they lied in affirming that Lamarck was the first evolutionist and cited other precursors of Darwin. In any case, even those whom Sierra calls "detractors" (Blanchard, Baer, and Perrier) admitted that natural and sexual selection were original with Darwin, even as Wallace himself affirmed. Sierra recognized that Haeckel exaggerated a bit and stated that Huxley was not a Darwinist but accepted the theory as the scientifically true one. The article closes with a eulogy of Darwin.[32]

---

[30] "Un nuevo libro de texto en la Escuela Preparatoria," *La Libertad*, January 26, 1878. I adhere to the common opinion that the author of this article was Justo Sierra, although I suspect that it may, in fact, have been written by his brother Santiago.

[31] "¿Su ignorancia?" *La Voz de México*, January 26, 1878. In this article the Catholics praise Darwin's work on coral reefs, geology, and other research prior to the extravagances of the *Origin of Species*. They accuse him of plagiarizing Lamarck's theory.

[32] Santiago Sierra, "Confesión paladina."

That same day *La Voz* began with sarcasm and jokes. The paper styled Darwin's theory ridiculous and added that, for that matter, there were some in Mexico who believed that man descended from the axolotl (a Mexican salamander and, here, an ill-intentioned reference to Justo Sierra). They referred to an incident when Darwin received a doctorate from Cambridge, when some prankster had a monkey climb down from the roof carrying a sign saying: "A link in the chain." They argued, further, that no one had ever seen mutations and closed with the assertion that the theory of evolution is degrading to man.[33]

La Voz's attack upon the Darwinist newspaper continued in the issue of January 29, in which two articles were published. The first was a reply to Santiago Sierra's article of January 27, containing various quotations in which Darwin affirms that there is no doubt that we descend from a branch of the simian trunk.[34] In the other, it is reported that some students from the Preparatory School had presented themselves in the paper's editorial offices to say that they had initiated a debate in the lecture room over Darwinism and had triumphed over the "ridiculous theories." They promised to let no chance slip by to combat evolution in the classrooms.[35] The same day a joke appeared in *La Libertad* to the effect that a parrot had confirmed that the staff of *La Voz* were descended from parakeets and not from monkeys.[36]

The polemic entered a satirical phase when in the January 30 issue of *La Voz* there was a reference to a long-tailed ape that escaped from his owner and went disrespectfully to rub elbows with the slanderers of *La Libertad*, who hired him to write articles on natural selection. Further, *La Voz* informed the Darwinians, so that they might not lack details of the genealogy that makes man descend from monkeys, the French translation of *The Descent of Man* was available in the Bouret bookstores at 8.50 pesos. This notice was of no use to the general reading public,

[33] "La Universidad de Cambridge y Carlos Darwin," *La Voz de México*, January 27, 1878.
[34] "Los redactores de *La Libertad* no han leído a Darwin," *La Voz de México*, January 29, 1878. The quotations are from *La Descendance de l'homme* (Paris: Reinwald, 1873).
[35] "La Escuela Preparatoria," *La Voz de México*, January 29, 1878.
[36] "Un perroquet," *La Libertad*, January 29, 1878.

"who will not waste their money and time reading the extravaganzas of the English scholar."[37]

The same day, *La Libertad* printed an article by Santiago Sierra concerning the passages from Darwin cited in *La Voz*. Sierra poked fun at *La Voz*'s misinterpretation of Darwin and refuted its construal of the passages in question. In another article, *La Libertad* applauded "with both hands" the anti-Darwin students of the Preparatory School because "our triumphs have all been achieved in the realm of controversy."[38]

On January 31, *La Voz* mocked *La Libertad* for applauding with two hands, since "Darwin's tailed brother" should have applauded with four. They added that debate should continue in the classrooms.[39] That day the Darwinists complained about *La Voz*'s jokes, alleging that they had expected their opponents to recur to satire once they had run out of rational arguments and further claiming that the monkey with the tail that had recently arrived at their offices had definitely escaped from the opposing paper.[40] The truth is that *La Libertad* had begun the barrage of jests with the parrot, as was pointed out in the February 1 issue of *La Voz*.[41] The same day *La Libertad* returned to serious argument in Darwin's favor with a long quotation and accused the writers of *La Voz*, with good reason this time, of not understanding what they were told.[42]

On February 2, *La Voz* continued in a sarcastic vein, however, by publishing a burlesque poem against Darwin by the Spanish poet Gaspar Núñez de Arce. And, furthermore, they reiterated that students of the Preparatory School were taught that man was descended from the salamander.[43] Against this last blast, *La Libertad* complained and stated that a professor of that institution taught the doctrine without taking sides.[44]

The debate returned to a more serious plane when *La Voz* amply

---

[37] "Un simio," *La Voz de México*, January 30, 1878; "A *La Libertad*," ibid.

[38] "*La Voz de México* versus Darwin," *La Libertad*, January 30, 1878; "También nosotros," ibid.

[39] "Ya somos dos," *La Voz de México*, January 31, 1878; also, "*La Libertad* y su Darwin," ibid.

[40] "Lo habíamos previsto," *La Libertad*, January 31, 1878; "Atrocidad," ibid.

[41] "Lo habíamos previsto," *La Voz de México*, February 1, 1878.

[42] "La cubierta peluda," *La Libertad*, February 1, 1878.

[43] "A Darwin," *La Voz de México*, February 2, 1878.

[44] *La Libertad*, February 3, 1878.

refuted Justo Sierra's article of January 26.[45] On this occasion the author of the history textbook returned once more to the debate in an article that ended the polemic. This time Sierra began by energetically combating the Catholic version of man's origin. He labeled as lacking in seriousness the Biblical doctrines of paradise and man made from clay: the historian cannot begin any serious text with such stories.

On the question of the origin of the human species, there is no room for doubt between the clay fable or the Darwinian hypothesis, because at least the latter is based exclusively on scientific facts, whereas the former is completely unsupported on scientific grounds. The Biblical cosmogony is, in the last analysis, a belief in the spontaneous generation of all species, and science has already relegated to the remotest origins of life on the planet a process which is not admissible in the special creation of any organized being among those who inhabit the world. Darwinism abstains from probing deeply into the great mystery of the origin of life, but given life in rudimentary beings it formulates a theory which, whether or not it be a true explanation of facts, is plausible and rests on considerations of a scientific nature, which observation and experiment will be charged to refute or support; it is, moreover, the only theory worthy of the name which exists concerning universal ontogeny. The historian should so expound it, just as the professor in the Preparatory School has done, not without pointing out that some eminent scholars are not in agreement with it, and that on the other hand many support it, illustrate it, and try to complete it. That is the truth, ignored only by those who are unaware that Darwin already counts among his proselytes the most eminent naturalists of England, Germany, Holland, Switzerland, Russia, Italy, and even in France and Spain where Cuvier still holds sway transformist doctrine is imposing itself rapidly and has snatched from its adversaries a multitude of confessions and desertions.[46]

Here the polemic ended.

It is easy to see that there is no great difference between this dispute with Catholicism and many other similar ones that arose for the same reason in other times and places. The controversy with Catholic society in no sense modified in essence the Darwinian impulse in Mexico. After the death of Barreda in 1881, Justo Sierra was the unquestioned leader

---

[45] "Un nuevo libro de texto en la Preparatoria y *La Libertad*," *La Voz de México*, February 5, 1878.

[46] Justo Sierra, "El Compendio de Historia y *La Voz de México*," *La Libertad*, February 7, 1878.

of Mexican education. His historical textbooks with their basic Darwinian themes were reprinted many times, and many generations of students at the Preparatory School were exposed to them.[47] In this manner, evolutionary ideas were introduced, however timidly, into official Mexican educational policy from 1878 on.

Sierra provides a good example of the utilization of Darwinian ideas in a wide variety of contexts. In that same year of 1878 he used them to refute a theory of criminality in Mexico,[48] and a year later he upheld the idea (naturally supported by Spencer) that the Darwinian model was applicable to sociology, stating that society as well as all other organisms is "subject to the laws of evolution." He quoted Huxley and spoke of evolutionary ideas as recognized scientific advances. Also he used evolutionary concepts to warn Mexico that, inasmuch as nations acted like organisms and Mexico had stronger neighbors, the country found herself destined "to be a proof of Darwin's theory, and in the struggle for survival we find that all the probabilities are against us."[49] In fact, Sierra even employed Darwinism in theatrical criticism: "If at any time Darwin's theories have been completely submerged in a psychological point of view, it is in the play of Sellés; the laws of heredity don't exist here."[50]

DARWINIAN BIOLOGY AND ANTHROPOLOGY

We have now examined the beginnings of Darwinism in Mexico in its connections with positivist philosophy, religion, and education, insofar as the data available to us have permitted. In regard to science, information is even more sparse. However, on the basis of our data it seems plausible to conclude that for many individuals Darwinism modified the direction of scientific research. It seems obvious that the greatest impact of Darwinism had to be upon anthropology, since Mexico was a country with great concentrations of Indians. However, impact on biology is also patent.

Dr. Ramón López y Muñoz was one scientist of the period already working with well-assimilated Darwinian ideas. In 1879 he published a

---

[47] See Mantecón, Contreras, and Osorio, *Bibliografía general*, pp. 74, 80–81.

[48] "Contestación a la carta del doctor Fenelón," *La Libertad*, October 12, 1878.

[49] "Positivismo político," *La Libertad*, September 3, 1879.

[50] "Las esculturas de carne: Drama de E. Sellés," *La Libertad*, June 5, 1878; June 6, 1878.

study entitled "Generación: Causa y condiciones de la sexualidad; ovo-
génesis y embriología," in which, as well as contributing some original
observations, he systematically reviewed Haeckel's theses.[51] In 1880 this
same author wrote an article entitled "La ley del hábito en biología," in
which he attempted to demonstrate the importance of habit, which he
studied in animals and in men. To emphasize his point, he noted: "It
suffices to say that the important law of 'habit or imitation,' common to
the animal kingdom, has led to the belief that considers custom as second nature and served Darwin with the name of 'adaptation to the environment' along with heredity and natural selection, to form his theory
about the origin of species."[52] Unfortunately, besides these two articles
we have no more notice of this Darwinian physician.

At this moment there appeared on the scene another person, this one
of great importance for the history of Mexican biology—Alfredo Dugès
(1826–1910). French by birth, he settled in Guanajuato, where he
trained several generations of doctors. His articles, along with those of
his brother Eugenio, fill many pages of almost all the scientific periodicals of the time.[53]

Dugès was a mature scientist who did not let himself be carried away
by enthusiasm for Darwin's theory. For the most part, his works are
simple reports in which the problem of evolution is never mentioned. But
this does not mean that he was unconcerned with it. It seems, rather,
that, although he sympathized with the new ideas, he preferred to
maintain an attitude of prudent scientific detachment. In 1882 he published a treatise entitled "Consideraciones sobre la clasificación natural
del hombre y de los monos," discussing the classification of man and the
apes and reviewing the various schemes proposed by Huxley, Haeckel,
Gervais, de Quatrefages, and others.[54] Dugès concluded:

[51] *Gaceta Médica de México* 14, no. 7 (April 1879): 121–128.

[52] Ibid. 15, no. 15 (August 1880): 333–345.

[53] See Robert M. Smith and Rozella B. Smith, *Early Foundations of Mexican Her-
petology* (Urbana: University of Illinois Press, 1969), pp. 13–23. In a review of this
work, Thomas F. Glick demonstrates that Maldonado-Koerdell's affirmation that Dugès
was silent concerning Darwinism is false, and he gives references by Dugès to Darwin.
He also doubts, with reason, that Darwin had scant influence in Mexico (*Isis* 61 [1970]:
551–552).

[54] *La Naturaleza* 6 (1882–1884): 280–283. This work has been discussed by Genovés,
"Darwin y la antropología," p. 32.

It is evident that the zoological characteristics which distinguish sloths and bats are of too great importance to authorize the inclusion of these divisions in the same group with bimanes and quadrumanes; but on the other hand one must push spiritual and religious considerations to their limits in order to segregate into a special kingdom a third group which analysis shows to have no essential differences from animals. Surely a monkey has few affinities with a fish or a tapeworm, yet no one would hesitate to declare the one as much an animal as the other. If therefore, man differs much less from the monkey than the monkey from an articulate or even a lower vertebrate, what reason is there to raise for man an altar over the rest of the animal kingdom and overlook his numerous ties to it?

Dugès himself was the author of a zoology text wherein he expounded Darwin's theory, raising objections to it of a scientific nature. In the book's prologue the reviewing commission, composed of Alfonso L. Herrera, José Ramírez, and Donaciano Cano y Alcacio, asserted: "Now that all biological problems are linked so closely with the question of the descent of organized beings, it is indispensable to initiate students in the principles of zoological philosophy, showing them the laws discovered by Darwin in his attempt to reconstruct a theory of living beings."[55]

In the body of the book, the entire forty-first chapter ("Trasformismo") is devoted to the problem of evolutionary theory, as Thomas F. Glick has pointed out. It begins with a quotation of an evolutionary nature from Hindu thought and with a list of transformists: Benoit de Maillet, Lamarck, Etienne Geoffroy Saint-Hilaire, Wallace, and Darwin, who is studied as the man who completed and perfected the theory. Dugès mentions five principal aspects of the theory: (*a*) present-day animals descend from a few primitive types; (*b*) they are modified principally by *selection*; (*c*) species are limitlessly variable (by crossing, external agents, and new habits); (*d*) those who acquire a favorable characteristic by vital competition or struggle for life destroy the others; and (*e*) evolution is progressive, although considerable time is needed for specific changes to occur. Dugès confessed he found the theory "seductive" and added that "if instead of resting upon a series of hypotheses

---

[55] Alfredo Dugès, *Elementos de zoología* (Mexico City: Secretaría de Fomento, 1884), pp. v–vi. Before this date Dugès had published several notes that I have been unable to consult for possible Darwinian references.

it can be supported with proven facts, there is no doubt that it will attract universal acceptance."[56]

Dugès criticized two elements of the theory: (*a*) limitless variability, which, he said, the facts did not confirm; for him, species varied within very restricted limits; (*b*) the fact that there are fossils of animals superior to those now living. The theory of descent also bothered him because forms transitional between species were not found, and he concluded that each form had appeared just as we know it, although we do not know how. Without mentioning him by name, he then refuted Haeckel, claiming that the famous law of ontogeny and phylogeny is based on superficial similarities. He said that Darwin had not proven the transformation of one species into another and accused him of talking of probabilities and suppositions in a science like zoology, which should be founded on the rigorous observation of the facts. But in sum he praised the Englishman for the hypothesis composed with talent and vast erudition. Noting the points in its favor, he concluded: "It is better to remain in philosophical doubt than to declare oneself either an absolute partisan or irreconcilable enemy of the theory."[57]

The measured and thoughtful stance of Dugès has its counterpart in the work of Vicente Riva Palacio (1832–1896), politician, novelist, historian, satirist, amateur naturalist, soldier, governor, cabinet minister, and (somewhat against his will) ambassador from Mexico to Spain. A man of great vitality, intelligence, and erudition, he was always a convinced liberal, and thus in 1884, the same year as the publication of Dugès's textbook, we find him in jail, where he passed the time in writing a good part of the history of the colonial epoch for the great work that he edited, *México a través de los siglos*.[58] When he was released from prison, he departed for Spain, where he published the second edition and died.

In this text of 1884, Riva Palacio employed a novel application of Darwin's theory. In a section dealing with the viceroyalty, analyzing the

56 Ibid., p. 224.

57 Ibid., pp. 223–229.

58 Vicente Riva Palacio, ed., *México a través de los siglos*, 5 vols. (Mexico City: Ballescá, 1884–1889; 2d ed., Barcelona: Espasa, 1888–1889). This work was directed by Riva Palacio, and such distinguished historians as Alfredo Chavero, José María Vigil, and others collaborated on it.

race and caste structure of the colony, he includes some truly peculiar considerations: "The indigenous race," he writes, "judged in accordance with the tenets of the evolutionary school, is undoubtedly in a state of bodily perfection and progress, superior to all other known races, even though its culture and civilization, as the conquest proved, were inferior to the civilized nations of Europe." And he adds: "Historians have only considered the Indians in their exterior aspects and by manifestations of their intelligence, but no one has yet undertaken the anthropological study of this race which, by reason of the clearest organic details which the first careful study will reveal, differs from the races studied to the present and demonstrates, following the accepted principle of correlations among living organisms, that there are characteristics which made it a truly exceptional race."[59]

The theory gets even more interesting. Basing his exposition on *The Descent of Man*, Riva Palacio begins by pointing out that Mexican Indians have no body hair and almost no facial hair, which modern naturalists (e.g., Darwin) view as annoying and harmful, thus demonstrating "progress in the constitution of the native race." A corollary is the perfection of the teeth—usually very good among Indians—whose evolution is shown in two facts: "the substitution of the eye or canine tooth by a molar and the lack of the last molar, known as the wisdom tooth." Riva Palacio continues in this vein with abundant quotes and examples from Darwin, concluding:

One can even go so far as to state, although without being able to prove it definitely, that the American races are autochthonous and in a state of advancement superior to that of other races, because if by progress we mean the accumulation of characteristics useful to the organism and necessary to sustain it in the struggle for existence, and the more or less complete disappearance of useless and harmful traits possessed by previous generations, then it is undoubtable that the Indians are in a more advanced stage of evolution. They conserve in a rudimentary state the same organs as do all other races, such as nipples on males, but they have lost the beard and body hair, the wisdom tooth, and have acquired a new molar in substitution for the canine which the most advanced European races still have in the vestigial state.[60]

[59] Riva Palacio, *México a través de los siglos*, II, 472–476.
[60] Ibid., p. 476.

Some persons objected to Riva Palacio's theory. Justo Sierra reviewed the work, and his cautious comment betrays the respect he felt for the author: "We must say, however, that there is a certain flavor of incomplete understanding in several chapters in the book containing ethnic and anthropological digressions and some hastiness in application."[61] We well know that Sierra was himself a distinguished Darwinist. Dr. Nicolás León reported that he had answered a questionnaire sent him by Riva Palacio, and asserted his priority in the study of Indian dental anomalies. In any case he did not agree that the canine had been replaced by a molar. He did not mention Darwin.[62] In 1896 an anonymous article also refuting the thesis appeared in the newspaper edited by Ramón Prida: "Some persons are preoccupied with the modern doctrines of the struggle for life and natural selection and, applying these to sociological theories, indicate that the Indian is inferior; defenseless in the tremendous struggle for survival which has the world as its stage and nations as actors, and they affirm that the Indian race is destined to disappear in the near future." The author mentions Riva Palacio's thesis and denies the absence of the canine tooth. He cites Spencer to the effect that Indians' teeth differ only in being very white.[63] It should be noted that none of the opponents of Riva Palacio's theory declared himself anti-Darwinian. We believe this is a good example of controversy in science, with application to research in Mexico.

Another evolutionary work published in Mexico that merits mention is one signed by the Sociedad de Estudios Clínicos de la Habana: "Lugar que ocupa la bacteriología en la categoría de las ciencias," published in the journal of the Mexican Academy of Medicine in 1890.[64]

The most active and well-known Darwinist in the last twenty years of the century was, without doubt, Alfonso L. Herrera (1868?–1924). A restless individual, with solid biological training, he came to exert de-

---

[61] Justo Sierra, "México a través de los siglos," *Revista Nacional de Ciencias y Letras* 2 (1889): 113.

[62] Nicolás León, "Anomalías y mutilaciones étnicas del sistema dentario entre los tarascos pre-colombianos," *Anales del Museo Michoacano* 3 (1890): 168–173. On Dr. Nicolás León, see Germán Somolinos d'Ardois, *Historia y medicina: Figuras y hechos de la historiografía médica mexicana* (Mexico City: Universidad Nacional Autónoma de México, 1957).

[63] *El Universal*, October 14, 1896.

[64] *El Estudio* 3 (1890): 353–357.

cisive influence on several generations of youth.[65] Herrera was a convinced Darwinian, a researcher and teacher in the natural sciences, in whose career can be demonstrated the Darwinian impact in the new scientific orientation. There is not room here to analyze all of Herrera's evolutionary research but some examples will be provided. In an essay of 1891, "Nota relativa a las causas que producen atrofia de los pelos: Refutación de un argumento de M. de Quatrefages," Herrera used Darwin and Wallace to combat the scarcely Darwinian ideas of the French author.[66] The following year he published another study, "Medios de defensa en los animales," in which he made contributions to evolutionary theory and showed a great capacity for systematization.[67] In his 1895 study on the fauna of Lake Texcoco, he made clear the necessity for studying sexual dimorphism, sexual selection, and sexual atavism in Mexican fauna.[68] The discussion was based on Darwin. In another article,"Filosofía comparada: El animal y el salvaje," he discussed theoretical considerations concerning the origin of man and called attention to the real closeness between primitive man and animals.[69] This type of citation could be greatly extended; further detail may be gleaned from the work of Enrique Beltrán.[70] Worth mentioning here, however, is a work that Herrera coauthored with Dr. Daniel Vergara Lope, which won a silver medal from the Smithsonian Institution and which abounds in quotations from Darwin and evolutionary theses.[71]

Herrera's most theoretical work and that which best demonstrates his

[65] Enrique Beltrán, "Alfonso L. Herrera (1868–1968): Primera figura de la biología mexicana," *Revista de la Sociedad Mexicana de Historia Natural* 29 (December 1968): 37–92. In this article Beltrán presents a balanced evaluation, motivated by his esteem for Herrera.

[66] *Anales del Museo Nacional de México*, 1st ser. 4 (1887). Signed in March, 1891.

[67] *Memorias de la Sociedad Científica Antonio Alzate* 6 (1892–1893): 251.

[68] Alfonso L. Herrera, "Fauna del lago de Texcoco: Notas acerca de zoología del lago de Texcoco y sus alrededores," in *Estudios referentes a la desecación del lago de Texcoco*, ed. Fernando Altamirano (Mexico City: 1895), pp. 41–62.

[69] *Memorias de la Sociedad Científica Antonio Alzate* 9 (1895–1896): 77–96. This is not Herrera's only anthropological work; see also the two works cited by Genovés (note 3 above).

[70] Beltrán, "Alfonso L. Herrera."

[71] Alfonso L. Herrera, *La Vie sur les hauts plateaux: Influence de la pression barométrique sur la constitution et le développement des êtres organisés; traitement climatérique de la tuberculose* (Mexico City: I. Escalante, 1899).

uncommon knowledge of Darwinian ideas is the *Recueil des lois de la biologie générale*.[72] We do not know why he had to publish it in French, and, although it is possible that pressure was brought to bear on him, it is unlikely, inasmuch as he showed himself an open partisan of Darwinism in all his studies. The book is written in the manner of a text or catechism. It begins, not with reflections on the value of Darwinism or the importance of laws in biology, but rather with a direct statement of what he calls great laws: laws of chronology, unity, cellular life, differentiation, variability, adaptation, selection, distribution, struggle for life, and evolution. Each law has secondary laws, which he enumerates. Each is enunciated as if there were no doubt in its affirmation. We believe that this work of Herrera is the most important of a long list of Darwinian writings in Mexico, and it represents something like a synthesis of the evolutionist movement in the country. Although Herrera continued to work in the first decades of this century with somewhat tangential notions, his influence on the development of biological research and education is indisputable.

Thus Darwinism produced changes in scientific orientation and in general thought. But two areas seem to have been pre-eminently affected: sociology and anthropology, often interrelated. The indigenous problem motivated interest among scholars who adopted clearly Darwinian positions. The notions of adaptation, natural selection, and the struggle for existence were translated into the social and anthropological domain by many thinkers of the epoch, whether with regard to the whole social organism, or only to the indigenous substratum of the population.[73] That Darwinian anthropology had great importance has been shown in several cases (Riva Palacio, Sierra); here can be added another.

At the Congress of Americanists that took place in Mexico City in 1895, the representative of Instituto Médico Nacional, José Ramírez, insisted on the thesis that American Indians were autochthonous. He did not cite Riva Palacio's arguments but introduced others. For Ramírez, three facts were basic: "(1) Plant life in America is as perfectly developed

72 Dedicated to his teacher Dugès; see note 9 above.

73 See González Navarro, *Sociología e historia en México*; without making special mention of Darwinism, he shows with great clarity the Darwinist-Spencerian thought of many authors.

as in the Old World; (2) The same is true of animal life; (3) No traces of the cultivated plants or domesticated animals of the Ancient World have been found." The Darwinian core is found in the second point:

Paleontologists have shown us that groups which in Europe or Asia are still undergoing natural evolution have already disappeared from America, leaving behind their fossil remains like a page of their most ancient history; such was the case with the horse, the bull, and the elephant. The group of quadrumanes, precursors of man, is represented by multiple forms which demonstrate that the environment has been favorable to their variation. Finally, we reach man and at the moment of the discovery of America, what did the bold adventurers of the conquest find? Multiple races whose ethnic or sociological characteristics established profound differences which were perceptible even to the first Spaniards who encountered them.

He concluded: "Indeed, gentlemen, can it be admitted given these basic facts, that the Animal Kingdom ceased evolving with the quadrumanic group? That is to say, that man could not have evolved spontaneously in America? For my part, I do not believe that, up to this time, there has been established a perfect phylogeny of any American race which traces its origin to the Old World. Philological, architectural, and sociological analogies are secondary and even of no value, compared with anatomical or ethnological ones and I repeat that these last-mentioned have not yet been established." In a manner distinct from that of Riva Palacio, but with the same anthropological preoccupations (similar to those of Florentino Ameghino in Argentina), Ramírez believed that evolution in America must have followed the same lines as in the Old World and thus produced man as the last step in the evolutionary process.[74]

Darwinism had still not received total recognition by the end of the century. In 1895 there was yet another serious attack by Agustín Aragón, a disciple of Gabino Barreda, in an article purporting to be a positivist critique of the survival of the fittest.[75] With abundant references to Darwin, Wallace, Haeckel, and Spencer, Aragón heralded the downfall of

---

[74] José Ramírez, "Las leyes biológicas permiten asegurar que las razas primitivas de América son autóctonas," *Congreso Internacional de Americanistas: Actas de la Undécima Reunión* (Mexico City: F. Díaz de León, 1895), pp. 360–363.

[75] Agustín Aragón, "Apreciación positiva de la lucha por la existencia," *Memorias de la Sociedad Científica Antonio Alzate* 9 (1895–1896): 145–161.

Darwinism for not having fulfilled the conditions of positivist science. His intent was to expose "the absurdity of the bases given up to now to the law of the struggle for life and the consequent necessity that such a struggle, given that it is a phenomenon rigorously proved by observation and experiment, rest on incontestable bases in order that it might be elevated to the status of a scientific truth." Further along he adds: "How can we accept the consequences which are deduced from the so-called struggle for existence, when it is not even demonstrated that such a struggle exists, whether it be considered as a true struggle or as the result of competition among plants and animals. In no way will we accept those results so long as the transformists do not begin with truthful postulates." In Aragón we still have an example of recalcitrant positivism, in the style of Barreda.

But scientific research continued independently of such attacks. Many other personalities made evolutionary ideas their own and worked along the course charted by them. To close this exposition, there is the case of Dr. Jesús Sánchez, who, at the end of the century, showed clearly the same anthropological preoccupations discussed above. In 1898–1899 he published two articles on the relationship between anthropology and medicine, in which he claimed that the Indians of Mexico were slowly being extinguished because they were less well prepared than Europeans for the struggle for life. In the second article he summarized Riva Palacio's thesis concerning the Indians and declared that the supposed substitution of the canine tooth was no more than the effect of attrition caused by a special diet. For his part, he observed simian characteristics in the native races. He added, as well, a nice study of hermaphroditism, citing Darwin.[76]

Sánchez sustained a brief polemic with Porfirio Parra (of the positivist controversy) about definitions of physiology and biology, in which Sánchez showed his broad biological knowledge by reducing the terms to their strictest scientific application, while the Spencerian Parra attempted to make the matter a philosophical topic.[77] Here we again note that some

---

[76] Jesús Sánchez, "Relaciones de la antropología y la medicina," *Gaceta Médica de México* 35, no. 10 (May 1898): 193–206; 36, no. 6 (March 1899): 112–122.

[77] Porfirio Parra, "Biología y Fisiología," *Gaceta Médica de México* 36, no. 18 (September 1899); Jesús Sánchez, "Fisiología y Biología," ibid. 36, no. 24 (December

positivists continued offering resistance, ever weakening, to Darwinism.

In the first few years of the twentieth century Mexican Darwinism still counted Herrera as its principal proponent. The various university institutes and the Office of Biological Studies (with Herrera at the head) began to systematize biological research along the lines marked out in the previous century.

CONCLUSIONS

1. Darwinism was introduced into Mexico with a slight delay. The clearest early manifestations of the evolution controversy that we have thus far found occurred in the decade of the 1870's.

2. Probably the reason for this delay was that Darwinism was known more from *The Descent of Man* (1871) than from the *Origin of Species* (1859), given that the impact of the former was greater in Catholic countries, as well as in those with scant scientific tradition.

3. The Darwinian revolution was not limited to the scientific camp. The liberal triumph in Mexico permitted the extension of evolutionary ideas to many varied areas.

4. The indigenous problem in Mexico and the general situation of the country caused a situation in which applications of Darwinian models to other social disciplines were not infrequent in the political writings of the epoch. The Sierra brothers and the persons associated with them— all Spencerians—are an example.

5. Anthropology is the science that shows the greatest traces of the impact of Darwin.

6. There is evidence that Darwinism entered official education toward the end of the 1870's and during the 1880's (Sierra and Dugès).

7. As in other countries, opposition originated from two sources: Catholicism and Comptian positivism. Catholic opposition, though by no means weak, was diluted by other burning issues of the time also originating in the liberal victory, a circumstance that lessened its negative impact on Darwinism.

8. Positivist opposition, headed by Barreda, represented without any

---

1899): 618–624. See also the report of the session of June 13, 1900, ibid. 37, no. 14 (1900): 267–268.

doubt a greater threat, inasmuch as for many decades the majority of Mexican educators were positivists.

9. In the sciences, too, there existed among certain people a form of opposition manifested by lukewarm refutations or by simply regarding the problem as nonexistent.

10. In spite of the lack of a structured scientific tradition in the research nuclei of the day, there were no small number of Darwinian scientists who were able to carry out positive work.

THE ISLAMIC WORLD

NAJM A. BEZIRGAN

The title, "The Islamic World," is, in a sense, misleading: I restrict my-self to the part of this world that is referred to as the Arab countries. My reason for this is that the significance and proportions of the controversy on the theory of evolution are greater in the Arab Islamic world than in other parts. The title is also appropriate because the controversy took place within a purely Islamic context, although the men who first ush-ered Darwinism into the Islamic world were Christian secular thinkers.

Although the European intellectual penetration of the Islamic coun-tries began in the early eighteenth century, the real impact of this pene-tration was not felt until the second half of the nineteenth century. Even then, the reception of Western ideas was anything but consistent. For reasons that are well known now, political ideas were, by and large, re-ceived with some measure of enthusiasm, while the scientific ideas that had the slightest implication of contradicting the Islamic religious dogma met considerable opposition.

In 1876, Ya'qub Sarruf, one of the pioneers of scientific thinking in the Arab world, published an article in the well-known and influential periodical *al-Muqtataf*[1] on the subject of the "rotation of the earth." In it he stated that "the rotation of the earth on its axis and its rotation around the sun have become well known and evident to any healthy

[1] *Al-Muqtataf* [Selection] was first established in Beirut in June, 1876, by Fuad Sarreuf and Faris Nimr, then moved in 1885 to Egypt, where it continued to be pub-lished until 1952. *Al-Muqtataf* is considered instrumental in introducing Western ideas into the Arab world.

mind that reads and thinks."[2] There was a clamor against the article, and it was by no means confined to the Muslim clergy. The archbishop of Antioch took a leading part in decrying the theory and refuting it.

Such was the intellectual milieu in the Islamic world when a young Christian physician by the name of Shibli Shumayyil (1850–1917) first introduced Darwinism in a more or less systematic form. Shumayyil had studied medicine at the Syrian Protestant College in Beirut and later on in Paris before settling in Egypt, where he started the controversy and remained at the very center of it until his death.

It seems that Shumayyil was already familiar with Darwin while still a student in the medical school. His thesis is entitled "The Influence of Nature, Environment and Climate on Man and Animal," but it is not clear from his writing how he became acquainted with the theory of evolution. We only know that Darwinism was not entirely unknown in his college, since in 1882 an instructor in natural sciences had referred to the theory of evolution with some approval in his commencement address. The reference was received with alarm by his associates and the administration of the college, and the instructor had to submit his resignation in the face of opposition from the board of trustees. The president of the board expressed the opinion "that neither the Board of Management, the Faculty, nor the Board of Trustees would be willing to have anything that favours what is called 'Darwinism' talked of or taught in the college."[3]

Shumayyil himself referred to this incident and explained that the reaction on the part of the college administration was in defense of Christianity.[4] It is, however, quite possible that the college administrators were not only defending Christianity but also trying to avoid giving the impression that the college tolerated doctrines frowned upon by the Muslims.

Shumayyil was quite aware of his position as a pioneer in introducing the theory of evolution to the Arab world. In the introduction to his

[2] Quoted in *al-Tali'a* [Vanguard], July, 1969, p. 131. Translations are mine unless otherwise identified.

[3] Albert Hourani, *Arabic Thought in the Liberal Age* (Oxford: Oxford University Press, 1967), p. 248.

[4] Stephen B. L. Penrose, *That They May Have Life* (New York: Trustees of the American University of Beirut, 1941), pp. 33–34.

book, "The philosophy of evolution and progress," he says: ". . . and when I began spreading the principles of this doctrine of evolution, it had no followers, and there was nothing written on it in the Arabic language. In fact, even its supporters in the West could be counted on the fingers."[5] The book was essentially a translation of Ludwig Büchner's commentary on Darwin, which had first appeared in the 1880's as a series of articles in the influential Cairo journal *al-Muqtataf*. These articles were published in book form with a long introduction in Cairo in 1910.

Although Shumayyil was a scientist by training, his interest in Darwinism was basically social and philosophical. His understanding and adoption of the evolution theory remained in the context of the materialist philosophy he advocated throughout his life. "My purpose in this treatise," he writes, "is to establish the materialist philosophy on solid scientific grounds in order to eliminate the misconceptions many people had in mind in those days that it is a base philosophy."[6] In the introduction he talks at length about social and political justice and launches a violent attack on despotism and monarchy. He introduces Büchner to his Arabic readers as "the foremost exponent of Darwinian philosophy." Yet later on he criticizes Büchner scathingly, saying he "could not escape from the effect of imaginary dreams, often forgot his materialism and made certain meaningless poetic statements."[7]

Büchner, however, was not among the European philosophers who dominated the intellectual scene in the Arab world in the second half of the nineteenth century. These included the Enlightenment philosophers, particularly Montesquieu, Rousseau, and Voltaire, and such nineteenth-century French thinkers as Auguste Comte, Ernest Renan, and Gustave Le Bon and, among English philosophers, John Stuart Mill, Herbert Spencer, Darwin, and T. H. Huxley. There were also a number of admirers of Nietzsche and Schopenhauer. (These European schools of thought, however, did not have corresponding currents of thought among Muslim and Arab intellectuals.) Often the same intellectuals echoed the ideas of more than one philosopher, no matter how frequently their respective ideas were contradictory.

[5] Shibli Shumayyil, *Falsafat al-Nushu wa al-Irtiqa'* (Cairo: al-Muqtataf Press, 1910), p. 25.
[6] Ibid., p. D.
[7] Ibid., p. 5.

Shumayyil was unique among his contemporaries, for he was very careful not to confuse Darwin with Schopenhauer or Nietzsche, as many Arab thinkers did. These two philosophers are described by Shumayyil as "dreaming philosophers." "Though their advocacy of evolution is based on Darwinism," he says, "it is incomplete as far as the details go," and, he adds, "it is bad dreams, not scientific . . . and would only lead to the degeneration of the society."[8]

Shumayyil's interest in evolution went beyond Darwinism: he was more concerned with the philosophical question of the origin of the universe. He explains this in the introduction of his book and argues that he did not entitle his book "Darwinian theory," but rather "The philosophy of evolution and progress." For, he says, "I did not confine myself to a mere account of the evolution of living organisms, but meant the theory as applied to nature in its entirety, inorganic, plant and animal, and the relationship of one to the other, by showing that the whole is interrelated in all its forms and action."[9]

There is no evidence to show that Shumayyil ever read Friedrich Engels, but his views on the laws of nature are very similar to those of Engels, particularly when he attempts to formulate a dialectic of nature, with its three laws—the law of transformation of quantity into quality, the law of the negation of the negation, and the law of interpretation of the opposites.[10]

Throughout the introduction, Shumayyil reiterates the relativity of knowledge and the primacy and eternity of matter. Form, energy, and motion are characteristics of nature that may manifest themselves but are potentially there in the essence of matter. This belief in the materiality of the world could lead to only one conclusion, namely, the denial of the principle of creation from nothing, which had been held and defended by all Muslim philosophers.

Shumayyil also advocated an extreme reductionism. "The natural sciences," he wrote, "are the source of all sciences and they must be so and they must precede all other sciences."[11] Since "the future is for science," traditional disciplines came under heavy attack from him. Philosophy,

8 Ibid., p. 158.
9 Ibid., p. 357.
10 Ibid., p. 1.
11 Ibid., pp. 36, 41.

theology, linguistic science, and law are worthless. Naturally, literature topped his list, but paradoxically the attack on it was couched in poetry. Progress, he held, is feasible only through natural science, and, when he was asked to formulate a program for this purpose, he responded by suggesting the abolition of law school and the tearing up of economic and philosophical books. Instead, he suggested establishing schools for natural sciences and mathematics.[12]

Shumayyil was an ardent advocate of social and political change and reform throughout the Islamic world. In his attack on supernaturalism and idealism he was of course echoing Büchner and the German materialists. Yet this materialist reformer argued with a great deal of enthusiasm against the budding feminist movement in Egypt. He was adamant in his belief in the natural inferiority of women, and to prove this he maintained that man's brain is heavier than that of woman and that "the more advanced humanity was, the more inferior a position women occupied," and that they became "equal or superior to man only in primitive societies."[13] The superiority of man, he concluded, is a requirement for progress, and vice versa.[14]

The reaction to Shumayyil's materialistic outlook and his advocacy of the theory of evolution was in many ways similar to the opposition that confronted Darwinism from churchmen in the West. The initial outbursts on the part of the clergy, both Muslim and Christian, were characterized by attempts to vilify the theory and its author. The main opponent, however, was not a member of the clergy: he was, rather, an enlightened Muslim thinker who himself was a pioneer in the modernist movement in the nineteenth century.

This was Sayyid Jamal al-Din al-Afghani (1839–1897), the most important political thinker of his era in the entire Muslim world. But, before discussing his anti-Darwinist polemics, two late nineteenth-century Christian Arab anti-Darwinists should be mentioned briefly, both of whom attempted to refute the theory on what they believed to be scientific grounds.

The theory of evolution, up to the end of the nineteenth century, was

---

[12] Ibid., p. 3.
[13] *Al-Tali'a*, July, 1969, p. 140.
[14] Quoted in *al-Tali'a*, July, 1969, p. 141.

known in Arabic only through fragmentary translations of Darwin, Spencer, Büchner, and others.[15] Both opponents, however, showed not insignificant knowledge of Darwin. The first, Father Jirgis Faraj, published in 1880 a book in which he tried to refute the principle of natural selection. The argument in the book is in the form of a dialogue between two characters: a "humanist" and an "apist." Toward the end of the dialogue, the humanist succeeds in wrenching an admission from the apist to the effect that the principle of natural selection is based on mere chance and probability. Having done this, the humanist triumphantly declares that chance enters the picture only when something has to take different forms, but never in matters determined by necessity, such as the essence of man or the law of gravitation.[16] The implication is that, if natural selection does in fact take place, then man is immune to it.

The second refutation, published in Lebanon in 1886, bears the title "The absolute certainty in the refutation of Darwin's falsity"; it was written by Ibrahim Hourani, a Christian by faith and a philologist by profession. Here again the theory of evolution is rejected on the ground that it is based on chance and guesswork. Hourani's "scientific" refutation does not merit detailed treatment here, for it is an echo of the theological objection that was so common in the West. The interesting point is that Hourani divides the evolutionists into three groups, or rather "sects": the theists (*Ilahiyya*), the skeptics (*al-La'adriyya*), and the materialists (*al-Mu'atilla*), using terms that were common in medieval Islamic philosophy and theology. Among the materialists he mentions Büchner, Ernst Haeckel, and Johann Gottlieb Fichte, while describing Spencer, Huxley, and John Tyndall as skeptics. The theists are further divided by Hourani into two groups: the first, to which Darwin belonged, believed that man is a descendant of the ape, and the second, people like A. R. Wallace, believed that God created man as man.

The two main objections Hourani raises against the theory of evolution are, first, that the missing link has not been proved and might not

15 The first five chapters of Darwin's *Origin of Species* were translated by Ismail Madhar and published in 1918 in Cairo. Four more chapters were added to a second edition in 1928. A complete translation of the *Origin* appeared in Cairo only in 1964.

16 Quoted by al-Aqqad, *Al-Insan Fi al-Quran* [Man in the Koran] (Cairo: al-Hilal Press, 1962), p. 114.

exist at all, and, second, that Darwinian theory would lead us to think that evolution took place on earth while it was still a burning mass, which Hourani thinks is impossible.[17]

Turning to Büchner, Hourani argues that creation *ex nihilo* is an unscientific principle, for there is no evidence to show that atoms ever existed. If they were in fact the causes of the universe, then everything would be eternal. Furthermore, he adds, how can atoms come together to form things, if they have not been created?[18] (Incidentally, these were the favorite arguments against atomism in medieval Islamic theology.)

Both of these men lacked significant religious or intellectual standing, and it seems that neither succeeded in tipping the scale for or against Darwinism. The reason that neither of them succeeded in creating an anti-Darwinist movement of any significance was partly that they were not noted religious or political leaders, but mostly, I think, that they were not Muslims. The task was left to al-Afghani.

Al-Afghani's onslaught on Darwinism was couched in an extremely violent and derisive rhetoric. But it was very effective in creating a strong opposition to the theory of evolution all over the Islamic world. The book that contains his polemics against Darwinism is called "The truth about the Neicheri sect and an explanation of Neicheris." Originally published in Persian in 1881 and translated into Arabic as *Al-Radd 'ala al-Dahriyyin* [Refutation of the materialists], it became known under its Arabic name.

The drift of the arguments in this book is rather religious, emphasizing that the essence of Islam is belief in a transcendent God, in a universe created by him. But the religio-philosophical refutation of Darwinism and defense of Islam as religion were not the main motives behind writing the book. As Nikki Keddie points out, it was not "for all its expedient character so much a defence of religion as it has usually been taken to be."[19] Al-Afghani's book was written for the purpose of political agitation for pan-Islamic solidarity against the West, forging an

---

[17] Quoted in *Al-Fikr al-Arabi Fi Mi'at 'Aam* [Arabic thought in a hundred years] (Beirut: American University of Beirut, 1967), p. 385.

[18] Ibid., p. 359.

[19] Nikki Keddie, *An Islamic Response to Imperialism* (California: University of California Press, 1968), p. 23, n. 33.

ideology that would serve to unite all Muslim peoples. The cornerstone of this ideology was that Islamic religion was not in any way contradictory to reason.

The political character of al-Afghani's attack on Darwin and Darwinism is borne out by the fact that the theory of evolution in his time hardly posed an immediate danger to Islamic beliefs. After all, Shumayyil was not a Muslim, and his critique of religion was not directed particularly against Islam as such, but against what had become of Islam, a view al-Afghani would not dispute. What prompted al-Afghani to write the "Refutation" was the political stance the advocates of materialism had taken. They were, he thought, champions of close cooperation with the West, particularly with Britain. Sayyid Ahmad Khan in India (1817–1898), the Tanzimat people in Turkey, and the Shumayyil school in Egypt were all for complete Westernization at a time when anti-Western sentiment was gaining increasing support. Sayyid Ahmad Khan, who coined the word "Neicheri" from the English "nature" after a visit to England in 1870, preached a doctrine similar to that of Shumayyil, albeit less elaborate. He taught that all Islamic principles and beliefs, including the Koran, must be interpreted in accordance with reason and nature.[20]

Instead of arguing against Ahmad Khan's views, al-Afghani launched a bitter attack against him, in an attempt to ruin his political reputation. He was abandoning Islam, al-Afghani wrote, in order to ingratiate himself with the English: he was a hoax, a British agent who was "sowing disunion among the Muslims and scattering their unity."[21]

Behind all this vituperation, there is an element of truth so far as the attitude of secularists toward the West is concerned. Shumayyil's attitude in this regard is rather curious for a man of his stature. He held that "Egypt under British occupation developed its irrigation system and agriculture; the life of the peasants became richer . . . and its finances better organized . . . to the extent it gained the confidence of the world and freedom prevailed."[22] These may all be true, except for the prevailing of freedom under the Earl of Cromer's occupation.

Bearing this background in mind, we now turn to al-Afghani's dis-

20 Hourani, *Arabic Thought*, p. 124.
21 *Al-Urwa al-Wathqu* [The strongest link], Cairo, August 28, 1884.
22 Quoted in *al-Tali'a*, July, 1969, p. 143.

cussion of Darwinism. To begin with, all his arguments against evolution are traditional and conventional, based on the then widespread confusions about and misunderstandings of Darwinism. It is evident from the section on the theory of evolution in his "Refutation" that al-Afghani had no firsthand knowledge of it, and there is no indication that he had even read Darwin's *Origin of Species*. The "Refutation" is an anti-materialist manifesto, extending in its range from the Greek materialism to Darwinism. The discussion takes the form of questions al-Afghani poses for Darwin and then volunteers to answer for him.

The theory of evolution is summarized as follows: "One group of materialists decided that the germs of all species, especially animals, are identical, that there is no difference between them and that the species also have no essential distinction. Therefore, they said, those germs transferred from one species to another and changed from one form to another through the demands of time and place, according to need and moved by external forces."[23]

The Origin of Species is described as ". . . a book stating that man descends from the ape and that in the course of successive centuries as a result of external impulses, he changed until he reached the stage of the orangutan. From that form he rose to the earliest human degree, which was the race of cannibals and other Negroes. Then some men rose and reached a position of a higher plane than that of the Negroes, the plane of Caucasian man."[24]

The discussion then continues by raising a number of rather imaginary and simplistic questions, which, al-Afghani says, would have left Darwin flabbergasted. Darwin supposedly would have bitten his tongue had he been asked about the variety of structure, length, leaves, and flowers of plants that have been in the forests of India from the olden times. Or, he "could not have raised his head from the sea of perplexity, had he been asked to explain the variation among the animals of different forms that live in one zone and whose existence in other zones would be difficult."[25]

The "ape business" is harped on incessantly: "Only the imperfect resemblance between man and monkey has cast this unfortunate man into

[23] Quoted by Keddie, *Islamic Response*, p. 135 (Keddie's translation).
[24] Ibid., p. 135.
[25] Ibid., p. 136.

the desert of fantasies, and in order to control his heart, he has clung to a few vain fancies." And finally this: "He reports that one society used to cut off the tails of their dogs, and after that had persevered for several centuries their dogs began to be born by nature without tails. He apparently is saying that since there is no longer need for a tail, nature refrained from giving it. Is this wretch deaf to the fact that the Arabs and Jews for several thousand years have practiced circumcision, and despite this and until now not one of them has been born circumcised?"[26]

How can one explain these absurd utterances about Darwinism from a man who was a champion of reason and enlightenment? One of his close friends and disciples, Muhammad Abdu, said that al-Afghani wrote the "Refutation" while he was in a state of passionate anger against the advocates of complete Westernization.[27] This seems the only possible explanation for the deliberate attempt to caricature Darwin and Darwinism. For only a few years later al-Afghani returned to the subject to discuss it less passionately and more systematically. In *Khatirat Jamal al-Din al-Afghani* [The ideas of al-Afghani], which was published in the thirties by Muhammad al-Makhzumi and attributed to al-Afghani, we find a chapter devoted to the "struggle for existence." Here "force" is taken as the greatest law of nature, applicable to the inanimate, the animal, and the human world. Natural selection is a principle he admits, not only in plants and animals, but also in the realm of thought. "Ideas constantly regenerate and new characters and new categories emerge from the old ones,"[28] not because of nature, but as a result of acquisition.

He admits "that the principle of natural selection was known and made use of long before Darwin in the pre-Islamic and Islamic culture either in selecting wines or in eugenics."[29]

He is now prepared to accept the theory of evolution with some reservations, arguing that Muslim thinkers preceded Darwin in advocating it.[30] He quotes poets and alchemists to prove this point of view. Al-

[26] Ibid.
[27] Quoted by Muhammad Amara, *The Collected Works of al-Afghani* (in Arabic) (Cairo: al-Katib al-Arabi Publishing House, 1969), p. 107.
[28] Ibid., p. 252.
[29] Ibid., p. 250.
[30] Ibid., p. 251.

though he is prepared to accept the evolution of inorganic into plant matter, and plant into animal, he retains his original rejection of the evolution of man from ape. The disagreement centers on the question of the "soul" or "breath" of life, which is created by God.[31]

Al-Afghani's general attitude to Darwin at this stage is rather sympathetic. He draws a distinction between Darwin and his followers, particularly the materialists, Büchner and his translator Shumayyil. Darwin in *Khatirat* is often referred to as "the sage" instead of "the wretch" or "this individual." Al-Afghani claims that, when Darwin reached the essential question, he admitted that all living things on this earth come from one primary species into which the Creator breathed life.[32]

Nevertheless, there is no harm, he argues, in limiting the living things to a few species from which many others branch, provided that the evolutionists prove the possibility of self-creation; otherwise the claim would make no sense.[33]

This transformation in al-Afghani's thought is significant, for, despite his continued criticism of Büchner and Shumayyil, he at last admits that there is no matter without force. What brought about this transformation is not quite clear. One explanation offered by Muhammad Amara is that al-Afghani was becoming more and more aware of the socialistic ideas of his time. For, while the "Refutation" teems with polemics against secularism, nihilism, and communism, the *Khatirat* is characterized by tacit acceptance of socialism. Büchner's concept of force is turned by al-Afghani into a social concept.

Al-Afghani's basic idea rests upon the presumption that it is possible to reach a reconciliation between science and Islamic religion, since the latter is essentially a rational religion. In this he was only elaborating the ideas of many of his predecessors. Rifā'a al-Tahtawi in Egypt (1801–1873), a great admirer of Western civilization, consistently exhorted the Muslims to adopt Western ideas and methods. In Tunisia, Khayr al-Din (d. 1889) also argued for the adoption of Western institutions and practices, in the face of powerful opposition by many of his contemporaries. Similar movements had been initiated in Morocco, Iraq, and Iran.

[31] Ibid., p. 252.
[32] Ibid.
[33] Ibid., p. 443.

None of these men, however, was as well aware as was al-Afghani of the danger of adopting Western ideas and of their impact, not only on religion or religious institutions, but also on the future of Islamic countries vis-à-vis Europe. Advocating Western ideas, while at the same time warning of their dangers, was the characteristic tendency of al-Afghani's thought. But toward the end of his life he seems to have become increasingly aware that the materialist ideas that had been encroaching upon the Muslim consciousness could no longer be dismissed as mere atheism. Thus, the only alternative left for him was to assume that these ideas were in fact part and parcel of Islamic civilization and that there was no conflict between them and Koranic teachings.

The message was well received by his followers and disciples. The task before them now was to give the Koran a new scientific interpretation. The Koran, according to Abdul Rahman al-Kawakibi (1843–1903), for instance, had stated the theory of evolution long before Darwin: the verse "And certainly we create man of an extract of clay" was quoted as evidence. Muhammad Abdu (1849–1905), al-Afghani's close friend and associate, wrote a major commentary on the Koran in which he claimed that the doctrine of struggle for survival is fully expressed in verse ii, 251: "Had God not repelled some of the people by means of others, the earth would have been corrupted."[34]

Outside the domain of religion, Darwinism had very little influence. *The Origin of Species* was still neither available in Arabic in a complete translation nor widely discussed. But the ideas of Comte, Mill, and Spencer were sweeping the educated Islamic world: they were more convenient for the Muslim reformers than the ideas of Büchner and Haeckel.

With the beginning of the present century, there was a revival of interest in Darwinism, particularly after the publication of Shumayyil's articles in book form. A number of books were written on Darwin and the theory of evolution in the first two decades of this century, among them Salama Musa's "The advent of superman" and "The theory of evolution and the origin of life." There was also an attempt by Jurji Zaidan to apply the theory of evolution to the development of the Arabic language.

[34] Quoted by Malcolm Kerr, *Islamic Reform* (Berkeley: University of California Press, 1966), p. 30.

The theological impetus behind the interest in Darwinism toward the end of the nineteenth century was already waning, and with the advent of the twentieth century more recent ideas were beginning to capture the imagination of the Muslim thinkers. The new interest in such philosophers as Nietzsche and Schopenhauer and Spencer overshadowed and cut short interest in Darwin and the theory of evolution.

DARWINISM AND HISTORIOGRAPHY

DAVID L. HULL

As the papers in this volume clearly show, Darwinism was many things to many people. It was rank materialism, an atheistic attack on the Christian faith, unadulterated positivism, a death blow to teleology. Simultaneously it was irresponsible speculation, an outrage against positivistic science, a rebirth of teleology, proof of the beneficent hand of God, a Christian plot to subvert the Muslim faith. It was also an intellectual weapon to use against entrenched aristocracies, a justification for laissez-faire economic policies, an excuse for the powerful to subjugate the weak, and a foundation for Marxian economic theory. Only incidentally, it seems, was it a scientific theory about the evolution of species by chance variation and natural selection.

Historians of science traditionally have written internal histories, emphasizing the scientific achievements of great scientists, their theories, the data available, rival theories, wrong turns, happy accidents, and similar factors. These internal histories have also tended to be rather elitist, concentrating on those scientists who were right or else wrong in big ways, exhibiting what Michele Aldrich calls the "good guys and the bad guys" syndrome. Rarely do we hear anything of the majority of scientists in these histories, the scientists who failed to make any important discoveries, who jumped on the wrong bandwagon, who did not push their priorities—in short, the vast majority of scientists. As M. J. S. Hodge argues, a thorough understanding of science and scientific change requires not only an investigation into the socioeconomic background of

NOTE. I wish to thank Michele Aldrich, Jon Hodge, Phillip Sloan, and Ernst Mayr for suggesting improvements in this manuscript; it was prepared in part under NSF grant GS 3102.

the period but also a careful conceptual analysis of the ideas at issue and a detailed discussion of the personalities of the men and women involved. Only a knowledge of all three provides adequate understanding of the course of science.

Even though the authors included in this volume realize that much work remains to be done, so much work that one is nearly immobilized at its prospect, the Darwinian revolution has been investigated by scientists, historians, and philosophers more thoroughly than almost any other chapter in the history of science. In addition, more references than usual are made in this literature to the intellectual setting of the period, the social ramifications of evolutionary theory and the conglomerate cluster of ideas known as Darwinism, the uses and abuses of the theory, and so on, perhaps because these issues figured more prominently in the controversy surrounding the *Origin of Species* than they did in similar episodes in science. As Ilse Bulhof remarks, it is hard to imagine a similar response to the theory of relativity.[1] By and large, the intent of the papers in this anthology is to balance internal and external factors in the story of the impact of "Darwinism" around the world, eliminating elitist biases as much as possible. But it should be kept in mind that the internal-external distinction itself assumes the existence of both sets of factors. There must be some internal history for external history to be external to.

TRANSLATION AND TRANSMISSION

Many of the interrelations pointed out by the authors in this volume are fascinating, even in such comparatively pedestrian matters as translation. Darwin would have preferred that his works be translated by noted scientists. In Germany he got his wish when H. G. Bronn, a second choice to Albert von Kölliker, agreed to translate the *Origin*; in France he did not. In fact, no important French scientist could be found to translate the *Origin*, or any of Darwin's later works for that matter. Instead Darwin had to settle initially for Clémence Royer. In both Bronn and Royer, Darwin got more than he bargained for. Bronn not only

---

[1] I doubt however that the difference between evolutionary theory and relativity theory that Bulhof mentions has much to do with Darwin's being in touch and Einstein out of touch with the general dispositions of their respective ages. Rather it stems most probably from the extreme difficulty in even misunderstanding relativity theory. Evolutionary theory is so easy almost anyone can misunderstand it.

omitted passages that he found offensive but also added a highly critical appendix to his translation. Royer turned out to be an enthusiastic evolutionist, but, for her, evolution was progressive. Not only did she preface her translation with an anticlerical diatribe, an indulgence that Darwin carefully denied himself in his public utterances, but she also gave her translation a title that emphasized the progressive nature of evolution, a notion that Darwin questioned increasingly as the years went by. But, even in his earliest publications, Darwin did not emphasize the progressive nature of evolution to the extent that so many of his followers did.

Royer's translation encouraged the predilection for the notion of progress among Darwin's nineteenth-century readers, especially in places like Spain and Mexico, where French was the language of intellectual discourse. As Thomas Glick tells us, Spanish-speaking participants in the evolutionary debate preferred to read Darwin's works in French even after Spanish translations were available. As one might expect, evolution was interpreted by its proponents in Spain and Mexico as being progressive: "Unchangeable laws of nature strive toward incessant perfection in everything" (Glick). But the European intellectual community did not need Royer's translation to encourage them to view evolution progressively. Even without it, evolution still would have been interpreted in the Western world as being progressive, since the notion of progress was pandemic at the time. In the United States, where translations were unnecessary, Asa Gray insisted on arguing for progressive evolution in spite of Darwin's increasing skepticism. And, as Edward Pfeifer has pointed out, Neo-Lamarckian versions of evolution came to predominate in the States. In the Netherlands, "Darwin confirmed the Modernists' optimism and faith in progress, for now it had been proved by science that everything obeys the law of perpetual progress" (Bulhof). Man would continue to develop into a higher and higher being.

Although the first Russian translation of the *Origin* did not appear until 1864, Darwin's *Variation of Animals and Plants under Domestication* could not have fared better. It was translated by a young scientist, Vladimir Kovalevsky, and appeared before the English original. The first volume of the *Descent of Man* appeared in Russian in the same year as the first English edition. The depressing feature of this success story is that it was largely redundant, since nearly all intellectual Russians could read English. As always, the rich get richer. In Mexico and the Islamic world, the poor could not have been poorer. In a survey made in 1959, Manuel Maldonado-Koerdell could find only two references to Darwin's works in Mexican publications between 1870 and 1900. Roberto Moreno's own investigations led him to conclude that the "explosion" of Darwinism did not occur in Mexico until the 1870's. In Islam only fragmentary translations of evolutionary ideas were known until the end of the nineteenth century. Although the Darwinian controversy in the Islamic world reached its peak in Egypt, a complete translation of the *Origin* did not appear in Cairo until 1964!

Of course, there was much more involved in the transmission of evolutionary theory than translations of the *Origin of Species*. In many instances, Darwin was a comparatively minor figure in the movement termed "Darwinism" in this volume. The importance of T. H. Huxley, Ernst Haeckel, and Herbert Spencer in spreading the evolutionary gospel is well known, but less familiar names appear with equal frequency, for example, those of Carl Vogt and Ludwig Büchner. Earlier, Büchner had translated Sir Charles Lyell's *Antiquity of Man*, and Vogt had translated Robert Chambers's *Vestiges of the Natural History of Creation*. In France, Vogt was the best-known foreign expositor of Darwinism. When Shibli Shumayyil introduced the theory of evolution into Islam, it was through a translation of Büchner's *The Philosophy of Evolution and Progress*. Both Vogt and Büchner had a widespread underground circulation in Russia. The list could be extended almost indefinitely, but, if the papers in this anthology do anything, they show how many writers in addition to Darwin figured in the spread of the broad social movement termed Darwinism.

PHILOSOPHY AND RELIGION

Darwin did not formulate evolutionary theory for the express purpose of helping to foster atheistic materialism, but that was one of its more significant consequences. It is no coincidence that almost all the early proponents of Darwinism were atheistic materialists—or their near relatives. A. R. Wallace and Asa Gray were major exceptions. Among Western intellectuals, religious and philosophical considerations far outweighed matters more directly concerned with science in determining their *initial* reaction to evolutionary theory. (Their eventual acceptance or rejection of the theory is another story.) The state of affairs that Ilse Bulhof describes in the Netherlands is fairly typical. Deists throughout the Western world had reached a polite reconciliation with religion. Since God was the author of all natural law, science could not possibly come into conflict with true religion. All that had to be added to evolution was some sort of inner force or teleological principle to make evolution compatible with or even supportive of the Christian vision of man.

Dutch freethinkers (and radicals around the world) were attracted to Darwinism, but for reasons directly opposite to those of the deists. The freethinkers saw Darwinism as a materialistic refutation of religion, and they were for it. Orthodox Protestant and Catholic thinkers agreed with the radicals' interpretation of Darwinism and hence were opposed. The Protestant reaction to evolution in the United States is too well known to repeat here. Judging from my own experience in teaching the products of Christian homes, evolution is still viewed in the Midwest as an atheistic lie. In Roman Catholic countries like Spain and Mexico, the reception of evolutionary theory was also largely negative. St. George Jackson Mivart, an early convert to the Catholic Church and later to evolution, if not natural selection, provides a minor embellishment on the Catholic reaction in England. For a while it seemed as if Mivart might convince the Vatican that a teleological version of evolution could be reconciled with Catholic dogma, but eventually Pius IX rejected Mivart's proposals and put his stamp of approval on one of the most scurrilous anti-Darwinian attacks to appear in the French language. Incidentally, Mivart was excommunicated from the Catholic Church shortly before he died, not for his evolutionary writings, but for a paper entitled "Happiness in Hell."

At first the Eastern Orthodox Church in Russia ignored Darwinism, since it, unlike the Church of Rome, was highly mystical and not at all concerned with such earthly matters as the evolution of organic species, but eventually even it was roused to oppose Darwinism. Strangely enough, in Islam, Darwinism was viewed as a Christian heresy, since the Darwinians tended to be Christians and their opponents Muslims.

In the midst of all this talk of religious objections to evolutionary theory, however, there is some need to emphasize that religious arguments were all but excluded from the public debate between scientists. Perhaps religious convictions motivated certain scientists in their initial choice of sides, but these convictions were rarely voiced openly by scientists as being relevant to their eventual acceptance or rejection of evolutionary theory. This reticence on the part of scientists in the second half of the nineteenth century probably tells us more about the scientific etiquette of the day than about the considerations that were actually operative in their decisions.

A second, equally significant, factor in the reception of evolutionary theory is related to both science and philosophy—the reputed failure of Darwin to adhere to proper scientific method. Regardless of which side one might be on, the issue was "positivism," and, like "Darwinism," it meant many things to many people. The term had its origin in the philosophical writings of Auguste Comte, who saw a progression in the intellectual development of man from the theological stage through the metaphysical stage to the final positivistic or scientific stage. In this final stage of development, phenomena are to be explained in terms of observation, hypotheses, and experimentation. Similar exhortations about the value of scientific method can be found in the writings of Claude Bernard, John Herschel, William Whewell, and John Stuart Mill. On the notion of science expounded in these popular treatises, Darwin consistently came up wanting.

Paradoxically, Darwin was attacked on one side for being too positivistic and on the other for not being positivistic enough. He was too hard-nosed for those who insisted on the ultimate mystery of human nature and too fuzzy-headed for those who wanted science to be nothing but the recording of facts, preferably under experimental conditions. In many cases, the methodological objections to Darwin's work were just so much smoke screen to cover the critic's real objections to the content of his theory. In other cases, especially for experimentalists like Louis Pasteur and Bernard, it was a way of avoiding uncomfortable decisions. Just when Pasteur thought that he had laid to rest the ancient belief in spontaneous generation of extant forms of life, along came a theory that seemed to entail the spontaneous generation of very simple extinct forms somewhere in the foggy past. As Robert Stebbins points out, the very silence of such prestigious members of the scientific community on evolutionary theory was condemnation enough.

As I have argued elsewhere, at least some of the methodological objections to evolutionary theory were well founded.[2] For example, long after the evolution of species was accepted by all knowledgeable scientists, considerable reticence remained concerning Darwin's mechanisms, and for good reason. The evidence for the efficacy of natural selection did not approach that for the existence of evolution. The reader should be reminded that, to the extent that Darwin and his theory had anything to do with the "Darwinism" that was sweeping the Western world, it was evolution and not Darwin's mechanisms that was being accepted.

POLITICS, ECONOMICS, AND RACISM

The authors in this volume have also shown the central role played by economic and political factors in the reception of evolutionary theory. Europe was feeling the full impact of the industrial revolution and was busily laying the groundwork for World War I and the Russian Revolution. The pivotal events in the period as far as Darwinism was concerned seem to have been the unsuccessful revolution in Germany in 1848, the Crimean War of 1854–1856, and the Franco-Prussian War of 1870–1871. The disillusionment caused by the failure of the radical movement in the German National Assembly of 1848 resulted in numerous German radical reformers leaving the country, including Karl Marx, Friedrich Engels, and several scientists who would later become champions of evolutionary theory: e.g., Carl Vogt, Hermann Burmeister, and Fritz Müller.

In 1854, Russia went to war with Turkey, France, England, and Sardinia, ostensibly to retain control of the holy places in Palestine. Alexander Vucinich credits Russia's defeat in 1856 as one of the factors that led to a period of national awakening and a critical reassessment, culminating in the Great Reforms that emancipated the serfs. Throughout Europe, evolutionary theory was well received by liberals and radicals alike as undermining the static political order. If species could evolve, so could political systems. The theory was especially welcomed by Marx, who had settled in London in 1849. Marx had replaced the traditional justification of social reform in terms of natural rights by the action of inevitable laws of nature. Although he and Engels privately ridiculed Darwin's naïve re-

---

[2] David L. Hull, *Darwin and His Critics: The Reception of Darwin's Theory of Evolution by the Scientific Community* (Cambridge: Harvard University Press, 1973).

liance on Malthusian economics, Marx was not above sending Darwin a copy of *Das Kapital* in 1867, asking permission to dedicate the book to him. Darwin, of course, declined, perhaps because he was repulsed by the ideas that he had heard it contained. He himself never read the copy presented to him, because the pages were never cut. More likely he declined because of his general aversion to political causes and public affairs. He probably would have declined having the charter of the League of Nations dedicated to him.

Apparently the Franco-Prussian War jarred those who believed in the continuing improvement of mankind. The brutality of the war and the cruelty of the famine that resulted from the German siege of Paris cast a shadow over the high opinion that European intellectuals, especially the French, had manufactured of themselves.

The United States had just endured a bloody civil war, initiated at least in part over the issue of slavery. Evolutionary theory had a special significance in America. The status of blacks was no academic issue for them. Actually, two distinct, though related, questions were involved—polygenism and the relative superiority of the various races of man. In the first instance, the question was whether man as a biological species was monogenic or polygenic. Whether one was a special creationist or an evolutionist, the question was, did the various races of mankind have a single origin or several? Traditional Christian teaching favored a single origin, but such a belief did not make Christians any less bigoted than adherents of other religious faiths.

By and large, biologists at the time opted for the monogenic origin of all species, including *Homo sapiens*. For special creationists that meant that each species was specially created, man included, and that the various races of man developed from this original creation. For evolutionists it meant that man evolved from a single immediately ancestral species and that the races of mankind were later productions. Only a few reputable scientists, such as Louis Agassiz (a special creationist) and Carl Vogt (an evolutionist) argued for a polygenic origin for mankind. Nor was the idea especially popular among political Darwinists. As James Rogers tells us, the Russian revolutionary intelligentsia saw a close connection between the emancipation of the slaves in America and the freeing of the serfs in Russia. They were not at all pleased when such evolutionists as Vogt and Varthalomew Zaitsev came out in favor of polygenism.

The second issue was the relative superiority of the various races, regardless of their origin. With rare exceptions (and Charles Lyell was one of these exceptions), the general attitude among Europeans of the day, scientists and nonscientists alike, was that Caucasians of European extraction were vastly superior to all other segments of mankind. In the struggle for existence, they would surely win out, except perhaps in extremely hot climates. In countries, like Mexico and the United States, still inhabited by large numbers of aborigines, the issue again was not of purely academic interest. Moral compunctions aside, the conclusion reached by nearly all those concerned was that Europeans would inevitably extinguish the inferior races, perhaps not by slaughtering them wholesale, but by competing more successfully than they for the necessities of life. Even Wallace observed that, inevitably, higher, more "intellectual and moral" races must displace the lower and more degraded races (Hodge). The Dutch naturalist Pieter Harting was even more specific. The Franco-Prussian War was just another example "in which the lower [the Celtic race] succumbs to be replaced by the higher [the Germanic race]" (Pieter Harting, quoted by Bulhof).

For those who might feel some moral qualms about such higher forms of human beings exterminating lower forms, there was always the consolation that the process would result in man's continuing improvement. Instead of challenging the notion of progress, the Franco-Prussian War by a perverse logic only confirmed it. After all, man's brain had already undergone noticeable development since the Middle Ages. Of course, there was always the possibility that the inferior races might prove to be better adapted to the harsh city life that was emerging in the nineteenth century and in time outbreed Anglicans and other higher forms of life. The human race might degenerate as well as progress. Michele Aldrich informs me that, in the United States, at least, both a polygenesist and a degradationist view of man were popular.

UNIVERSITIES AND THE SCIENTIFIC ESTABLISHMENT

The most notable feature of the papers in this volume is the understanding that they bring to the role of universities and the scientific establishment in the reception of evolutionary theory. Today we tend to forget that universities in 1859 were very different from modern centers of higher education and that the nature of these universities varied considerably from country to country. We also tend to forget that science as a separate and powerful discipline with its own sources of financial support is a relatively new phenomenon. Men who occupied themselves with what we now call science have been around for a long time, but they tended either to be of independent means or else to make their living in other professions, frequently becoming clergymen. Nor was science a very significant part of university education prior to the nineteenth century. All that was in the process of changing when Darwin published his *Origin of Species*. Museums were thriving. Parliament had forced reforms on Cambridge and Oxford, reforms that included the establishment of chairs in natural science. Scientists were engaged in conspicuous public projects like laying the trans-Atlantic cable. Science as we know it today was in the midst of being formed. One would think that such developments would have affected the reception of evolutionary theory, and they did.

In Russia the stodgy Academy of Sciences was being replaced by universities as the seat of scientific research. As Vucinich informs us, the Academy of Sciences was made up of aging scholars, many of foreign birth, who had long ceased to do any original research. During the twenty years after the publication of the *Origin*, it did not produce a single Darwinian scientist of consequence. The situation in the newly emerging universities, however, was not all that much better. The three leading evolutionists at the time in Russia were I. I. Mechnikov and two brothers, A. O. and Vladimir Kovalevsky. Both Mechnikov and A. O. Kovalevsky were forced to accept teaching positions in the provinces, not so surprising for young academics, but, even after they had made international reputations for themselves, they remained at provincial universities instead of obtaining positions at more prestigious universities in Moscow or St. Petersburg. Eventually Mechnikov joined the Pasteur Institute in Paris, but A. O. Kovalevsky remained in provincial universities throughout his career. His brother fared worse. Though Vladimir Kovalevsky had greater personal contact with Darwin than any other Russian scientist and had received recognition outside Russia for his work in paleontology, he never obtained a teaching position in Russia or elsewhere.

The situation was somewhat better in Spain—for a while. Several Darwinians obtained university positions in Spanish universities after passage of the Free Education Act of 1868 but lost them in 1875 after the restoration. In the United States, Darwinism became associated with the attempt to wrest control of the universities from the clergy. It helped the cause of evolution, of course, to have a champion like Asa Gray already firmly established at Harvard. Another factor that helped evolutionists in America was the presence of more than one or two respectable universities. Perhaps an evolutionist might be denied a position at Harvard, but there were other alternatives. Milestones in the career of Darwinism in the United States are clear. Soon after the appearance of the *Origin*, William Barton Rogers debated the great Agassiz on evolution—and won! In 1872 Gray was elected president of the American Association for the Advancement of Science. In 1874 J. D. Dana, surpassed only by Agassiz in his opposition to evolutionary theory, announced his conversion. Finally, in 1880 when the editor of *Popular Science Monthly* challenged the editor of the Presbyterian *Observer* to name two working naturalists in the United States who were not evolutionists, the best that the Presbyterian editor could manage was one aged American geologist and a Canadian.

In the Netherlands, the conversion of scientists to evolution and the rise of evolutionists in the academic hierarchy was even more rapid. By the 1860's most professors of natural history in the Netherlands were Darwinists. (Once again I should remind the reader that a scientist could be a Darwinist without accepting natural selection as the major evolutionary mechanism.) Bulhof attributes the success of Darwinism to the domination of Dutch universities by the liberal bourgeoisie, a segment of society that was already well disposed toward ideas of evolutionary progress.

In Germany, evolutionists began their careers in small, remote universities, but, unlike their counterparts in Russia, many succeeded in moving up to more prestigious universities as their reputations grew. French evolutionists climbed the academic ladder too, but more slowly and more cautiously. In most European countries, the gap between the one or two leading universities, usually located in the capital, and the provincial universities was large, but nowhere was it larger than in France. If a man failed to obtain an appointment at the Sorbonne, there was hardly a second choice. An evolutionist could obtain a respectable teaching position in France, but, unlike his fellow laborers in Germany, he had to exercise extreme discretion. In 1888 a chair of "évolution des êtres organisés" was established at the Sorbonne, but the evolution that was taught was surely not Darwin's version. In France, it still is not.

A detailed study of the effect that the particular organization of universities and systems of education in the various countries had on the reception of evolutionary theory would prove to be extremely valuable, not because evolutionary theory is more important than any other scientific theory, but because it is an excellent example of a startling new theory with broad ramifications. It would help to know what sorts of educational structures are most receptive to new ideas. The educational system in France was (and still is) highly authoritarian and centralized, in England highly centralized if not quite so authoritarian, in Germany authoritarian but not so centralized, and in the United States, by comparison, neither authoritarian nor centralized—though it is still too authoritarian and centralized for the tastes of most students. What effects do these and other factors have on the growth of knowledge?

THE CONSPIRACY OF SILENCE

Frederick Burkhardt has done us all a great service by setting out a detailed account of the role that learned societies and journals played in the reception of evolutionary theory in England and Scotland. The amazing thing about the story that Burkhardt tells is that it is repeated in country after country, a story that Huxley aptly termed a "conspiracy of silence." The older and more prestigious a society and its journal was, the more profound and impenetrable was the silence. For example, "between 1859 and 1870 no paper in either the *Proceedings* or the *Philosophical Transactions* of the Royal Society had as its subject a discussion of Darwin's theory of the origin of species" (Burkhardt). The silence was just as pervasive in the French and Russian Academies of Science. Numerous bulletins of various scientific societies barely mentioned Darwin or evolutionary theory, including the bulletins of zoology and geology in France. Similarly, the Moscow Society of Naturalists and the Russian Geographical Society greeted Darwin with total silence. Occasionally a journal would review the *Origin* and then not mention it again for years; e.g., the Dutch journal *Wetenschappelijke Bladen*. Several authors in this anthology have investigated possible correlations between field of inquiry and reaction to evolutionary theory, but no international pattern seems discernible. For example, English botanists were very receptive; German botanists were largely opposed.

Alvar Ellegård's *Darwin and the General Reader* (1958) bears out Burkhardt's finding that evolutionary theory received its fullest coverage in newer, less prestigious scientific journals and in the popular press. In England, evolutionary theory was followed most closely and completely in the unlikely *British and Foreign Medico-Chirurgical Review*, due in no small part to the efforts of one of its earlier editors, W. B. Carpenter. During the first half of the nineteenth century, numerous societies and journals were started to encourage a wider understanding and appreciation of science among the general public, a role not fulfilled by such exclusive organizations as the Royal Society of London and the various national academies of science. In England it was the British Association for the Advancement of Science that took the lead in popularizing science. It also served as a forum for the debate over evolutionary theory. The American Association for the Advancement of Science was unable to do the same in the United States, since it did not meet during the war. America had more pressing battles on its hands. In Russia the Society of Admirers of Natural Science, Anthropology, and Ethnography attempted to popularize science, publishing its proceedings in Russian instead of the traditional French or German. In the Netherlands the *Album der Natuur* was founded "to spread the knowledge of nature among civilized readers of all ranks of society" (Bulhof), but, when it failed to give ample coverage to evolutionary theory, *Isis* was begun to champion the Darwinian cause. Darwinism was discussed extensively in the pages of the *American Journal of Science*, thanks to the efforts of Asa Gray and the cooperation of its editor, J. D. Dana, who was still opposed to evolutionary theory at the time. Francis Bowen, of bee-cell fame, published several attacks on Darwin by Mivart and rebuttals by Chauncey Wright in his *North American Review*, not to mention one intemperate paper by himself. But it was the newly formed *Popular Science Monthly* that proved to be the most enthusiastic organ of Darwinism. Nor were the articles that appeared in the more popular journals especially inferior to those in the more learned journals. As Burkhardt mentions, men like Huxley wrote for both.

Perhaps there was a conspiracy of silence among the more prestigious scientific organs, but the conspiracy failed. All facets of evolution and the wide variety of scientific, philosophical, political, and social issues associated with it were thoroughly aired. Spokesmen of the societies and journals that tended to ignore evolutionary theory justified their actions on the grounds that it was too speculative and controversial. Discussing it would only engender discord among scientists and damage the reputation of science among the general public. The presentation of facts that bore on such speculative ideas was acceptable; open discussion of such ideas was not. It is difficult to decide how sincere these protestations actually were. Certainly the philosophies of science popular in the day emphasized the superiority of facts over hypotheses. One fact was worth a thousand theories. And the silent treatment was not restricted just to new, controversial ideas like evolution. Special creationism, the received doctrine at the time, was ignored just as studiously in such places as the Royal Society. Before the appearance of the *Origin*, most scientists who had opinions on the subject were special creationists. Yet technical discussions of special creationism as a scientific doctrine were just as rare in the leading journals as were discussions of evolutionism.

The difference between special creation and evolution was that special creation was widely accepted among both scientists and the general public; evolution was not. The introduction of controversial topics into scientific meetings *does* engender discord. It is equally true that open disagreement among scientists *does* tend to damage the reputation of science among the general public. The policy that one adopts in the face of these facts depends upon how much one is willing to pay for harmony and public good will, since debate and discord seem to be the inevitable price of scientific change. I doubt that there is a law of nature to the effect that scientists must go for each other's throats when they disagree, but usually that is exactly what happens. The problem is even more pressing today than it was in the nineteenth century, now that science has become so highly institutionalized, research so expensive, and government support and control so pervasive.

One way of judging the official reaction of the scientific community to evolutionary theory is by the honors that it bestowed upon Darwin. In 1864 Darwin was awarded the prestigious Copley Medal of the Royal Society of London; in 1865 the Royal Society of Edinburgh elected him an honorary member; in 1867 the Russian Academy of Sciences and in 1871 the French Society of Anthropology followed suit. The French Academy

of Sciences held out as long as possible, not electing Darwin a corresponding member until 1878, after eight nominations. In nearly every case,
however, some attempt was made to dissociate Darwin from his theory
of evolution, especially the doctrine of natural selection. Darwin was being honored for his solid work on barnacles or earthworms, not his erroneous speculations on the origin of species.

EVIDENCE AND RATIONAL ARGUMENT

Finally, a word must be said in defense of evidence and rational argument. The authors in this volume have set out a strong case for the
relevance of socio-politico-economico-idiosyncratic considerations on the
initial acceptance or rejection of evolutionary theory by the people involved. The correlations are too great to be accidental. They have also
shown how diverse these considerations were in the second half of the
nineteenth century in the various countries investigated. But, as diverse
as these circumstances were, eventually most of the scientists came to
have one thing in common—the acceptance of evolution. And they were
joined in this acceptance by large segments of the general public. Though
it took much longer, scientists and even large segments of the general
public came to accept natural selection as playing a major role in evolution. I am not so naïvely optimistic as to believe that essentially coherent
and cogent theories like Darwin's always eventually prevail. Yet the correlation between the amount of evidence, the cogency of argument, the
explanatory power of evolutionary theory, and its *eventual* acceptance is
also too great to be accidental. Maybe rational argument and the amassing of evidence is not the most efficient way of changing people's minds,
but it is one way and the way to which science is committed.

RELIGION AND DARWINISM
Varieties of Catholic Reaction

HARRY W. PAUL

> For one who believes in transformism, most of the questions
> which naturally present themselves to the human mind change
> their meaning; some, indeed, cease to have any meaning.*

Friedrich Heer noted as one of the themes of European intellectual history that "The secular philosophies and intellectual systems of our day are theologies of laymen. As surrogate theologies of nonconformist circles, they are intrinsically committed to a discussion with the theologies of orthodoxy."[1] This shrewd judgment can be used to explain the arguments and dialogue that came out of the long and frequently furious interaction between Darwinism and religion. In spite of disagreement over "substantive issues," it could be argued, for example, that "the universe over which Darwin and the natural theologians fought was really the same universe."[2] This is the basis of Walter F. Cannon's paradox: "Darwinism as the triumph 'of a Christian way of picturing the world over the other ways available to scientists,'" especially Charles Lyell's uniformitarianism.

By the time that Darwin's scientific outlook was being formed, Christian natural theology had adopted or assisted in the adoption of several concepts

---

* Félix Le Dantec, as cited by Roger Martin du Gard, *Jean Barois* (New York: Bobbs-Merrill, 1969), p. 85.

[1] Friedrich Heer, *The Intellectual History of Europe*, trans. Jonathan Steinberg, 2 vols. (New York: Doubleday, 1968), I, xi.

[2] John Lyon, "Immediate Reactions to Darwin: The English Catholic Press' First Reviews of the 'Origin of the Species,'" *Church History* 41, no. 1 (March 1972): 90.

which were quite alien to Lyellian uniformitarianism. That laws of nature were more than mere observed regularities; that matter could not organize itself and had to be organized by an outside force; that organic beings were marvellously adapted to their environments; that the concept of "purpose" had meaning in biology; that each organ had its specific function for its own species; that the past was historical (not cyclical); and that historical development in the past was a record of progress culminating in man, were all concepts which Darwin found in the intellectual milieu of his youth.[3]

But Darwin did worse than poach in the sacred grove. "For he showed how the universe which was developed under the aegis of Christianity could run quite as well without the Christian God at all." The clash became inevitable when "Darwin gave the world of natural science an historical view. And the ideas which led him to do so were in great measure theological in origin. They 'were of importance in shaping Charles Darwin's thought, especially because he did not recognize them as theological.' "[4] And so the Darwinian concept of progress, for example, can be regarded as "a secularization of ideas which have their origin in scripture."[5] Although Darwinism per se was never anathematized by the Roman lions of orthodoxy, many of the ideas with which Darwinism was associated were condemned, especially in history and philosophy, the two secular areas Rome regarded as most dangerous. In a certain sense the condemnation of modernism in 1907 was the use of the ultimate weapon of orthodoxy in its fight against the intrusion of secular values into theology, philosophy, and history, areas in which it had traditionally exercised control or strong influence in matters it considered vital to its interests.[6] The secular city of the nineteenth-century Darwinians was built upon substantial if transformed heavenly foundations.[7]

[3] Ibid., p. 90 (a summary of Walter F. Cannon, "The Bases of Darwin's Achievement: A Revaluation," *Victorian Studies* 5 [December 1961]: 127–128).

[4] Lyon, "Immediate Reactions to Darwin," pp. 90–91. Lyon finds that Cannon's thesis applies to the Catholic reviewers of Darwin for the *Rambler* and the *Dublin Review*. "What really divides Christian from would-be theist here are questions regarding the nature of miracles, the nature of causation and the meaning of chance, and the problem of evil" (p. 91).

[5] Ibid., p. 90.

[6] See Emile Poulat, *Histoire, dogme et critique dans la crise moderniste* (Tournai: Casterman, 1962).

[7] Cf. Peter L. Berger and Thomas Luckman, *The Social Construction of Reality: A*

But little can be gained from imitating the debate over the architecture of the heavenly city of the eighteenth-century *philosophes* or that over the sources of the secularized values of Marx, "the last of the schoolmen." Although, in the quarrels arising out of Darwinian arguments, there is the persistence of certain perennial problems, especially in philosophy, startlingly new solutions were given in the nineteenth century. There existed enough unity in the intellectual community to permit a debate over certain fundamental questions. The old questions are interesting and important because they elicit new answers every generation and sometimes stimulate debate within the same generation. It is precisely this aspect of the interaction between religion and Darwinism, rather than a sort of structuralist study of an identity of ostensible opposites, that has occupied the attention of the authors of the papers in this volume.

All missed the profound insight of Cannon: that "Darwin's views of the randomness of nature until corrected, of the irreversible and purposeful processes of nature, and of species as entities to be examined through their relation to the environment, are ideas made available by natural theology. . . . The ecological mechanisms of natural selection are not Christian, but they operate in a universe already accepted and defined by the natural theologians."[8] But it was not the debt of Darwinism to natural theology that struck the protagonists in the heat of battle, but rather the fundamental opposition between the groups involved. Let us briefly consider the nature of this opposition.

In contrast to seeing a close relation between evolution and natural theology, it might be argued that "any evolutionary theory clashed with the *exegetical* conceptions of the majority of Christian theologians of the middle of the nineteenth century."[9] This remark needs careful qualification. The long-range reaction of theologians to biological evolution, as it

---

Treatise in the Sociology of Knowledge (New York: Doubleday, 1966–1967), p. 112: "Theology is paradigmatic for the later philosophical and scientific conceptualizations of the cosmos. . . . Modern science is an extreme step . . . in the secularization and sophistication of universe-maintenance." See also Peter L. Berger, *The Sacred Canopy: Elements of a Sociological Theory of Religion* (New York: Doubleday, 1967–1969).

[8] Cannon, "The Bases of Darwin's Achievement," p. 133.

[9] Reijer Hooykaas, *Natural Law and Divine Miracle: The Principle of Uniformity in Geology, Biology and Theology* (Leiden: Brill, 1962–1963), p. 148, n. 1.

became increasingly accepted by scientists, was that there was no conflict between evolution and scripture. Most arguments by Catholic intellectuals against evolution were taken from science. Of course, beliefs in a Creator of the universe and in the immediate creation of the soul by God were maintained as essential beliefs, unamenable to modification by changing fashions in science. Conveniently, these items were of little concern to biology, although some biologists took great pleasure in antagonizing believers by covering their philosophical *obiter dicta* with a patina of science. The application of evolutionary concepts to theology, to the history of the church, and especially to dogma was quite another matter. Toward the end of the nineteenth century and especially in the first decade of the twentieth century, orthodoxy was kept quite busy condemning books and articles of clerical and lay intellectuals tainted by evolutionary assumptions. In the beginning of the Darwinian epoch, of course, it was common to find intellectuals, and not only among the clergy, who had difficulty in making any move toward acceptance of evolution because of its ostensible disagreement, especially in its implications, with traditional interpretation of scripture. But, on an official religious level in most of Western Europe, there was a conspicuously cautious attitude with respect to condemnation in a technical sense. A considerable number of past mistakes by religious authorities probably worked in favor of restraint in this matter. The Galileo case, which was completely reworked in the nineteenth century on the basis of newly available official documents, was constantly put forward by writers as proof of the need for caution in such matters. Regardless of the justification that might be advanced for the action against Galileo, in the end the church had lost more than it had gained and had put a powerful propaganda weapon in the hands of its enemies. Such a mistake in the nineteenth century would have been an even greater catastrophe. Yet the clash of some features of Darwinism with the traditional religious *Weltanschauung* could not be ignored.

Only on the western and eastern fringes of the European intellectual community were there condemnations, in a technical sense, of Darwinian ideas by the religious-political establishment. In 1876 Gregorio Chil's natural history of Las Palmas, with its "picture of the Quaternary epoch, during which the simian mammalian form was modified until it developed into man" (Glick), was banned and a serious attempt made to implement the decision on a parish level. "Most of the refutations of Chil's evolutionary arguments were based on scripture" (Glick). Such "inane posturings of the orthodox" (Glick), along with restrictions on scientific societies, led the intelligentsia to regard Spain as a model of a society held in thrall by a clergy of antidiluvian mentality. A harsh but not untypical judgment was given by J.-M. Guardia in 1890. "C'est en Espagne qu'il faut aller pour étudier cliniquement et sur place ces maladies chroniques de la conscience et de la raison, qui . . . constituent une diathèse sinon une cachexie nationale. . . . La vérité est qu'il n'y a point de philosophie espagnole, et que l'Espagne peut tout au plus compter une demidouzaine de penseurs originaux, ayant spéculé aussi librement qu'ils pouvaient, sous un régime mortel à la philosophie et à sa soeur, la science."[10] The temporary banning of the *Origin* in Russia in 1866 was the result of an immediate political-social scare and a general association of Darwinism with radicalism in a climate where "the politically conservative and religious criticisms of Darwinism were often inseparable" (Rogers), as in Spain.[11] But it was only in several states of the United States that political and legal barriers were erected comparable to and often more serious than those in Spain. (See Aldrich for bibliography on the United States.) In these cases a continued respect for scripture, combined with powerful social, political, and educational quarrels, produced situations without parallel in other countries. On a strictly intellectual level, at least, in the United States, "Biblical questions formed surprisingly little part in the controversy" (Pfeifer), especially before the appearance of the *Descent of Man*. And, even with the issue of the descent of man, the arguments considered by anti-Darwinian religious thinkers were chiefly scientific and philosophical. No doubt the absence of a direct confrontation in some situations was in large part due to a deliberate attempt by many

---

[10] J.-M. Guardia, "Philosophes espagnoles (J. Huarte)," *Revue Philosophique* 30 (1890): 250. "Spain is the place to go for a clinical, on-the-spot study of these chronic maladies of conscience and reason, which . . . constitute a national predisposition, if not a national cachexia. . . . The truth is that there is no Spanish philosophy and that Spain can count at most a half-dozen original thinkers, who have speculated as freely as they could, under a regime that is fatal to philosophy and to her sister, science."

[11] For an analysis of the "relatively rational and dispassionate attitude toward the Darwinian 'threat to morals and religion' in Russia," see George L. Kline, "Darwinism and the Russian Orthodox Church," in *Continuity and Change in Russian and Soviet Thought*, ed. Ernest J. Simmons (Cambridge: Harvard University Press, 1955), pp. 307–328.

scientists to avoid presenting their ideas as in conflict with scripture. Several years before the appearance of the *Origin*, the Dutch theistic natural historian Pieter Harting set an example by deliberately ignoring Genesis in his sympathetic treatment of the evolutionary hypothesis (Bulhof). But it was not unusual for an anticlerical or antireligious scientist to adopt for polemical purposes the view that adherence to the tenets of Christianity, especially its Catholic variety, meant ipso facto a belief in the literal truth of Genesis. This was matched by the belief of some extreme clericals in the equation "Darwinism equals materialism equals atheism." These are the extreme fringes of the quarrel and do not impair the validity of the argument that specifically biblical questions were not usually an essential part of the encounter.

In contrast to the power Catholicism was able to exert against Darwinism in Spain, it was practically impotent in Italy. Neither could the Italian Catholic intellectual establishment draw upon a repertory of anti-Darwinism arguments from the Italian scientific establishment, as was done in France. As in France under the Third Republic and as was the case sporadically in Spain, the advent of Darwinism in Italy provided a source of ideology for the anticlerical movement. Although Darwinism enjoyed a number of close connections with the English source, the peculiarities of the Italian situation set Darwinism in Italy apart from other situations. Italy was in the forefront in recognizing Darwin, electing him to various academies and societies and awarding him the famous Bressa prize in 1875. Among his admirers and critical followers were many of the leading figures of the Italian scientific community, including Federico Delpino, Giovanni Canestrini, Michele Lessona, Lorenzo Camerano, Enrico Giglioli, Paolo Mantegazza, Enrico Morselli, Nicolaus Kleinenberg, and Giovanni Mengozzi. Mario Cermenati explained the low intensity of Italian anti-Darwinism on the basis of a combination of scientific and religious factors. In contrast with France, he proudly noted the absence of the Cuvier authority syndrome in Italy and, with some exaggeration, also pointed to the consequent existence of a liberty of research in Italy that did not exist in France.[12] It is true that the power of the French anti-

---

[12] Mario Cermenati, "Nel cinquantenario dell' *Origine delle specie,*" *Nuova antologia,* 5th ser. 145 (February 16, 1910): 601–632.

Darwinian scientific establishment, especially as centralized in the Academy of Sciences, ensured that any professor who wanted to ensure maximum upward mobility in the academic system was an anti-Darwinian. This situation was radically reversed after the 1870's with the emergence of a dynamic Darwinian biology led by Alfred Giard at Lille. As in Russia and in Germany, the conquest of the biological segment of the scientific community moved necessarily from the periphery to the center. It also had to wait until the weight of political power was thrown definitively in favor of Darwinism. The process of triumph was one of subtle interaction between the adoption by biologists of the most fruitful paradigm for research and the adoption by the political establishment of a scientistic ideology that contained a strong Darwinistic component. In Italy the translation of the *Origin* (1864) was given an impeccable scientific presentation by Giovanni Canestrini and Leonardo Salimbeni, which avoided the type of situation that arose from the presentation of Darwinism in France by Clémence Royer as a new scientific basis for a secularistic *Weltanschauung*.

Paradoxically, Darwinism was promoted in Italy under government auspices when Filippo De Filippi, a scientist and practicing Catholic, was minister of education. De Filippi had preceded Darwin in applying Darwinism to man in his famous public lecture "L'uomo e le scimie" in 1864 at Torino. Cermenati considered this work as seminal in the introduction of Darwinism into Italy.[13] Although De Filippi argued in favor of the anatomical similarity of man and the other primates, he still retained the traditional idea of the existence of a separate human order based on the unique intellectual and religious faculties of man as well as his special position in the world. This concession did not save him from denunciations by the excited clergy. Why then was there an absence of general public excitement over the clash of scientific and clerical views? There was the serious problem posed by the nearly universal acceptance of Darwinism by the Italian scientific community, which forced the opposed clerical intelligentsia to draw upon foreign critics of Darwin and thus lose the advantage implicit in setting up a national paradigm against a

---

[13] Cermenati, "Nel cinquantenario," p. 622. Also important in the diffusion of Darwinism in Italy were Achille Quadri's *Note alla teoria darwiniana* (Bologna, 1869) and Giovanni Canestrini's *La teoria di Darwin criticamente esposta*, 2d ed. (Milan: Dumolard, 1887; first published, 1880).

foreign one. Cermenati also argued for the validity of Jacob Burckhardt's hypothesis as an explanation of the low-key nature of the debate over Darwinism in Italy. Descended from a Renaissance prototype, the modern Italian is not easily excited. *In partibus infidelium*, Catholicism was reduced to a veneer over a basic paganism and unbelief. There existed an aversion to the hierarchy and a distaste for attacks on modernity. The ravings of the clergy could have little impact on a people whose religion is best characterized as *sans souci*.[14] In view of an evident intellectual virility in late nineteenth- and early twentieth-century Italian Catholicism, which manifested itself, as in France, on the level of higher education and in the growth of organizations and journals, the arguments of Cermenati must themselves be viewed as part of the complex of anticlerical arguments common in Catholic Europe. Even Ernst Haeckel recognized the education and sophistication of the Italian Catholics as compared with the more primitive reaction of German Catholics. Haeckel's remark is really unjust to German Catholics, among whom could be found extremely sophisticated reactions to Darwinism, such as that of the well-known Jesuit biologist Erich Wasmann. As a general explanation, of course, it is reasonable to accept Cermenati's arguments that the favorable receptivity of the scientific community and the general indifference to ecclesiastical objections to Darwinism are the chief factors explaining the quick spread of Darwinism in Italy.

La Civiltà cattolica, a Jesuit periodical founded in 1856, quasi-official source of the parameters of orthodoxy, was in the forefront of the attack on Darwinism. Giovanni Battista Pianciani's articles criticizing Lamarck appeared in *La Civiltà cattolica* in 1860, and when they appeared in book form in 1863 they contained criticism of Darwin. A massive marshaling of evidence against Darwinism, which also contained an exposé of its main tenets, was done in a long series of articles by Beniamino Palomba

---

14 A variety of this explanatory model based upon the notion of Mediterranean clarity of mind appears in Benedetto Croce's distinction between the historiography of the Renaissance and that of the Enlightenment: "The rationalism of the Renaissance was especially the work of the Italian genius, so well balanced, so careful to avoid excesses, so accommodating, so artistic; enlightenment, which was especially the work of the French genius, was radical, consequent, apt to run into extremes, logistical" (Croce, *History: Its Theory and Practice*, trans. Douglas Ainslie [New York: Russell and Russell, 1960], p. 263).

in 1871–1872. Much of the series was concerned with attacking the materialistic interpolations to Darwinism by Ludwig Büchner, Carl Vogt, and Haeckel. In this regard the articles differed little from the reactions of Catholics elsewhere in Europe. Louis Agassiz, "one of the best naturalists of our time," Armand de Quatrefages, and G.-R. Meignan were among Palomba's favorite sources of anti-Darwinian arguments. By 1872 the translation of *The Descent of Man* by Michele Lessona had turned Palomba's attention to the animal origin of man. This argument, already made by De Filippi in 1864, had been restated by Canestrini in 1867: "In the universal existence of a rudimentary tail in man, I see a new proof in favor of the view that man has an animal origin."[15] The continuing concern of *La Civiltà cattolica* with Darwinist arguments on the origin of man was evident in a series of articles by Pietro Caterini in 1879, in which he restated scientific objections to the Darwinist position, emphasized the concept of the unity of the human species, and insisted on the distinctive characteristics of man.[16] Another article in the same year made it clear that the scientific aspect of the evolutionary debate was closely associated in the Catholic mind with Clémence Royer, who was quoted as providing a paradigm of Darwinist thinking.[17] On a quasi-official level, Catholic theologians had rejected Darwinism, or at least its distinctive features.

Having rejected the new scientific dogmas that the scientific community was in the process of accepting, the clergy were left in a precarious position. Over the preceding centuries the religious community had advanced as one of the guarantees of the validity of their theodicy the ostensible accord, or at least the absence of significant clash, between religious and scientific knowledge concerning the natural world. Although a few bold clerical minds moved rapidly to seek a readjustment of their theodicy to the new biological theories, most of the clergy understandably did not. Certainly, the fact that the new biology was used as one of the

---

[15] Giovanni Canestrini, "Caratteri rudimentali in ordine all'origine dell'uomo," *Annuario della Società dei Naturalisti* 2 (1867), cited in Beniamino Palomba, "Gli argomenti dei trasformisti," *La Civiltà cattolica*, 8th ser. 5 (February 3, 1872): 410.

[16] Pietro Caterini, "La scienza e l'uomo bestia," *La Civiltà cattolica*, 10th ser. 11 (1879): 294.

[17] "La scienza e la genealogia trasformistica," *La Civiltà cattolica*, 10th ser. 11 (1879): 182.

foundations of the new secularistic *Weltanschauung* partly explains this refusal. But another and largely unarticulated factor was shock at the realization that the very foundations of the scientific view of the natural world could be so radically altered. Unconsciously, perhaps, the clerical intelligentsia moved to find a safer foundation for its theodicy. Not that it rejected any support from science. At first it stubbornly held on to the old scientific dogmas. This became gradually more and more difficult as the generation of anti-Darwinian scientists died and the new biological community accepted, however critically, more and more Darwinistic dogmas. The Thomistic system of thought, adopted by Catholicism as insurance against threats resulting from a change of scientific paradigms by part of the scientific community in an area directly related to the intellectual foundations of its theodicy, showed that a tradition of centuries of interaction between science and theology could not be so easily overthrown. As was made clear by Leo XIII's encyclical letter *Aeterni Patris* (1879), prescribing the restoration of Thomism in Catholic schools, the guardians of doctrine were anxious to avoid any suspicion "of being opposed to the advance and development of natural science." The physical sciences would "find very great assistance in the restoration of the ancient philosophy."[18] Yet the adoption of Thomism by many Catholic intellectuals meant an anti-Darwinian position in biology, with specific Thomistic teachings being substituted for Darwinian ones. Zeferino Cardinal González made use of "the Thomist concept of species and criticized Darwin for converting the logical and ideal relations of the classical concept of species into genealogical ones" (Glick). An accommodation could be reached by ingenious equations like that of Juan González de Arintero, who accepted evolution within classes and equated "the Thomist concept of *species* with the modern biological *class*" (Glick). Zeferino González's reaction was more typical, especially among theologians.

The second volume of the periodical organ of the Accademia Romana di S. Tommaso d'Aquino contained a study, "Della vita," by Father Mazzella, in which the opposition of the official interpretation of Thomism to Darwinism was unequivocally stated. A clear implication of authority derived from the pope's patronage of the academy. Mazzella re-

---

[18] Etienne Gilson, ed., *The Church Speaks to the Modern World: The Social Teachings of Leo XIII* (New York: Doubleday, 1954), pp. 49–50.

jected the system of evolution that accepted the existence of prototypes (*groupes primitifs d'êtres vivants,* "primitive groups of living beings") as its point of departure. According to Mazzella, God made life, created matter and living beings separately, distributed them in different species, and divided them into varied races. The distinctive character of the species is essential and unchangeable; the principle of variety of races is accidental and changeable but always in keeping with its specific character. The individuals of different races but of the same species reproduce, keeping the essential traits of the group to which they belong—this is the great law of the generation of living beings. It is impossible to create a new species. The first generators were created by God without an intermediary. Man stands outside evolution. Coming directly from God, the soul is the substantial form of the individual, the specific principle of the human being. By the simplicity of its essence, it is above any transformation, and because of its spirituality it has no possible relation to matter or a generating principle. Darwinism is a simple scaffolding of absurd hypotheses; the biological science of Saint Thomas rests on the data of the most exact observation.

The true answer to the question of the origin of living beings is in Saint Thomas; it was stated earlier in the first chapter of Genesis. For scientific support of the largely obsolescent scientific part of his views, Mazzella, like Palomba, was forced to fall back on the dwindling number of anti-Darwinian scientists, chiefly Agassiz and de Quatrefages.[19] It was only after the neovitalistic movement made its impact that Thomism was able to integrate its philosophical and theological positions with a segment of biological theory that was paradigmatic for some contemporary research.

In France, Catholic critics of Darwinism continued to utilize the ebbing anti-Darwinian current that was still very much evident in the 1880's. In this way Catholic rejection of Darwinism could present an appearance of being in agreement with part of the scientific community. Since it was

---

[19] See Abbé O. Rey, "Etudes de biologie: Saint Thomas d'Aquin et le transformisme à l'Académie de Saint Thomas à Rome," *Annales de Philosophie Chrétienne,* n.s. 8 (1883): 69–85. For a modern example of "Roman" hostility toward Darwinism and evolution in general, see Ernesto Cardinal Ruffini, *La teoria dell'evoluzione seconda la scienza e la fede* (Rome: Ed. Orbis Catholicus, 1948).

vital to avoid rejecting a purely scientific hypothesis for religious reasons, scientific criticism of Darwinism was always given the place of honor in the complex of factors advanced as an explanation of the rejection. The possibility of maintaining that a rejection of Darwinism was in agreement with respectable contemporary scientific opinion is superbly illustrated in the influential work *La Vie et l'évolution des espèces* by the Jesuit Albert Farges.[20] The volume on evolution was the third of a series of eight *Etudes philosophiques* written "to popularize the theories of Aristotle and Saint Thomas and their agreement with the sciences." The *Etudes* were adopted in a great number of European seminaries and translated into Spanish, Italian, and German. Leo XIII wrote a letter to give his approval. The Roman Academy of Saint Thomas, the Roman College, and many Catholic universities were unstinting in praising the vast effort of this "complete course of philosophy." Farges, a "docteur en philosophie et en théologie," had been head of the Séminaire de Saint-Sulpice and the Institut Catholique de Paris and became the Supérieur du Séminaire de l'Institut Catholique d'Angers. Farges thought that the radical vice of Darwinism was its replacement of a guiding intelligence and a providential will in explaining the progressive development and harmonious spread of the world of living beings by the accidental competition of blind causes. This is a general observation presented in a page or so. Although he also paid attention to the possibility of conflict with Scripture, he gave most attention to the philosophical and especially the scientific deficiencies of Darwinism. But the difficulty of finding anti-Darwinian scientists is evident, for he could draw upon only Emile Blanchard and Charles Contejean, the latter of whom accepted evolution for antireligious reasons in spite of its scientific deficiencies! Extreme Darwinists were easier game. The Haeckelian dogma that ontogeny recapitulates phylogeny could be rejected on the authority of Henri Milne-Edwards, Edmond Perrier, and Karl von Baer. Most anti-Darwinian arguments had to be drawn from Pierre Flourens, de Quatrefages, Agassiz, and Joachim Barrande, with occasional references to Georges Cuvier. Finally, in spite of some criticism of some of his Darwinist phraseology, Farges accepted Gaudry's "passive evolution under the hand of the Cre-

---

20 Albert Farges, *La Vie et l'évolution des espèces avec une thèse sur l'évolution étendue au corps de l'homme* (Paris: Letouzey et Ané, 1894; 6th ed., 1902).

ator," which could be restricted to the geological periods of the formation of the world and thus reconciled with the great principle of the normal fixity of the species. Farges accepted the Marquis de Nadaillac's conclusion at the international scientific congress of Catholics in 1891: Gaudry's work in paleontology had put beyond doubt that many transitions had occurred between species in geological time, and this proved the existence of an *enchaînement*, although the similarities did not prove ascent or descent of species. Classifications would have to be revised and probably simplified.[21] In the present state of scientific knowledge, this was the only concession necessary. It was comforting to make it on the basis of work done by a Catholic scientist.

In a general philosophical sense, Darwinism led more to a recrudescence of old quarrels than to new debates. Nowhere was this more evident than in the case of the exchange over materialism. Darwinism was immediately pressed into service as providing scientific proof of the validity of materialist principles. The fact that the more fervid Darwinians, like Büchner, were materialists led to antimaterialistic reactions against Darwinism in the Islamic world (Bezirgan) as well as in Europe. In Mexico, Darwinism could serve as "a weapon against theological cosmogonies" (Moreno). In Spain, Haeckel could be regarded as having created "a total system which explained the evolutionary development of the cosmos. Anti-Darwinists did not fail to get the message and uniformly branded Haeckel as the chief promoter of anti-Christian materialism" (Glick). It is now difficult to appreciate the alarm caused in religious circles by the fertilizing of materialism by Darwinism. It had been difficult enough to fight the battle on a solely philosophical front. The added weight and prestige of one of the more dynamic branches of science might, it seemed, prove fatal to the religious cause. Salvation from physics and the philosophy of science, as well as help from neovitalism in biology, was several decades off. With some exaggeration, Alfred North Whitehead makes clear the seriousness of the general situation. "The eighteenth and nineteenth centuries accepted as their natural philosophy a certain circle of concepts which were as rigid and definite as those of the philosophy of the middle ages, and were accepted with as little criti-

---

[21] Ibid., pp. 242–278.

cal research. I will call this natural philosophy 'materialism.' Not only were men of science materialists, but also adherents of all schools of philosophy. The idealists only differed from the philosophic materialists on question of the alignment of nature in reference to mind. But no one had any doubt that the philosophy of nature considered in itself was . . . materialism." The great advantage of materialism as a competitor over any other theory is that it "has all the completeness of the thought of the middle ages, which had a complete answer to everything, be it in heaven or in hell or in nature."[22] It was a competing *Weltanschauung*, especially when embodied in a set of dogmas and endowed with disciples and propaganda organs.

One of the leading French clerical intellectuals, Msgr. G.-R. Meignan, Bishop of Châlons-sur-Marne and later Cardinal-Archbishop of Tours, viewed the materialistic aspect of Darwinism as a great threat to Christian humanism.[23] The materialist could fall back on science to show that man was descended from "an arboreal primate who in the long course of Tertiary time had descended to the ground and achieved some dexterity in the manipulation of stones."[24] Rather than any new philosophical premise, the real danger of the Darwinist-materialist symbiosis was that now man could be catalogued scientifically according to the tenets of materialism. Man had lost the place of honor assigned to him in the creation and recorded in biblical anthropology. Meignan noted with alarm the Darwinian definition of man given in a contemporary organ of materialism: "L'homme, animal mammifère, de l'ordre des primates, famille des bimanes, nez saillant, oreille nue."[25] Philosophically, the work of Darwin, especially as interpreted by the materialists, had "left man only one of innumerable creatures evolving through the play of secondary forces and it had divested him of his mythological and supernatural trap-

---

[22] Alfred North Whitehead, *The Concept of Nature* (paperback ed., Cambridge: Cambridge University Press, 1964), pp. 70, 73 (first published, 1920).

[23] G.-R. Meignan, *Le Monde et l'homme primitif selon la Bible*, 3d ed. (Paris and Brussels: Palmé, 1879), a revised version of his lectures in the Faculty of Theology at the Sorbonne.

[24] Loren Eiseley, *Darwin's Century: Evolution and the Men Who Discovered It* (New York: Doubleday, 1958), p. 195.

[25] Meignan, *Le Monde*, p. 178. "Man, mammalian animal, of the order of Primates, family of Bimana, projecting nose, naked ear."

pings."[26] The logical conclusion was brutally put by Haeckel: belief in the immortality of the soul and belief in God can be relegated to the rubbish heap of "outworn articles of the Christian faith."[27]

A rejection of biblical anthropology and its concomitant "Christian humanism" and an avoidance of materialism were possible at the cost of adopting, like T. H. Huxley, "a scepticism in which no proof of God or the soul could be admitted" (Pfeifer). Another escape was in Haeckel's "Monistic Religion," which changed as his conceptions of monism changed and passed through four stages: mechanism-pantheism à la Spinoza and Bruno; a "nature-philosophy which united idealism and materialism, and spirit and matter," resulting from the introduction of hylozoism into pantheism; between 1890 and 1904, "Monism as a link uniting science and religion," which featured the replacement of any deity by the Monistic Law of Substance; and the final stage (1904–1919), which saw "the reemergence of specific references to God in Haeckel's Monism, with a new formulation of the Monistic Religion, tinged with vitalistic elements and markedly similar to Schelling's Identity-philosophy."[28]

A third alternative was available for those who wished to accept evolution but not abandon the comforts of the faith of their fathers. Essentially, this was the position of St. George Mivart and the Catholic clergy M.-D. Leroy and John Zahm, who attempted to save the uniqueness of man by attributing his special faculties and characteristics to direct divine intervention.[29] After nearly a century of squabbles, condemnations, and prudence, this position was pretty much adopted officially by Pius XII in the encyclical *Humani Generis* (1950), but with a reassertion of the authority of the church in the intellectual domains traditionally reserved for its ultimate jurisdiction.

[26] Eiseley, *Darwin's Century*, p. 196.

[27] Ernst Haeckel, *The Riddle of the Universe*, trans. Joseph McCabe (1901); summarized by Vance Randolph (New York: Vanguard, 1926), p. 56. Contrary to widely accepted opinion, Haeckel's monism was not "an extension of Büchner's popular materialism." See Niles R. Holt, "Ernst Haeckel's Monistic Religion," *Journal of the History of Ideas* 32 (1971): 265–280.

[28] Holt, "Ernst Haeckel's Monistic Religion," pp. 267–268.

[29] To avoid ambiguity on the issue of man, the second and revised edition of Leroy's book had its title changed from *L'Evolution des espèces organiques* (Paris: Perrin, 1887) to *L'Evolution restreinte aux espèces organiques* (Lyon: Delhomme et Briquet, 1891); it was put on the Index in 1895.

Few issues were more confused than the teleological implications of Darwinism. It has been made clear that in most countries there were a considerable number of Darwinists who "insisted that teleology be rooted out of biology since the only true science is mechanistic. They wanted to reduce all explanation to some form of causal determinism." But this "conflict between causality and teleology was ultimately a superficial one." Even August Weismann admitted that "causality pursued far enough is indistinguishable from teleology" (Montgomery). Christians could take some comfort in the thought that "such prestigious figures as Owen, Sedgwick, Herschel, and William Thomson continued to proclaim the need for design as an explanatory principle in biology" (Burkhardt). But Darwin was eventually irritated by Asa Gray's evolutionary teleology because he could not see the evidence of design in nature, which, Gray insisted, not without "petite malice," was shown by Darwin's work (Pfeifer).[30] That Christian intellectuals were not at all happy with the claim that the design argument had not been destroyed is seen in the caution with which the famous *Revue des Questions Scientifiques*, organ of the Catholic Société Scientifique de Bruxelles, presented an article by Zahm favorable to evolution and claiming that teleology had been modified but not destroyed, since God operated through secondary causes rather than directly. An editorial note pointed out that until then the journal had published essentially antievolutionary articles and that *La Civiltà cattolica* had taken an antievolutionary line while attempting to veneer its opinion with orthodoxy by claiming that the teachings of Saint Augustine, Saint Thomas Aquinas, and Francisco Suárez were not favorable to evolution.[31] The theological opponents of Darwinism who thought that the teleological cause had suffered a serious blow were justified in thinking so. First of all, powerful and dynamic groups of Darwinists everywhere hailed the elimination of teleology from biology and thought as a result of Darwin's work. Secondly, from a general philosophical viewpoint, it can be argued that this was not an unreasonable interpretation, regardless of Darwin's ambiguity on the subject.

---

[30] See Hooykaas, *Natural Law and Divine Miracle*, pp. 212–215.

[31] John Zahm, "Evolution et théologie," *Revue des Questions Scientifiques* 43 (1898): 403–419. This had been a paper at the International Scientific Congress of Catholics in 1897.

The new biology thought of life as resembling matter and unlike mind in being wholly devoid of conscious purpose; Darwin talked freely of selection, and constantly used language implying teleology in organic nature, but he never for a moment thought of nature as a conscious agent deliberately trying experiments and aware of the ends which she was pursuing; if he had troubled to think out the philosophy underlying his biology he would have arrived at something like Schopenhauer's conception of the evolutionary process as the self-expression of a blind will, a creative and directive force utterly devoid of consciousness and of the moral attributes which consciousness bestows on the will of man; and it is such ideas which we find floating everywhere in the atmosphere of Darwin's contemporaries, such as Tennyson.[32]

The ethical and religious consequences of the end of cosmic teleology would be horrendous.

If my body comes from brutes, my soul uncertain, or a fable,
Why not bask amid the senses while the sun of morning shines,
I, the finer brute rejoicing in my hounds, and in my stable,
Youth and Health, and birth and wealth, and choice of women and of wines?[33]

In 1910 the Jesuit biologist Robert de Sinéty, looking back on a half century of Darwinism, noted that the philosophical poverty of Darwinism about which his contemporaries complained had been nearly a guarantee of its success two or three decades earlier.[34] Sinéty made a Procrustean division: before Darwin, the preoccupation had been with the final causes of biological phenomena; after Darwin, the only interest was in the efficient determinism connecting phenomena. Distinguished biologists like Giard, Sinéty's master, a pioneer of Darwinist-inspired studies in France, claimed that this was the distinctive characteristic of Darwinian theory. Natural selection and adaptation ended all interest in the purpose of an organ; it had become antiscientific to define an organ by its function; the scientist no longer had the right to admire the marvelous

[32] R. G. Collingwood, *The Idea of Nature* (New York: Oxford University Press, 1960), p. 135 (first published, 1945).

[33] Alfred Tennyson, "By an Evolutionist" (1889), in Tennyson's *Poems and Plays* (London: Oxford University Press, 1965), p. 811.

[34] Robert de Sinéty, "Un Demi-Siècle de Darwinisme," *Revue des Questions Scientifiques* 67 (1910): 5–38, 480–513. Sinéty's doctoral thesis, guided by Giard, had been called "Recherches sur la biologie et l'anatomie des phasmes" (Université de Paris, 1901).

order and harmony of an animal body abstracted from the physico-chemical conditions that produced it. All teleological commentary had been made tautological. Such an intimate union existed between Darwinism and antifinalism that teleophobia became one of the most important allies of Darwinism. Yves Delage, another leading evolutionary biologist, was delighted that the "theory of causality," which was closely allied to evolution, "compelled him [man] to reject the too facile explanations offered by teleological systems and to consider causal explanations as the only satisfactory ones."[35] The Christian reaction was summed up by Nikolai Danilevsky's thesis: "Will without motive, power without design, thought opposed to reason, would be admirable in explaining chaos, but would render little aid in accounting for anything else" (Rogers). W. B. Carpenter's view provided a prototype of the uneasy accommodation that "the higher idea of Creative Design" could reach with Darwinism. "That the 'accidents' of Natural Selection should have *produced* that orderly succession [of forms] is to my mind inconceivable; I cannot but believe that its evolution was part of the original Creative Design; and that the operation of natural selection has been simply to limit the survivorship, among the entire range of forms that have successively come into existence, to those which were suited to maintain that existence at each period" (cited in Hodge). But this was essentially a negation of one of the chief tenets of Darwinism, and, since the antiteleological ingredient of Darwinism provided one of its chief attractions for many converts to Darwinism, no real compromise was possible.

The emphasis upon the teleological character of an organism that is characteristic of modern organismic biology should not really be viewed as a triumph for any sort of extrinsic teleology. As Richard Bevan Braithwaite, George Summerhoff, and Ernest Nagel have argued, "teleological language is not nonsense, and not necessarily anthropomorphic"; neither does a teleological statement "presuppose a metaphysical doctrine of final causes; . . . it has legitimate scientific uses. . . . No nonphysical agents or causes need be active in teleological behavior . . . In fact, the teleological system is simply a special case of physical system in the ordinary sense of that term, and can be described, qua teleological, in a vocabulary suited

[35] Yves Delage and Marie Goldsmith, *The Theories of Evolution*, trans. André Tridon (New York: Huebsch, 1912), p. 6.

for the description of nonteleological systems."[36] In modern biology, teleology is more likely to lead to a J. S. Haldane than to a Pierre Teilhard de Chardin.

Writing near the end of the long clash of church and state and in the final phase of a long anticlerical wave in France, Sinéty emphasized the important role played by antireligious factors in favor of Darwinism. In the first volume of the *Revue Scientifique* in 1881, the materialist and atheistic scientist Charles Contejean undertook a devastating review of the proofs of *transformisme* and concluded that he remained an evolutionist because it is the only way of obviating recourse to miracles. Such argumentation was similar to the later admission of Delage in his *L'Hérédité et les grands problèmes de biologie générale* that he was an evolutionist for purely philosophical reasons. Such admissions led Farges to conclude that a system forced to fall back on such arguments was very sick.[37] An alliance developed between Darwinism and the anticlerical "party." Varieties of Darwinism became the quasi-official teaching of the university, the "ideology" of those who wanted the best appointments. The fact that Darwinism was an eminently respectable scientific doctrine enhanced its value from the viewpoint of the republic, which would be scientific or would not be at all. Giard celebrated "the tendency of science to replace gradually the role hitherto enjoyed by religion," and Paul Bert, Félix Le Dantec, Marcelin Berthelot, and Paul Painlevé shared this sentiment.[38] Similar situations developed in other countries. "The reception of Darwinism in Spain was linked to a social and political event of the first order—the Revolution of 1868" (Glick)—and to a movement of educational reform whose ideology contained a significant anticlerical component. French Darwinists were also generally supporters of the new republic and devoted to educational reform. In Russia, Darwinism arrived during the national awakening after the Crimean War and was adopted by the "rebellious intelligentsia" partly because "it supported the material-

[36] Morton Beckner, *The Biological Way of Thought* (Berkeley: University of California Press, 1968), pp. 132–133 (first published, 1959).

[37] Farges, *La Vie et l'évolution des espèces*, p. 276.

[38] Alfred Giard, "Le Laboratoire du Portal, les grandes et les petites stations maritimes," *Bulletin Scientifique de la France et de la Belgique* 20 (1889): 298. For the interaction between scientism and politics, see Harry W. Paul, "The Crucifix and the Crucible: Catholic Scientists in the Third Republic," *The Catholic Historical Review* 58, no. 2 (July 1972): 195–219.

istic view totally opposed to the spiritualistic and irrationalist bent of autocratic ideology" (Vucinich). "The conservative and religious rejection of Darwinism meant that Darwinism acquired among Russian social thinkers a social and political context reflecting only the radical side of the political spectrum. . . . They [the young Russian Radicals] found in Darwinism the keystone to the structure of their materialistic philosophy" (Rogers). In contrast, most Dutch scientists "had no radical inclinations" (Bulhof). But generally, as in England, "the Darwinian cause was just another progressive cause to be championed or resisted according to one's attitude in general. . . . Darwin joined Tom Paine, Ben Franklin, Shelley, and others on the roster of heroes adopted by the small but growing bands of organized secularists, positivists, and freethinkers" (Hodge, original version). It may not be too much of a distortion to argue that in the Continental cases, where religion, science, and politics mingled, the views of Karl Mannheim have a limited applicability. There was a tendency for Darwinians to overcome their "homelessness" and avoid "failure" by a "direct affiliation" with parties, if not always with "classes." In a more general sense, it might be argued that one sees the passage of Darwinism from a stage of utopian thinking into a stage of ideological thinking.[39] But Darwinism, or its utopian and ideological spin-offs, supplied only one element of the new secularist set of ideas used to criticize various establishments in the nineteenth and twentieth centuries. It must also be kept in mind that the scientific debates were carried on in isolation from social and political quarrels. Even in Russia the scientific revisionist debate over Darwinism, "although deeply touched by the philosophical controversy over the meaning of materialism, was much more detached [than the conservative attack on Darwinism] from the political struggle" (Rogers).

The antireligious exploitation of Darwinism had a serious biophilosophical consequence, for it was the long association of the idea of finality with the idea of God that was the source of most of the hostility it evoked. Etienne Gilson makes the point in his recent study of the idea of finality as "a philosophical inevitability" and as "a constant of biophi-

---

[39] See Karl Mannheim, *Ideology and Utopia: An Introduction to the Sociology of Knowledge* (New York: Harcourt, Brace, and World, 1964), pp. 159–160 (first published, 1936).

losophy or philosophy of life." "Soit par hostilité contre la notion de Dieu, soit par désir de protéger l'explication scientifique contre toute contamination théologique, fût-ce de théologie naturelle, soit enfin par une alliance de ces deux motifs, les représentants de ce que l'on peut nommer le scientisme s'accordent à proscrire aujourd'hui la notion de finalité."[40] Teleological explanation may be at one of two levels: first, "explanations of the behavior of individual natural units, e.g., an organism," and, second, "explanations of systems of interacting natural units, such as organisms in a particular environment or even the universe as a whole."[41] It may be that in their eagerness to get rid of the cosmic teleology, the antiteleological group unintentionally jettisoned the first as well. "In short, when Darwinism evicted the watchmaker of Paley's famous watch, it threw out as well the *telos* of the watch itself. But without a *terminus ad quem* of development, without a *terminus ad quem* for our understanding of the organization of a living system, of an organism, of an organ, of an organelle, one has no univocal concept of adaptation, of the adjustment of these means to that end."[42] In getting rid of the cosmic *telos* imposed upon Aristotelian nature by Judaic-

---

[40] Etienne Gilson, *D'Aristote à Darwin et retour: Essais sur quelques constantes de la biophilosophie* (Paris: J. Vrin, 1971), p. 9. "Whether from hostility to the notion of God, from a desire to protect scientific explanation from any teleological contamination, even that of natural theology, or from a combination of these two motives, the representatives of what can be called scientism agree, now, in proscribing the idea of finality." See especially pp. 133–147, "Evolution et théologie," on the paradoxical role of Darwin in associating morphology with teleology, although, as Huxley pointed out, it was a new teleology. But was not new presbyter but old priest writ large? "Une téléonomie immanente au vivant et analogue à une connaissance ne diffère de la finalité classique que par le nom" (ibid., p. 141, n. 9: "A teleonomy immanent in the living being and analogous to consciousness differs from classical finality only in name"). For the argument that teleonomy can be substituted for finality, see Jacques Monod, *Le Hasard et la nécessité: Essai sur la philosophie naturelle de la biologie moderne* (Paris: Le Seuil, 1970). For criticism of the philosophical parts of the work of Monod and of François Jacob's *La Logique du vivant: Une Histoire de l'hérédité* (Paris: Gallimard, 1970), see Madeleine Barthélémy-Madaule, *L'Idéologie du hasard et de la nécessité* (Paris: Le Seuil, 1972), as well as François Russo's review in *Le Monde*, August 6–7, 1972.

[41] For this distinction between intrinsic and extrinsic teleology, see the *New Catholic Encyclopedia* (1967), XIII, 979. See p. 981 for comment on the point raised by Gilson about the link between teleology and God.

[42] Marjorie Grene, "Aristotle and Modern Biology," *Journal of the History of Ideas* 33, no. 3 (July–September 1972): 407–408.

Christian thought and Neo-Platonism, the antiteleological Darwinians produced an epistemological merry-go-round.

Why do we keep going round this merry-go-round? "Adaptation" is a means-end concept. Yet if all adaptations are for no specifiable end except survival, one keeps falling back into a universal necessity which is in turn reducible to the same old tautology. Stretch it out: it's teleology. Collapse it one level, it's necessity. Collapse it further, it's tautology. What is lacking to stabilize this endless vacillation? This brings me at last to my second major Aristotelian concept: *eidos*. For what is lacking in the modern concept of adaptation is precisely the definite *telos*, which in Aristotle is the mature form of the species.[43]

Contrary to what Marjorie Grene says, modern biology did concern itself with precisely the Aristotelian concepts she thinks related to present-day "investigation and knowledge of living nature." As Montgomery shows, the German "idealist" biologists "rejected natural selection and sought to substitute some internal developmental principle as the primary motive force of evolution. . . . Biological form could be explained independently of its historical adaptation to the external environment. . . . By attributing form to an inner developmental law, the idealists attempted to save the idea of type and give it an autonomous position in biology, independent of descent." The idealists were not cosmic teleologists, of course; "most idealists granted a certain limited role to natural selection; and some of them were second to none of the Darwinists in their insistence on mechanistic causality." It was the Darwinist innovation of "the introduction of historical modes of explanation for the observable phenomena of living nature" that the idealist biologists rejected (Montgomery).

As R. G. Collingwood noted, along with the development of the view of nature "as attempting, like a human cattlebreeder, to produce always new and improved forms," there developed the view that life was "like mind and unlike matter in developing itself through an historic process and orienting itself through the process not at random but in a determinate direction, towards the production of organisms more fitted to survive in the given environment. . . . This theory implied the philosophical conception of a life-force at once immanent and transcendent in relation

---

43 Ibid., p. 407.

to each and every living organism; immanent as existing only as embodied in these organisms, transcendent as seeking to realize itself not merely in the survival of the individual organisms, nor merely in the perpetuation of their specific type, but as always able and always trying to find for itself a more adequate realization in a new type."[44] The complexity of the disagreements between the proponents of "idealism" and the proponents of "mechanism," along with the possibility of one person's embracing components of both models or even of one person's moving from one to the other model, leads one to appreciate the profundity of Henri Bergson's observation, "teleology is only mechanism turned upside down—*un mécanisme au rebours*."[45] As might be expected, the criticism of Darwinism made by Johannes Reinke, Albert Fleischmann, and Gustav Wolff, among others, was picked up and exploited by its philosophical and theological opponents. The case of Hans Driesch (1867–1941) is the most striking illustration of this trend.

Driesch, student of Haeckel, Wilhelm His, and Wilhelm Roux, moved from a position of mathematicism-mechanicism to a vitalist position between 1891 and 1895.[46] Under serious attack were Darwinism as a *general* theory of descendance and especially Darwinian selection. Evolutionary doctrine remained, of course, the guiding theory of nearly all biologists. Although the quarrels had a distinctly German flavor, the repercussions were clearly felt in the rest of Europe. Driesch's views were particularly attractive to Catholics. The revival of Thomism had greatly increased interest in Aristotle and therefore in specifically Aristotelian biological concepts. Driesch showed his admiration for the genius of Aristotle by adopting the word *entelechy* to describe the agent-outside-the-machine he introduced as "a regulator of organic development," which could not be reduced to terms of physics.[47] In living matter there is a characteristic internal finality, different from that of machines. This partial *rapprochement* with the Thomists incurred the displeasure of part of the biological community. Adolph Wagner thought that it marked a regression to dependence on a sort of demiurge

---

[44] Collingwood, *The Idea of Nature*, p. 135.

[45] Quoted in ibid., p. 138.

[46] Jane Oppenheimer, "Driesch," in *Dictionary of Scientific Biography* (1970–), IV, 188.

[47] Ibid.

or other principle hostile to science. Others were delighted with the emergence of this new symbiosis of science and philosophy that had seemed so remote a decade or so before. In spite of Driesch's "Kantian agnosticism" and "solipsist idealism," the general reaction of the Catholic intelligentsia was favorable.

When the first volume of Driesch's 1907 Gifford lectures was translated into French, the author of the preface was Jacques Maritain, then professor of philosophy at the Institut Catholique de Paris.[48] During his forays into biology at Heidelberg (1906–1907), Maritain had known Driesch and had given a technical analysis of new developments in German biology in the Catholic *Revue de Philosophie*. Maritain hailed Driesch's restoration of natural philosophy in the Aristotelian and scholastic sense, especially in the sciences of life. "The reign of pure phenomena and brute fact, . . . three centuries of *mathématisme*" had come to an end. Basic metaphysical teachings on man now found complementary support in at least one segment of contemporary science. In Maritain's opinion, only Thomist thought was "capable of assimilating in a living and progressive synthesis the materials accumulated by science."[49]

An excellent touchstone by which to judge the change in the attitudes of Catholic groups toward evolution is the series of five international scientific congresses of Catholics held between 1888 and 1900. At the first congress a proposal was put forward in the anthropology section declaring it to be a duty for Catholics to oppose evolutionary theory because it is contrary to the faith and to Scripture. A lack of enthusiasm greeted the idea, and the section did not approve the motion. In the natural sciences section, Paul Maisonneuve, professor at the Catholic faculty of sciences in Angers,[50] warned that the issue was a purely scientific one, that some Catholic scientists had adopted evolutionary ideas, and that,

[48] Hans Driesch, *The Science and Philosophy of the Organism* (London: Black, 1908); idem, *La Philosophie de l'organisme*, trans. Max Kollman (Paris: Rivière, 1921).

[49] Jacques Maritain, "Le Néo-vitalisme en Allemagne," *Revue de Philosophie* 17 (September–October 1910): 417–441. See also F. Donau, S.J., "Un Vitaliste idéaliste: Hans Driesch," *Revue des Questions Scientifiques* 67 (1910): 426–453.

[50] On science and the new Catholic Institutes created in 1875, see Harry W. Paul, "Science and the Catholic Institutes in Nineteenth-Century France," *Societas—A Review of Social History* 1, no. 4 (1971): 271–285.

although he was not an evolutionist, the theory was a scientific possibility. Science has its rights, and evolution should be freely discussed. No collective opinion on evolution came out of the congress, but the anthropology section passed a motion approving the organization of a Catholic fight against atheistic anthropology.[51] This is not surprising in view of the antireligious orientation of Paul Broca's Ecole d'Anthropologie de Paris. Orthodoxy viewed the "science of man" as one of the more dangerous sciences of the nineteenth century. Under Isabella II the Spanish government prevented the establishment of an anthropological society in Madrid. Many of those associated with the French Ecole d'Anthropologie were, like Eugène Dally, Gabriel de Mortillet, André Lefèvre, and Jules Soury, among the chief bêtes noires of the clergy. Disputes over the attitude of the congress toward evolution, which eventually included strong criticism by the Jesuit *Etudes* and a defense by the *Revue des Questions Scientifiques*, led Msgr. Maurice d'Hulst, rector of the Institut Catholique de Paris and one of the key organizers of the congress, to emphasize that a majority of the participants had been opposed to Darwinism and that no unorthodox position had been taken.[52] But Catholics had come to discriminate between the exploitation of evolution against religion, including its materialistic interpolations, and the evolutionary hypothesis as a scientific tool. To oppose all evolutionary theory was to fail to distinguish between an extremist and unacceptable interpretation and the simple evolutionary theory adopted by religious men, among whom were included Catholic scientists, like St. George Mivart, d'Omalius d'Halloy, Charles Naudin, the Marquis Gaston de Saporta, and Albert Gaudry. The charge of uselessness, hurled at evolutionary theory by many of its opponents, was palpably absurd.[53]

In the congress of 1891 the leading star of the anthropology section, a prehistorian, the Marquis de Nadaillac, argued that the trend in contemporary anthropological research was antievolutionary. This was chal-

---

[51] See *Compte-rendu du congrès scientifique des catholiques* (Paris: Bureau du des Annales de philosophie chrétienne, 1889), pp. 606–609.

[52] Msgr. Maurice d'Hulst, *Le Congrès scientifique international des catholiques— communication présentée à l'assemblée générale des catholiques le 17 mai 1888* (Paris, 1888).

[53] See Jean d'Estienne, "Le Transformisme et la discussion libre," *Revue des Questions Scientifiques* 25 (1889): 76–141, 373–420.

lenged by Maisonneuve, who had also delivered the paper by Nadaillac. Another attempt to secure Catholic scientific unity on evolution was beaten back, with d'Hulst again taking the initiative in preventing a general antievolutionary stand. But, in closing the congress, the bishop of Angers, Charles Emile Freppel, who had earlier spoken eloquently on male mammary glands, proclaimed his opposition to all evolutionary and transformist hypotheses. He added, of course, that science enjoys complete liberty as long as it says nothing contrary to divine revelation and the teaching of the church.[54] Not much change had occurred since the congress of 1888.

In the congress of 1894, at Brussels, Nadaillac changed his emphasis from the antievolutionary trend in anthropology to an antimaterialistic trend. Abbé Guillemet boldly asserted the irrelevance of the venerable marquis's banal antievolutionary arguments. The anthropology section then enthusiastically passed a resolution calling for the encouragement of the study of the role of evolution in secondary causes in bringing the physical world into its present state. A startling change had taken place in the climate of opinion of the congress as compared with that of 1888. Abbé Nicolas Boulay, botanist and professor of the Catholic faculty of sciences at Lille, a critic of evolution, anticlimactically expressed his alarm over the enthusiasm of the younger scientific generation for evolution. At the fourth congress, held at the five-year-old Catholic university in Fribourg, Switzerland, Boulay showed the influence of the results of new research by calling for a revision of the popular Catholic chronology of the human species as derived from Genesis. Mortillet's chronology was revised, but no mention was made of his notorious anticlericalism. This is indicative of the sophistication that had become characteristic of Catholic scientific circles and evident in Catholic publications, such as the *Revue des Questions Scientifiques*, the *Annales de Philosophie Chrétienne*, and, later, the *Revue du Clergé Français*. An absence of the old type of dispute was conspicuous in the fifth congress, in Munich (1900). In one of the rare direct references to Darwinism in the science sections, J. Boiteux simply stated his acceptance of the Darwinian hypothesis that all zoological species descended from some prototypes, whose origin lay

[54] See *Compte-rendu du congrès scientifique des catholiques tenu à Paris . . . 1891* (Paris: A. Picard, 1891), pp. 212–224.

in an act of the Creator. He did not hesitate to speculate on the probable evolutionary link between man and animal. Scientific views that had seemed "heretical" a few years ago could now be safely proclaimed from the most respectable pedestals of "Catholic science."[55]

Catholicism did not necessarily entail rejection of evolution or even of Darwinism. Teilhard de Chardin had many intellectual ancestors among nineteenth-century Catholics. As Henri Begouen pointed out, a good deal of the support for evolutionary concepts was cautious or even clandestine, especially among the clergy, in part due to the hostility of the powerful intransigents in the hierarchy to evolutionary ideas.[56] A powerful factor in reducing opposition to evolution was probably Denys Cochin's *L'Evolution et la vie*.[57] Having done research in the laboratories of Paul Schützenberger and of Louis Pasteur before veering off into a conservative Catholic political career, Cochin was capable of speaking authoritatively in scientific matters. Rejecting the idea of materialistic universal evolution, Cochin argued for the scientific solidity of a good many features of Darwinism. Cochin admired Darwin, the conservative gentleman, who, neither atheist nor materialist, was not a German pedant proclaiming a new theory of matter or a deliberate democratic atheist proclaiming, in the new French fashion, that evolutionary theory contained the new principles of 1789 for the animal kingdom. Darwinism could also easily accommodate the doctrine of final causes. The ideas of a struggle in nature, natural selection, and sexual selection seemed reasonable to Cochin. Cochin ended by recommending that Catholics read Darwin and Herbert Spencer for scientific confirmation of the glories of the creation and the wisdom of the Creator.

In his brief personalized account of Catholic evolutionary thought in

[55] See *Compte-rendu du congrès scientifique des catholiques tenu à Bruxelles . . . 1894* (Brussels: Société Belge de Librairie, 1895); *Compte-rendu . . . tenu à Fribourg (Suisse) . . . 1897* (Fribourg: Librairie de l'Oeuvre de St. Paul, 1897); *Akten des fünften internationalen Kongresses katholischer Gelehrten zur München . . . 1900* (Munich: Herder, 1901).

[56] Comte Henri Begouen, *Quelques Souvenirs sur le mouvement des idées transformistes dans les milieux catholiques suivi de la mentalité spiritualiste des premiers hommes* (Paris: Blond et Gay, 1945). John C. Greene, *Darwin and the Modern World View* (New York: Mentor, 1963), p. 29, uses Begouen in reference to the Catholic "Council of Altamira" (1925) and its consequences.

[57] Denys Cochin, *L'Evolution et la vie*, 3d ed. (Paris: Masson, 1888).

late nineteenth- and early twentieth-century France, Begouen noted a cooling-off of controversy in this period. Part of the explanation lay in the emergence of a new intellectual climate more favorable to religion. Catholicism itself was undergoing a dynamic intellectual renewal, in part due to the activities of the Catholic universities. The writings of Emile Boutroux, Henri Poincaré, and Pierre Duhem had limited the competence of science itself to severely circumscribed areas. Although some scientists like Berthelot, Delage, and Le Dantec kept up the attack on religion, it was easy for apologists to turn the tables and take the offensive with powerful ammunition from the sciences themselves. Of capital importance in the cooling-off was the widespread acceptance of the evolutionary hypothesis in prehistory, anthropology, and biology, even by Catholic scientists. It was clear that evolutionary doctrine, at least in the natural sciences, did not pose any threat to religion. Apologists were soon at work finding support for evolution in the writings of the church fathers, who can be used to support nearly any idea. A certain caution is evident in the works of most Catholic writers, partly due to the vigilance of opponents of evolution in quickly indicating to Rome the deviations of their coreligionists. The intellectual *Gleichschaltung* accompanying the condemnation of modernism accentuated this caution. So the quarrel continued well into the twentieth century, with both sides vying for the support of Rome against their opponents, a practice typical of the vicious infighting in modern Catholicism. But, when the centenary of Darwin's birth was celebrated at Cambridge University in 1909, the Catholic University of Louvain accepted an invitation to attend and sent Canon Henri de Dorlodot, director of the university's geological institute. Dorlodot's praises for the Newton of the organic world surprised many Catholics and elicited some criticism. A series of Dorlodot's lectures at Louvain was eventually printed, thus giving a sort of Catholic intellectual *nihil obstat* to the teachings of Darwin.[58]

But divisions in the Catholic world over Darwinism were just as sharp as ever; only the issues had changed. The development of idealistic and vitalistic trends led some Catholic intellectuals to conclude that, al-

[58] Henry [Henri] de Dorlodot, *Darwinism and Catholic Thought*, trans. Ernest Messenger (London: Burns, Oates, 1922). The first edition in French was printed in 1918 but dated 1913 in order to avoid German censorship.

though there might be no necessary clash of Darwinism with religious orthodoxy, the debate was irrelevant because of the scientific irrelevancy of Darwinism. At the same time, of course, there was a definite acceptance of Darwinian ideas in segments of the Catholic intellectual community, which has never been noted for unanimity on important ideas. Since only the remotest of parameters of orthodoxy existed, a great deal of disagreement could occur. This was clear in a series of studies on Darwinism that appeared in the *Revue de Philosophie* in 1910.[59] Agostino Gemelli harped on the bankruptcy of Darwinism in research on the nature and origin of living beings. In this area the dynamic research paradigm was vitalism.

Le darwinisme, qui ne devait être qu'une explication particulière du mécanisme de l'origine des espèces organiques, a étendu ses prétentions; il est devenu l'expression la plus authentique du mécanisme moderne, le boulevard le plus puissant du positivisme matérialiste, la base réputée inébranlable des sciences naturelles modernes. Et c'est là que sa chute a été la plus désastreuse, c'est là qu'il a le plus manqué à ses promesses. C'est sur ce terrain indûment envahi que les nouvelles recherches ont clairement montré l'inexactitude de sa solution. Pour nous, nous pouvons considérer les deux noms de vitalisme et de darwinisme comme deux noms antithétiques, et nous pouvons dire que le second n'a plus qu'une valeur historique.[60]

Maritain took much the same approach but was more radical in arguing

59 "Etudes sur le darwinisme," *Revue de Philosophie* 17 (September–October 1910): 215–441. The authors were well-known Catholic scientists and philosophers: Agostino Gemelli, "Darwinisme et vitalisme"; A. Briot, "Le Problème de la vie"; C. Torrend, "Le Transformisme et les derniers échelons du règne végétal"; Erich Wasmann, "La Vie psychique des animaux"; Henri Colin, "La Mutation"; Robert de Sinéty, "Mimétisme et darwinisme"; Max Kollmann, "Les Facteurs de l'évolution"; R. D., "La Loi biogénétique fondamentale"; J. Gérard, "Evolution, darwinisme, vitalisme: Etat de la controverse en Angleterre"; Jacques Maritain, "Le Néovitalisme en Allemagne et le darwinisme."

60 Gemelli, "Darwinisme et vitalisme," p. 249. "Darwinism, which was supposed to be simply a special explanation of the mechanism of the origin of organic species, has extended its claims; it has become the most authentic expression of modern mechanism, the most powerful avenue of materialist positivism, the supposedly unshakeable basis of modern natural science. And it is here that its fall has been most disastrous, here that it has most certainly failed to live up to its promises. It is on this improperly invaded terrain that new research has clearly showed the inaccuracy of its solution. For our part, we may consider the two names vitalism and Darwinism as antithetical, and we may say that the second has only historic importance."

that transformism in general was not so much denied as simply ignored by the Driesch school; the issue of evolution remained open, but it was a purely historical one. The German neovitalists had absolutely abandoned Darwinism. Vitalism interpreted the nature of life qualitatively, in contrast to the quantitative interpretation given by transformism. "Le vitalisme scientifique: 1° exclut formellement et totalement le *darwinisme*; 2° réduit formellement le transformisme au rôle d'hypothèse historique; et 3° est même par tendance, implicitement, antipathique à cette hypothèse."[61] It is clear that Catholic opinion was part of the division of opinion that existed in the scientific world itself.

Erich Wasmann, the German Jesuit biologist who had defended Darwinism against the attacks of Albert Fleischmann, merely emphasized the gulf between the other animals and man. "Chez l'homme seul la vie sensible se couronne et s'achève en vie intellectuelle."[62] An explicit acceptance of a limited type of evolution was stated by C. Torrend, of the Institute of Natural Sciences, Collège de Campolide, Lisbon. Species can and do change, although never beyond certain limits—an ass will never become a cow, nor a hen a turkey. The role of Providence was conspicuously retained.

Et si le naturaliste à une connaissance intime de la nature joint *un amour insatiable* de la vérité, s'il étudie ces merveilles de Providence et de finalité qu'il découvre à chaque pas dans la nature, et cette Providence bien plus merveilleuse qui conduit le roi de la Création à sa fin suprême, oh! je ne doute pas que la vie de ce naturaliste se transforme en une vie de bonheur et d'allégresse, en une vie intime avec son Créateur, de véritable intuition face à face, suivant l'expression du vénérable M. Fabre, le roi des Entomologistes,

---

[61] Maritain, "Le Néovitalisme en Allemagne et le darwinisme," p. 441. "Scientific vitalism: (1) formally and totally excludes *Darwinism*; (2) formally reduces transformism to the role of a historical hypothesis; and (3) is even, in tendency, implicitly antipathetic to this hypothesis."

[62] Wasmann, "La Vie psychique des animaux," p. 321. "Only in man is the life of the senses crowned and completed in the life of the intellect." Wasmann is an obvious exception to the exaggerated thesis of Hermann Josef Dörpinghaus, in *Darwins Theorie und der deutsche Vulgärmaterialismus im Urteil deutscher katholischer Zeitschriften zwischen 1854 und 1914* (Freiburg: E. Wasmann, 1969), that Catholic writers failed to separate the scientific part of evolution from the interpolations of atheistic popularizers. See review by Ralph H. Bowen in *The American Historical Review* 77 (April 1972): 537–538.

dont les sociétés savantes se plaisent à célébrer cette année-ci le cinquantième anniversaire de travaux entomologiques.[63]

But the fact that Torrend, writing in one of the leading Catholic journals of the time, published under the aegis of the philosophers and theologians of the Catholic Institute in Paris, made only limited use of the notoriously antievolutionist views of Henri Fabre is in itself an event worthy of note. The famous *Souvenirs entomologiques* of the "Virgile des insectes" were usually exploited by religious antievolutionists to provide scientific blasts against all evolutionary ideas. To call on Fabre as a scientific witness against atheism was in keeping with the antiscientistic move that had been underway since the late 1870's and in tune with the intellectually based religious revival of the period. It was intellectually respectable to use the idea of cosmic teleology as the only possible hypothesis to explain the enigmas that science had failed to solve. "Dieu, non seulement je crois en Lui, mais je Le vois. Sans Lui, tout est ténèbre; sans Lui je ne comprends rien dans la nature. L'athéisme n'est qu'une lubie des temps présents. Non, on m'ôtera la vie plutôt que de me faire perdre la croyance en Dieu."[64]

In spite of the growing accommodation between Catholic thought and varieties of evolutionary thought, even Catholic scientists did not usually accept all the implications of evolutionary thought present in the work of many essentially neutral but nonreligious scientists. Acceptance of Christian values meant that any evolutionary view accepted would at least have to avoid any dissonance with the basic elements of the Christian outlook. In reviewing Marcellin Boule's *Les Hommes fossiles* (1921), Teilhard de Chardin was able to endorse enthusiastically the scientific part of

[63] Torrend, "Le Transformisme dans les derniers échelons du règne végétal," pp. 311–312. "And if the naturalist joins *an insatiable love* of truth to an intimate acquaintance with nature, if he studies those wonders of Providence and of finality that he discovers at each step in nature, and that still more wonderful Providence that conducts the king of Creation to his supreme goal, oh! I cannot doubt that this naturalist's life will be transformed into a life of joy and happiness, a life of intimacy with his Creator, of true face-to-face intuition, to use the expression of the venerable M. Fabre, the king of Entomologists, whose fiftieth anniversary of entomological studies the learned societies are celebrating this year."

[64] Henri Fabre, *Souvenirs*, cited in ibid., p. 313. "Not only do I believe in God, but I see Him. Without Him, all is darkness; without Him I understand nothing in nature. Atheism is only a fad of the times. No, they will take my life before they will make me lose my belief in God."

Boule's work and recommend it to theologians and philosophers. The work had been written in a great spirit of conciliation and in absolute good faith. In view of the antireligious orientation of much preceding similar work, Teilhard noted this characteristic with great satisfaction. But Teilhard noted that, in his conclusions, Boule's statement of the unity of man with his history and the material world contained some expressions that could not enter unchanged into Christian thought. Without some explanations the book could not be given to all. Philosophers and theologians should try to put into orthodox language the teaching whose outlines, still encumbered with conjectures and hypotheses, seemed to conform to reality.

La lettre de la Bible nous montre le Créateur façonnant le corps de l'homme avec de la terre. L'observation consciencieuse du monde tend à nous faire apercevoir aujourd'hui que, par cette "terre," il faudrait entendre une substance élaborée lentement par la totalité des choses,—de sorte que l'homme, devrions nous dire, a été tiré non pas précisément d'un peu de matière amorphe, mais d'un effort prolongé de la "Terre" tout entière. Malgré les difficultés sérieuses qui nous empêchent encore de les concilier pleinement avec certaines représentations plus communément admises de la création, ces vues (familières à saint Grégoire de Nysse et à saint Augustin) ne doivent pas nous déconcerter. Petit à petit (sans que nous puissions encore dire dans quels termes exactement, mais sans que se perde une seule parcelle du donné, soit révélé, soit définitivement démontré), l'accord se fera, tout naturellement, entre la science et le dogme sur le terrain brûlant des origines humaines. Evitons, en attendant, de rejeter d'aucun côté, le moindre rayon de lumière. La foi a besoin de toute la vérité.[65]

[65] Pierre Teilhard de Chardin, Review of Marcellin Boule, *Les Hommes fossiles: Eléments de paléontologie humaine, Etudes* 166 (March 5–20, 1921): 570–577. "The letter of the Bible shows us the Creator fashioning the body of man from the earth. Conscious observation of the world tends to make us perceive today that for this 'earth' one must understand a substance slowly elaborated by the totality of things—so that man, we should say, was taken not exactly from a little bit of amorphous matter, but from a prolonged effort of the whole 'Earth.' In spite of the serious difficulties that still prevent us from reconciling them completely with certain more commonly accepted representations of the creation, these views (known to Saint Gregory of Nyssa and Saint Augustine) should not disconcert us. Little by little (without our being able yet to say in exactly what terms, but without losing a single particle of the data, whether it be revealed or definitively demonstrated), an agreement will be reached, quite naturally, between science and dogma on the burning issue of human origins. Let us avoid, in the meantime, rejecting the slightest ray of light from any direction."

Later Teilhard himself would try to give "a full and coherent account of the phenomenon of man" and work out the agreement that he thought necessary. "Religion and science are the two conjugated faces or phases of one and the same complete act of knowledge—the only one which can embrace the past and future of evolution so as to contemplate, measure and fulfill them."[66]

In spite of the revolution that has taken place in biblical interpretation since the nineteenth century and in spite of a certain reconciliation of Catholicism with evolutionary thought, certain areas of disagreement or at least difficulty remain, especially concerning man. It is possible to interpret the biblical account "as saying that God drew the human body from an animal organism which was transformed so as to receive a soul. It is possible that this transformation occurred before the infusion of the human soul." But, if one conceives of the animal organism as having been finally sufficiently perfected to receive the soul, one cannot remain orthodox in thinking that "this perfection . . . *required* the infusion of a human soul from purely immanent intramundane processes." The whole process must be recognized as directed by God. A Catholic cannot accept "the hypotheses of the purely animal origin of man." God played a key role in the formation of the body as well as the soul. Several varieties of religiously orthodox evolutionary belief are possible if one accepts the proposition that "body and soul are still formed by the immediate intervention of God."[67] Equally vital for Catholic thought is the venerable doctrine of the unity of the human race. It is impossible to square polygenesis with the dogma of original sin. *Humani Generis*, whatever its careful concessions in other areas, was explicit on this issue. "No Catholic can hold that after Adam there existed on this earth true men who did not take their origin through natural generation from him as from

---

Faith needs the whole truth." Boule was professor of paleontology at the Museum of Natural History and editor of *L'Anthropologie*.

66 Pierre Teilhard de Chardin, *The Phenomenon of Man* (New York: Harper, 1959), pp. 312–313. This work was finished in 1938 and published posthumously in 1955 as *Le phénomène humain* (Paris: Le Seuil).

67 All quotes from the Catholic theologian Robert W. Gleason, "A Note on Theology and Evolution," in *Darwin's Vision and Christian Perspectives*, ed. Walter J. Ong (New York: Macmillan, 1960), pp. 104–113.

the first parent of all, or that Adam is merely a symbol for a number of first parents. For it is unintelligible how such an opinion can be squared with what the sources of revealed truth and the documents of the Magisterium of the Church teach on original sin, which proceeds from sin actually committed by an individual Adam, and which, passed on to all by way of generation, is in everyone as his own."[68] There are certain areas where theology and science advance mutually exclusive hypotheses; in these rare cases orthodoxy demands that the Catholic safeguard dogma by choosing another "scientific" hypothesis.

In conformity with the present state of science and theology, the doctrine of evolution should be examined and discussed by experts in both fields, in so far as it deals with research on the origin of the human body, which it states to come from pre-existent organic matter (the Catholic faith obliges us to believe that souls were created directly by God). But this must be done in such a way that the arguments of the two opinions, that is, the one favorable and the other contrary to evolution, should be weighed and judged with all necessary seriousness, moderation, and restraint, and on condition that they are all ready to submit to the judgment of the Church, to which Christ has entrusted the office of interpreting authentically the Holy Scriptures and of defending the dogmas of the faith.[69]

Such a concession, however cautious, had it been made even two decades after the appearance of the *Origin*, would probably have saved a great deal of anguish on the part of those few intrepid clergy who admitted the validity of substantial parts of Darwinian arguments, and it might have placed a damper on the regurgitation of a great many sterile anti-Darwinian arguments soon shrouded in spurious orthodoxy.

[68] Cited in ibid., pp. 112–113.
[69] Cited in Greene, *Darwin and the Modern World View*, p. 24.

DARWINIAN AND "DARWINIAN" EVOLUTIONISM IN THE STUDY OF SOCIETY AND CULTURE

ANTHONY LEEDS

DARWINIAN, "DARWINIAN," AND NON-DARWINIAN MODELS OF EVOLUTION

My comment pertains to the distinction between Darwinian and "Darwinian" evolutionism in the domain of the social sciences and general history—that is, the domain of the study of society and culture and of their changes. The distinction—signaled throughout what follows by the presence or absence of the quotation marks—is that between Darwin's model of evolution, as expressed primarily in *The Origin of Species*, and all those models of evolution whose authors conceived them to be, or which other writers have asserted to be, Darwinian but which, on closer analysis, are demonstrably different from Darwinism. By extension, they are also different from the evolutionary models which are derived from Darwin's, albeit with necessary modification or with basic additions, such as that of the Mendelian process. They are distinct from these neo-Darwinian or "genetico-Darwinian" models because they still do not contain the central assumptions and viewpoints of Darwin's original model as carried down into the derivative models: they operate, as I shall show, with quite different sets of assumptions and outlooks.

The Darwinian model may briefly be described as follows. A living population of animals is designated a 'species' when it is marked off from other populations rather similar to itself if the first population cannot or can only poorly and under highly specific circumstances interbreed with the second.[1] Changes observed in such a population do not occur within an unchanging, ever-repeating range, fixed forever by the nature of the species itself (the fixity-of-species notion), nor are they immanently determined by the essence of the species even if these changes occur in an unfolding, nonrepetitive sequence (the Lamarckian version of the fixity of species). Changes observed are (a) events inherent in natural process; (b) phenomena which are nondirectional, i.e., random, in origin, even if, cumulatively, directional in retrospective view; (c) the basis of the appearance of new species;[2] (d) *statistical* results of the reproductive activity of the population of *individuals*; (e) *statistical* results of the elimination of individuals less suited and the survival of those more suited to external environmental conditions by "natural selection." The mechanism of sexual reproduction accounts for both the continuity of species characteristics *and their changes* through natural selection. In this view, no god is needed to direct things, even though one is not excluded necessarily; no corporative social body exists as an actor in the scenario of modification, although it may be possible to introduce organized aggregates, like ants, as cooperative bodies affecting evolution; the evolutionary process itself is not directed toward any goal except "improvement";[3] the world is basically ruled by chance rather

[1] See John S. Haller, Jr.'s treatment of hybrids in "The Species Problem: Nineteenth-Century Concepts of Racial Inferiority in the Origin of Man Controversy," *American Anthropologist* 72, no. 6 (1970): 1319–1329, esp. pp. 1320–1321, 1322, 1325. Discussion of hybrids revolved around efforts, virtually without basis in biological observation, to designate different stocks of mankind as *species*—largely as rationales for extant social and ideological stances. The fact that all stocks could interbreed made difficult the argument for their species status, hence the hybrid-sterility argument (which also flew in the face of facts). The species problem—in the context of prior monogenist-polygenist conflict—was crucial in the acceptance, modification, or rejection of Darwinism because of the social, ideological, and religious stances, threatened by Darwin's model, which were so important in the social philosophies and their views of societal evolution.

[2] See Haller's discussion (ibid., pp. 1326–1327) of Darwin's view of race as emerging species—a view that would accommodate polygenist and monogenist views of the origin of man.

[3] It is perhaps not clear in Darwin (and certainly not in Spencer) whether this "improvement" is to be seen as relative or absolute. In the former sense it would mean improvement relative to specifiable particular conditions subject to change; in the latter, a development which is absolutely better than all preceding it regardless of conditions. Darwin appears to have something of the latter in mind—and therefore also a teleological undertone of movement toward an all-purpose improvement in the over-all

than by design. The schema—a single, coherent explanatory system—accounted for a wider array of facts than any other conception except Christian creationism; dealt with certain aspects of reality—fossils, for example—much more simply than any other theory, including Christian creationism; and even permitted dealing with man-directed action—e.g., stockbreeding. The social evolutionism discussed or adverted to in the papers of this volume and discussed at length below is quite different from this model in almost every major respect.

It is important to note that the model is not made explicit in most of the papers in this book; the writers assume not only that it is known to the readers but also that it was not merely known but *held* by the writers they discuss. This is demonstrably false for a number of cases, e.g., the Mexican one discussed by Roberto Moreno—the Mexicans *said* they were Darwinian, but any close analysis of what they were actually assuming seems clearly strongly Lamarckian (see the quotations given in Moreno's article). What appears to me striking is how few of the figures discussed in these pages—with the exception of a small number of the Spanish, the Germans, and the English—held a Darwinian view at all. Mostly they assimilated a phrase or an aspect of Darwin's expression of his thought to their own understanding and thought, then, that they were Darwinians. The most striking case is that of the Russians, discussed in James Allen Rogers's paper, in which *not one* of the protagonists of his drama is remotely near the Darwinian model.

The problem of the effect of Darwin's works on the domain of the social sciences and general history is not discussed at all in a number of the papers of the present volume, merely touched upon in others, and discussed at length only by Thomas Glick, Moreno, Edward J. Pfeifer, and Rogers. Only Glick and Rogers provide an extensive discussion of the materials relevant to the goal of this comment: to distinguish between the models in which most thinking about the evolution of society

---

cumulative course of evolution. This sort of view is reflected also in Lamarck's passages on use and disuse as developers of immanent species characteristics. This kind of teleology is not essential to Darwin's model, which stands as well without it as with it, and has since been consistently eradicated to create a relativistic evolution seen as being "opportunistic" (see George Gaylord Simpson, *The Meaning of Evolution: A Study of the History of Life and of Its Significance to Man* [New Haven: Yale University Press, 1949], esp. pp. 160–186).

has been embedded and the model—the Darwinian—in which intellectual folklore alleges such thought to have been cast. I shall go well beyond the content of the papers in this book to elaborate the point, turning to my own field, anthropology, for my most salient examples, but referring to related fields as well.

In these papers, as elsewhere generally, the folklore of intellectual history and academic self-perception regarding Darwin and Darwinism, particularly but not exclusively with respect to the social sciences, is striking in two ways. First is the great overestimation of Darwin's contribution (as distinguished from the reaction to it; the strength of the latter has been mistaken for a measure of the former, with the result that Darwin is often made into a kind of Culture Hero rather than being seen in his proper place in the flow of ideas in the West).[4] Second is the very prevalent acceptance of the idea that evolutionistic views in the study of society emerged out of Darwin by application to sociocultural evolution of his vision regarding biological evolution and his adversions to "Man's" evolution.

PRECURSORS IN EVOLUTIONARY THOUGHT

With respect to the first point, it has been more than adequately demonstrated in a number of works[5] that virtually all the elements involved in Darwin's model of evolutionary process had been thought of before

---

[4] Arthur O. Lovejoy, in *The Great Chain of Being: A Study of the History of an Idea* (Cambridge: Harvard University Press, 1936), ch. 1, makes the point that ideas, to be fully understood, must be seen in the context of their times, in relation to other ideas of those times to which they were systematically linked, and in terms of the then-current axioms and logics by which they were linked. It is of the greatest urgency that one not read meanings retrospectively into past ideas, given latter-day wisdom (itself subject to vagaries of particularistic logics and axioms, prejudices and biases). This view is identical, I believe, with the basic methodological proposition of anthropological field work: one should try to see the society and culture as much as possible from the people's own contemporaneous point of view and to avoid imposing one's own conceptions and interpretations on them. I try here to observe these rules with respect to the historical situation under discussion. (In line with this ethnographic treatment, I take the Christian "God" not as axiomatically given but simply as a native statement about the universe. I refer to this deity as 'god' unless it is being spoken of by its proper name.)

[5] See Loren Eiseley, *Darwin's Century: Evolution and the Men Who Discovered It* (Garden City: Doubleday, 1958); John C. Greene, *The Death of Adam: Evolution and Its Impact on Western Thought* (Ames: Iowa State University Press, 1959).

him, especially by Georges Buffon, but also, to a lesser extent, by Erasmus Darwin and others, including (in his later years) Carolus Linnaeus. All of these were known to Darwin. Further, Buffon's evolutionism, in particular, is, in essence, quite as nontheistic as Darwin's—possibly more so —and quite as "naturalistic." What he seems to have failed to do is to put the elements into a single coherent theory, or to have discovered the mechanism linking the elements.

Although Lamarck's theory of evolution appears to have been more coherent than Buffon's and quite as internally consistent as Darwin's, this coherency did not induce such pervasive and violent reaction as did Darwin's (as Glick demonstrates for Spain, Rogers for Russia, and Robert E. Stebbins for France). The fact requires us to question why Lamarck's evolutionary theory—or, better, model—did not elicit a strength and severity of response remotely resembling the response to Darwin's and certainly did not create the same kind of folk beliefs that Darwin's evolutionism did in the history of science. The question is important because, first, some aspects of Lamarckian evolutionism—specifically, use and disuse—have never been adequately rejected, and, second, Lamarck's model—as I shall show below—was either itself a dominant one in sociocultural evolutionism until the middle of the twentieth century or had major features that characterized all the sociocultural evolutionisms till recently but are *not* present in Darwinian biological evolutionism.

The Lamarckian sociocultural evolutionism was even reinforced by the use-disuse conception, which continues to reappear both in sociocultural and in biological evolutionism. Though repeatedly pronounced dead, especially since the ouster and subsequent demise of the Soviet biologist Trofim Lysenko, biological Lamarckism continues to have a shadow existence and reappears now and again with a guttering flame.[6] A number of

---

[6] See, for example, "The Genetics Controversy," "I: The General Issues," by Marcel Prenant, and "II: Lysenko and the Issues in Genetics," by Jeanne Levy, *Science and Society* 13, no. 1 (Winter 1948–1949): 50–54, 55–78. See also Bernhard J. Stern, "Genetics Teaching and Lysenko," *Science and Society* 13, no. 2 (Spring 1949): 136–149. Stern demonstrates convergences between Lysenko's Lamarckism and Western genetics, including a discussion by the latter of possible external effects on cytoplasm. See, too, as a more recent example, Arthur Koestler, *The Case of the Midwife Toad* (New York: Random House, 1972), and its review by Stephen Jay Gould in *Science* 176 (1972): 623–625, entitled "Zealous Advocates."

observations and experimental results continue to suggest that use/disuse cannot be put aside so lightly, that such dismissal has more to do with ideology than with observation and scientific reasoning: e.g., the famous experiment with worms one end of which was treated with electric charges; the worms then cut in half regenerated as whole worms, but each former half, now whole, retained the effects of the electric charge. Other cases can be adduced with careful reading of the literature. For my part, I have never been able to understand the logic of denying possible chemical or electrical effects (in organisms constituted of chemical and electrical connections and charges) induced by the use of organs, if one at the same time allows effects of chemical and electrical stimuli on genes or cytoplasm from nonuse sources, e.g., X-rays, thalidomide, German measles, drugs of various sorts. This position, implying that the gene is a chemically and electrically closed monad, has always struck me as inherently untenable.

Further, it may even be that Baron Georges Cuvier's catastrophist theory, formulated to account for observed changes in the fossil content of geological sequences, was more coherent as a theory than Buffon's, although it was not properly an evolutionary theory, since it tried to account for the observed differentiating sequence essentially by denying natural process and attributing the change to the volition of a deity—the Christian God. His catastrophist theory could only stand as long as catastrophism as a general theoretical outlook in geology could stand. That outlook, and with it Cuvier's theory, died with the publication of Charles Lyell's monumental work on the principles of geology, just about the time that Cuvier himself died.

Finally, aside from a number of quite sophisticated and relatively coherent theories about biological succession and change which, for at least one hundred years before 1859, had existed for intellectuals to choose among, social or societal evolutionary thought *not* springing from Darwin pervaded the era. Characterized by the sort of Lamarckian features adverted to above, it was represented by a number of major thinkers or schools of thought: Spencer and the Social Darwinists; Marx, Engels, and the Marxists; the Scottish moral philosophers and jurists and their descendants; the antiquarians; the physiocrats and economists; and others. To some of these I return below.

Thus, in sum, not only was the whole era into which Darwin was born pervasively evolutionist in outlook—as one would expect with the eighteenth century's general secularization and temporalization;[7] not only was Darwin much less of an innovator of specific evolutionary ideas and concerns than the folk beliefs of intellectual history hold; but also the alternative evolutionary models—whether in biology or the social sciences—were not and, in the third quarter of the twentieth century, still have not been extensively supplanted by Darwinian evolutionism. Furthermore, the dominating evolutional models of society in the nineteenth century were, with minor exceptions, clearly not Darwinian; a few of the more important ones, like Herbert Spencer's, supposedly became Darwinian, although his is not so in fact; the rest were, in varying degrees, plainly Lamarckian, as Pfeifer cogently shows for American thought. Certainly with respect to the sociocultural world, Darwin and his influence have been vastly overrated, and I think there is a strong case for arguing that his specific contributions to biology have also been overstated in the histories.

In view of the preceding, two observations need to be made. First, alternative models of evolutionary process still have viability, *especially in the social sciences*, where Darwinian models struggle exiguously for marginal existence. One should note, in this connection, Rogers's discussion of the reception of Darwin in Russia, because not one single group that he discusses as responding to Darwinism is using a Darwinian model! Not even the Russian biological models of evolution can be fitted to Darwin's own model; they were, at best, "Darwinian." The fact of the great viability of alternative models in the social sciences suggests that the sociocultural world is not easily susceptible to Darwinian models at all, or at least is so only in limited ways.

Second, something special to Darwin's model—absent from other evolutionary models—must have fired the tremendous reaction it got. The reaction cannot be attributed to the model's evolutionism or even to its novelty as such, since virtually all the elements were not new and had figured in other evolutionary outlooks, from that of Buffon, through those of Erasmus Darwin, Thomas Malthus, Lamarck, and Cuvier, to that of Charles Lyell.

---

[7] Lovejoy gives a detailed and elegant account of these developments in *The Great Chain of Being*, esp. ch. 9, pp. 242–287.

What was involved appears to have been, on one hand, a degree of incontrovertibility of a dual and interlocking sort, resulting from the logical structure of Darwin's model, which linked all the analytical elements into a whole, and from the tight relationship between these logically linked elements and vast bodies of differentiated data—all tied together with a degree of simplicity and a minimum of axioms and suppositions. On the other hand was the fact that the entire evolutionary process could operate mechanically, without a god and as a natural process. The effect of the incontrovertibility was quite different in different national contexts but seems to me crucial in all the cases described in this book. The very power of the incontrovertibility once and for all destroyed the idea of the fixity of species (already under question for more than a century) and all the Christian world views, ontologies, metaphysics, and political power systems based on that notion. The threat and the actuality of such destruction are made poignantly sharp especially in Glick's paper. Darwin's theory, incontrovertibly pointing to a uniformly operating mechanism which certainly needed no god to keep it going, if, indeed, it required a god at all to start it, generally threatened the conceptions of the purposiveness and teleological nature of the universe—and most particularly of Man—characteristic of the Christian world and its secularized offspring, like Marxism and anarchism. Finalism, progress, advance, betterment, improvement, striving toward ultimate ends, dialectic self-fulfilment, total self-consciousness, moving toward the Last Judgment, millennialism—Christian or Marxist—none of these was necessary to the Darwinian model, although teleology was not entirely absent from it in its vestiges of Lamarckism and the ideas of "progress" and "improvement."

It is striking, by contrast, that in all these respects Lamarck's and Cuvier's models of evolution or succession and those of the sociocultural evolutionists were fundamentally different. All, although they allowed for changes, worked out compromises between change and fixity of species reasonably consonant with observation—Lamarck's with known geographical distributions and behavioral characteristics of species; Cuvier's with geological distributions of fossils; the sociocultural evolutionists' with known histories of institutions and cultural inventories, and with trait distributions.

Lamarck's model was consonant with a number of views prevalent in the eighteenth century[8] according to which changes in species were possible by virtue of their being immanent in the species as created in the beginning by a fecund god who attached perfectibility to the life of species on earth. The changes, therefore, were necessarily sequential (i.e., they had "stages"), linear, in some way teleological, and end-oriented; and they were almost necessarily progressive in the sense of "improving." The Lamarckian model dealt, furthermore, with a mechanism by which change from one immanent form of the species to the next could take place, namely, use and disuse of organs in adaptation to environment, a process observable in nature. Herein lay a rich body of observation, a vast array of data on species distribution and behavior, and a mechanism —use and disuse—which, on the face of it, appears quite plausible.

In sum, for Lamarck, the species was fixed but contained the seeds of its own transformation through predetermined, teleologically oriented stages.[9] This conception of species is fully consonant with that of the older man, Jean Baptiste René Robinet, who makes explicit his idea of "seed" or "germ":

All germs have individual differences; that is to say, their life, organization, animality, have *nuances* which distinguish each of them from all the others. There are no elements except the germs; all the elements are therefore heterogeneous. These elements are not simple beings. . . . Elements are composed of other elements. . . . There is no natural nor artificial process which can bring an element, or germ, to the last degree of possible division. Germs, as germs, are indestructible. They can be dissolved into other germs only after the completion or the beginning of their development; in the state of germ they admit of no division. In the resolution of a developed

---

[8] See Lovejoy, *The Great Chain of Being*, esp. chs. 8 and 9, particularly the passages on Robinet (pp. 269–283); note the evolutionistic quotes from Diderot, 1754 (pp. 274–275), and the fact that Robinet's views re-emerge in Lamarck. Robinet's magnum opus, *De la nature*, was written between 1761 and 1768; Lamarck was born in 1744; as a very young man he might well have been influenced by Robinet's views. One might suppose that the model of evolution that informed the thought of both men was *the* ethnic model of the times in France.

[9] I use the phrase "seeds of its own transformation" here as a deliberate allusion to Marx's "seeds of its own destruction" because I believe Marx's conception is a descendant of the kind of conception Robinet was expressing a century or so earlier. In Karl Marx, too, there is a kind of immanentism, teleology, and even finalism (to which I return later), *not* in the reality of the world so much as in the logic, of all possible logics, he chose to use to describe this world, that is, the dialectic.

germ into a multiplicity of other germs, there is no matter that dies. All of it remains alive; only its forms and combinations change. . . . This is to say that there is no destruction of anything in nature, but a continuous transformation. The idea of succession enters necessarily into the definition of Nature. Nature is the successive sum of phenomena which result from the development of the germs. . . . The series [of germs] is inexhaustible, whether read backward into the past or forward into the future. A germ which has begun to develop and has encountered an unsurmountable obstacle to the continuance of this development, does not retrogress to its original state. It struggles against this obstacle until its useless efforts bring about its dissolution, as its complete development would also naturally have done. . . . The existence of Nature is necessarily successive. . . . Germs created all together do not all develop together. The law of their generations, or manifestations, brings about these developments one after another. . . . Nature is always at work, always in travail, in the sense that she is always fashioning new developments, new generations.[10]

It should be noted, with respect to this sort of view, that much time was involved but was not *specified* in the form of rates—contrary to what Darwin's model eventually entailed, especially after the development of genetics.

Cuvier maintained the fixity of species and a creationist stance by a really most clever model, one quite plausible, given the general state of knowledge, the point of development of geology, and the condition of theology at the time. In his model, all species were fixed, but different species or sets of species were created (by the Christian god) at different times, prior sets being eliminated in great "natural" catastrophes, such as floods—for example, The Flood—presumably also willed by the god. The elimination of species did not imply that the god need have eliminated *all* species at each catastrophe; it left open the possibility that some, by his will, might have continued into the succeeding epoch. I think that, in our latter-day wisdom, we tend to pooh-pooh catastrophism as a simple-minded clinging to religious orthodoxy, while failing to see its correspondences to empirical observations. In Cuvier's case, geological nonconformities and disconformities indicated sharp differentiation of epochs. The evidence of such disparate sets of strata was strongly reinforced by the fact that the fossil content of each layer was

10 Quoted in Lovejoy, *The Great Chain of Being*, pp. 274–275.

widely but not totally disparate from that of other layers. Much of the fossil content was of fauna, especially mollusks and other shelled animals, confirming prior presence of water on a vast scale.

In this model, Cuvier was able to accommodate a god who still figured dominantly in Western ontogenies; the dogmas of creation;[11] the fixity of species; the fossil sequence; marine fossils in hill lands; the geological ordering, change, and time; and a classificatory scheme based on his immense knowledge of comparative anatomy, which did not force him to see descent with modification. Again, as with Lamarck, time had no units, or, at least, *rates* had no importance in the model. Finally, Cuvier's model, though not consonant with such calculations as Bishop James Ussher's, which gave to the world an age of only a few millennia, *was* consonant with the logical arguments concerning a fecund god whose creation of all possible species could have occurred, not at a single moment, but as continuous creation.[12]

Thus, both Lamarck and Cuvier developed models which either did not threaten orthodoxies or actually supported them. They did not eliminate Mind and replace it with Mechanism in a Newtonian sort of biological process—once wound up such a process would continue indefinitely—and therefore they did not challenge idealist views of Man by suggesting that Man and his societies operated by Mechanism rather than Mind, just as did the animals. They did not assimilate Man to man, and man to animals. Both models accommodated ideologically acceptable notions of essence, sequence, teleology, and progressivism. Yet both also dealt with changes of species, succession, time, geographical setting, form and structure of organisms, and, in Lamarck's case, with a plausible (and still not dismissed [see above]) mechanism of evolutionary transmutation—all of this together representing a materialist aspect of their evolutionism. Add to this Buffon's recognition of fecundity as delimiting species boundaries, the importance of geographical distributions, and other matters, and Linnaeus's recognition of temporal and processual implications of minute gradations within species, and one has virtually all of the elements of Darwin's evolutionary model present well before he was born.

[11] See Appendix A.
[12] See Lovejoy, *The Great Chain of Being*, esp. ch. 9.

The responses to Darwin, then, appear to be much more in terms of implications perceived (sometimes erroneously) in the model than in terms of the model itself and least of all in terms of anything radically new, from a strictly biological point of view, that it presented. I think the case histories in this volume, especially when they advert to "social sciences," clearly bear out this contention, particularly the case of Russia.

STUDENTS OF SOCIETY, RESPONSE TO DARWINISM, AND SOCIETAL EVOLUTIONISM

Turning to my second point regarding the intellectual historical folklore as to Darwin's importance as a transmitter of models for sociocultural evolution, the following quotations from A. I. Hallowell, otherwise one of the finest minds in anthropology and with a highly informed sensitivity to the history of the field, are characteristic: "In the intellectual climate of the post-Darwinian period, however, human evolution was conceptualized in much broader terms. The advent of Darwinism helped to define and shape the problems of modern psychology as it did those of anthropology." Again, further on he says: "It was also under the stimulus of Darwin's ideas as applied to man that historians, economists, sociologists, linguists, cultural anthropologists, and others began to apply evolutionary ideas to human institutions on a wide scale." He then goes on to show that a brief flurry of concern over the evolution of mind generated by some of Darwin's concerns died out rapidly, leading "to a *re*-creation of the old gap between man and the other primates which, it was once thought, the adoption of an evolutionary frame of reference would serve to bridge."[13] I will show that the "*re*-creation of the old gap" was simply the continuity and reassertion of the old evolutionary model which had existed prior to Darwin, continued through the initial florescence of Darwinism. J. W. Burrow, of course, shows this, but his views (*a*) are not those generally held by scientists in the various social disciplines, who tend to take a view somewhat like that in the quotations from Hallowell (which Hallowell himself modifies later in the article); (*b*) are not generally held by historians of science dealing with

---

[13] A. I. Hallowell, "Self, Society, and Culture in Phylogenetic Perspective," in *Evolution after Darwin*, ed. Sol Tax (Chicago: University of Chicago Press, 1960), vol. 2, *The Evolution of Man: Man, Culture, and Society*, pp. 309, 310, 314.

biological evolution when they talk about Darwin's effect on social thought; and (*c*) do not correspond to what the "social scientists" of the last and of this century say about themselves in respect of Darwinian influences (e.g., Spencer). The old evolutionary model has persisted ever since (see the discussion of Leslie A. White in text below).[14] It was the *prior* evolutionism which, as both Rogers's and Pfeifer's articles show, shaped certain elements of Darwinism to its own models, as in the case of Spencer, and created "Darwinism," "social Darwinism," social Spencerism, and social Lamarckism (to use Pfeifer's felicitous phrase).[15]

Such statements as Hallowell's (or, for example, Pfeifer's in this volume: "Darwin . . . had triumphed") seem to me simply historically false, in part dependent, as discussed above, on an erroneous weighting and post-hoc overevaluation of Darwin in general and particularly with respect to the social sciences. It is interesting to compare the entries for Buffon, Cuvier, Lamarck, and Darwin in the *Universal Pronouncing Dictionary of Biography and Mythology* by J. Thomas, published in 1873. The generally quite thorough and sophisticated entries include reference to one or more major sources of commentary on the life or work of the person. Buffon is given 62 lines of text, Cuvier 104, Lamarck 39. Darwin gets only 29, and the entry refers primarily to his literary style. Given the immense impression made by the earlier three, one wonders how devastating the effect of the *Origin* or the *Descent* had been in

---

[14] White has often contended that for the social sciences there is no "neo-evolutionism"—there is only one evolutionism, which he sees as a grand fusion of nineteenth-century societal evolutionism with Darwinian evolutionism. The fact is that his own evolutionism is purely of the nineteenth-century societal type that I discuss in the text, virtually without a trace of the Darwinian model. It is largely descended from Spencer and Durkheim. Burrow's book is, to my knowledge, scarcely known among anthropologists. It is, for example, not mentioned by Haller (see note 1 above), nor by Robert L. Carneiro in "Structure, Function, and Equilibrium in the Evolutionism of Herbert Spencer," *Journal of Anthropological Research* 29, no. 2 (1973): 77–95, nor by Elman R. Service in *Cultural Evolutionism: Theory in Practice* (New York: Holt, Rinehart, and Winston, 1971), though all three deal with evolutionism, Spencer, and social Darwinism, and Service deals with Tylor. See J. W. Burrow, *Evolution and Society: A Study in Victorian Social Theory* (Cambridge: Cambridge University Press, 1966).

[15] Spencer's shaping of Darwin's "natural selection" to his own conceptions of societal evolution and some of the consequences, in the form of social Spencerism, which arose from that are sharply set forth by James Allen Rogers in "Darwinism and Social Darwinism," *Journal of the History of Ideas* 33, no. 2 (1972): 265–280, esp. on pp. 277–280; note in particular p. 280.

America and in Scotland to have elicited primarily stylistic notes and a bare listing of his works (exclusive of the *Descent*).[16]

Darwinism did not become the generic model even for biological evolutionism till well on into the first quarter of the twentieth century; has not yet remotely become the generic model for sociocultural evolutionism (nor, indeed, I shall argue, does it give reason to expect that it will); and still encounters, in the biological sciences, vocal pockets of resistance with quite large audiences.[17]

With respect to the study of society, evolutionary models had fully emerged at least by the mid-eighteenth century and became one of the dominant modes of thought, certainly throughout Western Europe, by the turn of the nineteenth century. There were at least three and possibly four important centers of the emergence of well-formulated ideas of societal[18] evolution: Scotland, France, Germany, and Italy.

Represented by such figures as Adam Ferguson, David Hume, Andrew Millar, James Burnett Lord Monboddo, and Adam Smith, among others, and working in a tradition of institutional analysis already begun by the sixteenth-century George Buchanan and in part linked with Thomas Hobbes, John Locke, and Edward Gibbon in the England of the sixteenth, seventeenth, and eighteenth centuries respectively,[19] the Scots

---

16 J. Thomas, A.M., M.D., *Universal Pronouncing Dictionary of Biography and Mythology* (Philadelphia: J. B. Lippincott, 1873). Comments on Buffon by no less than Condorcet and Cuvier and on Lamarck by Cuvier and Saint-Hilaire are cited. Note that Chauncey Wright, reviewing Spencer's *Principles of Biology*, "stated that it marshaled the evidence for organic evolution more conclusively than Darwin's *Origin of Species*" (Carneiro, "Structure, Function, and Equilibrium," p. 81, n. 6, citing Wright, "Spencer's Biology," *The Nation* 2 [1866]: 724–725).

17 See Appendix A.

18 I use the term 'societal' because the notion of 'cultural' had not yet evolved; cultural evolution is a category of late nineteenth-century thought.

19 Note, for example, Ferguson's *Essays on the History of Civil Society*, 1767, and *History of the Progress and Termination of the Roman Republic*, 3 vols., 1783; Gibbon's *Decline and Fall of the Roman Empire*, 1776–1788; Hobbes's *Elementa Philosophica de Cive*, 1642, and his *Leviathan*, 1651; Hume's *Natural History of Religion*, 1755, and other works; Millar's *Origin of the Distinction of Ranks*, 1771, and *Historical View of the English Government from the Settlement of the Saxons to the Accession of the House of Stuart*, 1787; Monboddo's *Dissertation on the Origin and Progress of Language*, 1774, and *Ancient Metaphysics*, 1779; Adam Smith's works which have evolutional content; Locke's *Two Essays on Civil Government*, 1689.

developed an explicit conception of the evolution of society from pre-human (in Monboddo's thinking, monkey or monkeylike) beginnings, in terms both of institutions and of language and communication. It is interesting to note that the problem of independent invention (or "convergent evolution" or "parallelism") vs. diffusion, as processes of human cultural development, was already clearly formulated, *at latest*, by the latter part of the eighteenth century.[20] The Scottish conception of evolution itself and its institutional models were directly ancestral to a number of major streams of thought in the nineteenth and thence the twentieth centuries: from Adam Smith, through David Ricardo, to Karl Marx and Frederick Engels and thence to a world-wide descendancy of one kind of non-Darwinian evolutionists; from Hume,[21] perhaps, to the nineteenth-century trait-evolutionists (see below), like Edward B. Tylor; from Ferguson, Millar, and Monboddo to the institutionalists, like their direct successors John Ferguson McLennan, probably Henry Maine (whose book, *Ancient Law*, had, despite its publication date of 1861, been gestating long before Darwin's *Origin*), Frederic Seebohm, Frederick Maitland, William Robertson Smith, and to a lesser degree Tylor himself. One should not exclude the monumental figure of the American Lewis Henry Morgan, who as a lawyer was familiar with both the Anglo-Saxon and the Roman Law traditions and was also aware of the works of McLennan and Tylor (with whom he entered into controversy) and possibly the others of his period and of the previous century. Morgan's importance beyond the sphere beloved of anthropologists—kinship— need scarcely be emphasized, since his "periodizations" are still discussed and defended widely, functioning, as they do, in a far more inclusive ideology.[22]

The second great center was France. The generating milieu was, perhaps, the Encyclopedist movement, which conceived of the possibility of

[20] See Garland Cannot, "The Correspondence between Lord Monboddo and Sir William Jones," *American Anthropologist* 70, no. 3 (1968): 559–562.

[21] Hume developed Locke's description of human knowledge, an epistemic view whose consequences for models of society and its changes persist today; it is reductionist and atomistic, rather than institutional and holistic, though both writers also had the latter viewpoints in other phases of their work. It is interesting that the former were elaborated in Anglo-American social sciences other than anthropology.

[22] See Appendix B.

purposeful human directing of society and its changes (a form of evolutionary thought in itself). One may include here Montesquieu and Voltaire, whose influence spread throughout Europe, including, for the latter, Russia, and whose thought is intimately involved in conceptions of institutional change.

Related to this world view are several other developments of mid-eighteenth-century France: the socialists, such as Charles Fourier and Saint-Simon; the physiocrats, such as Turgot; and more specifically evolutionary thought culminating in Condorcet and Auguste Comte. Both the latter had explicit evolutionary schemes—the former in ten stages, the latter in the more traditional (Trinitarian!) three which also dominated most nineteenth-century sociocultural evolutionist thinking. The three-stage evolutionary scheme is still alive, through Morgan, in the Soviet Union.[23] The physiocrats were already formalizing early versions of developmental economic thought, clearly related to economic developments in their own society. That evolutionary thought regarding society was well crystallized and strategically placed for diffusion to large audiences of the elite can be inferred from the fact that the Encyclopedists were writing passages like the following from Diderot:

It seems that Nature has taken pleasure in varying the same mechanism in an infinity of different ways. She abandons one type (*genre*) of products only after having multiplied individuals in all possible modes. When one considers the animal kingdom and observes that, among the quadrupeds, there is not one of which the functions and the parts, above all the internal parts, are not entirely similar to those of another quadruped, would not one readily believe that Nature has done no more than lengthen, shorten, transform, multiply, or obliterate, certain organs? . . . When one sees the successive metamorphoses of the envelope of the prototype, whatever it may have been, approximate one another, from one to another kingdom, by insensible degrees, and people the confines of the two regions . . . with beings of uncertain and ambiguous kinds . . . —who would not feel persuaded to believe that there has never been but one primary being, prototype of all beings?[24]

---

[23] The ubiquity of classification into threes in the West should have long ago raised the suspicion that something other than empirical material and inductive reasoning was entering into the taxonomy; it seems to me clearly the imposition of the metaphysics of the Trinity on classifications to which it is entirely irrelevant.

[24] Quoted in Lovejoy, *The Great Chain of Being*, pp. 278–279.

The effect particularly of Auguste Comte on nineteenth-century social evolutionist thought of a clearly *non*-Darwinian nature is not to be underestimated. The direct inheritors and major proponents of Comtian views were, perhaps, Herbert Spencer and Emile Durkheim. The former had, of course, formulated wide-ranging social evolutionary views before Darwin published, and, in spite of much lip service to Darwin's influence, he maintained his own views with only minor substantive modifications.[25] The term 'superorganic' was Spencer's and represents a view of society, inherited through Comte from Jean-Jacques Rousseau (who also expressed evolutionary views), which postulates a structure of society that has an ontological priority over individuals and individual action— the antithesis of Hobbes's and Locke's positions—or at least emerges as an entity in its own right with its own properties and processes such that, on the whole, the individual is determined rather than determining. The Comtian-Spencerian view, in some formulations, becomes an organismic model (itself with roots in medieval conceptions still today preserved in terms like 'the body politic').[26] This same sort of model dominated Durkheim's thinking, was passed on to the British social anthropologists around the turn of the twentieth century, and persists, if in somewhat modified form, till today (see below). On the whole, the evolutionists of this heritage tend to be idealists, Spencer perhaps least so; achronic in their evolutionism (time has rather a mythic quality—"once upon a time there were simple societies . . ."); concerned more with delineations of the universal path of evolution of societies (homogeneity to heterogeneity; mechanical to organic solidarity) than with process; and partially concerned with stages. Both Spencer and Durkheim, like Comte, were imbued with a notion of progress or moral improvement, a conception which persisted well into the twentieth century among figures like L. T.

[25] See Rogers, "Darwinism and Social Darwinism," and Carneiro, "Structure, Function and Equilibrium," for extensive discussions of Spencer and the continuity of his elaborately formulated evolutionary world view from before *Origin* and on after its publication. With respect to the notion of "superorganic" discussed in the text, see Spencer's conception of society as an organism, as cited from his *Principles of Sociology* by Carneiro (ibid., pp. 90–91) and in "The Social Organism," *The Westminster Review* 73 (1860): 90–121 (one year after the *Origin*); see also the following footnote.

[26] See John of Salisbury, "The Body Social" (twelfth century), in *The Portable Medieval Reader*, ed. J. B. Ross and M. M. McLaughlin (New York: Viking, 1949), pp. 47–48.

Hobhouse, Morris Ginsberg, and, still later, Robert Redfield.[27] Thus, in them, we find forms of finalism, perfectibility, teleology, stages, immanence in societies of their evolutionary trajectories, and other aspects of and analogies to Lamarck or Lamarcklike evolutionism.

Durkheim's influence, of course, permeates much of social science still today. Figures as diverse as Leslie A. White, who sees himself as a materialist related to the Marxist stream of thought, and Claude Lévi-Strauss, who, in my view, is entirely an idealist of Cartesian descent, see themselves as descendants of Durkheim. White is proud of his superorganicism, which he sees as stemming from Durkheim's almost totally sociocentric approaches; Lévi-Strauss sees his structuralist approaches as, in important measure, derived from Durkheim's sociocentric structuralism.[28] The diversity of descendants claiming grandparentage in Durkheim increases when one includes British social anthropologists of the twentieth century like A. R. Radcliffe-Brown and E. E. Evans-Pritchard (who claims a separate descent from the Great Man from that of Radcliffe-Brown, under whom he did his doctoral work) and sociologists like Talcott Parsons.[29]

[27] See, for example, C. M. Williams, *A Review of the Systems of Ethics Founded on the Theory of Evolution* (New York: Macmillan, 1893); L. T. Hobhouse, *Morals in Evolution* (London, 1906); Edward Westermarck, *The Origin and Development of the Moral Ideas* (London, 1906); Morris Ginsberg, "Social Evolution," in *Darwinism and the Study of Society*, ed. Michael Banton (London: Tavistock, 1961), pp. 95–128 (esp. the sections "Moral Development as an Aspect of Rational Development" [pp. 112–114], "Development of Notion of Human Rights" [pp. 114–115], "Development of Religion" [pp. 117–118], and "Possibilities of Further Progress" [pp. 122–124]); Robert Redfield, *The Primitive World and Its Transformations* (Ithaca: Cornell University Press, 1953). Redfield makes the assumptions that choice is good and more choice is better and that, since sociocultural evolution leads to greater independence of the "parts" of society, there is more choice; hence there is evolutionary improvement. Note the notion of moral improvement in Theodosius Dobzhansky's *Mankind Evolving: The Evolution of the Human Species* (New Haven: Yale University Press, 1962), esp. in ch. 12, "The Road Traversed and the Road Ahead," the sections entitled "Evolution and Ethics," "Evolution, Values, and Wisdom," and "Man, the Center of the Universe," pp. 340–348.

[28] See Claude Lévi-Strauss, *The Scope of Anthropology* (London: Jonathan Cape, 1967).

[29] See Talcott Parsons, *The Structure of Social Action: A Study of Social Theory with Special Reference to a Group of Recent European Writers* (New York: Free Press, 1949; first published, 1937). The book deals primarily with Vilfredo Pareto, Durkheim, and Max Weber (a choice governed by Parsons's own derivation from their

As coda, it might be mentioned that the towering figure of French sociocultural evolutionism of the twentieth century, Pierre Teilhard de Chardin, preserves the Christian god, finalism, teleology, creationism, and other aspects of eighteenth- and nineteenth-century sociocultural evolutionism despite his training in physical and archaeological anthropology.

In sum, the French, like the Scottish and English, had an elaborate variety of sociocultural evolutionary models, ranging from programs of praxis to bring about societal change to divers ideas, both secular and religious (or some fusion of both), on the course of societal evolution. Some of these conceptions, like those of Diderot, Rousseau, Comte, and others, undoubtedly had much wider circulation throughout Europe and its intellectual circles than did those of the Scots. It is to be noted that many of these models, along with the Scottish ones and the German ones to be discussed below, are still central to our conceptions of sociocultural evolution, if in somewhat modified forms.

The third major center was Germany, where the evolutionist idea seems to have appeared somewhat later, possibly as a result of the French ferment. The main formulations come at the end of the century, although there are earlier intimations, as in some of Goethe's writings, particularly on the history of Italian art, which he saw in the context of a stream of institutional development.[30] Three major figures emerge, each of whom has left a heritage of evolutionist thought and scholarship of the greatest importance, still generating both philosophical thought and research activity within their general frameworks, but *not* in Darwinian molds. These are Johann Gottfried von Herder, Georg Wilhelm Friedrich Hegel, and Wilhelm von Humboldt.

---

sort of social science), especially Durkheim and his contribution to "a *single* body of systematic theoretical reasoning" (p. v, emphasis in original) initiated and elaborated by the three theorists cited. In contrast to the 120 or so pages given to Pareto, the 160 to Durkheim, and the 200 to Weber, Karl Marx gets 20, Engels is not mentioned, anthropologists like Franz Boas and E. G. Taylor are unlisted, Radcliffe-Brown has 1 page, and Bronislaw Malinowski has 5. See also Parsons's *Societies: Evolutionary and Comparative Perspectives* (Englewood Cliffs, N.J.: Prentice-Hall, 1966).

[30] Haller (see note 1 above) points out in "The Species Problem," p. 1320, that Goethe adhered to what might be called the transformist monogenist school developed in France by Lamarck.

To the first of these may, in large part, be attributed the major beginnings of interest in folklore. Folklore has always tended to be seen in an evolutionary interpretation because of the peculiarities, from urban intellectual viewpoints, of the content and form of the materials which are allegedly "archaic," "residual," and "primitive." Herder's interests included philology, and the evolutionist perspective—already expressed with respect to language by people such as Monboddo, William Jones, and some of the French—was, naturally, not lacking here either. Herder's interest in the origins and development of language[31] was not the first expression of this theme but was one of the great monuments of synthetic thought which led to a flowering of comparative and evolutionary philology, detailed linguistic study and collection (as by the von Humboldts), and further expressions of the evolutionary principle throughout the nineteenth century by people like the brothers Jacob and Wilhelm Grimm, Friedrich Max Müller, Andrew Lang, Edward B. Tylor, and eventually James George Frazer. During the course of that century, the folkloric and philological interests were diffused to other spheres of inquiry, such as religion, magic, and science, also treated evolutionarily in the framework of predominating models to which I return below. Finally, between 1784 and 1791, Herder wrote on the problem of a philosophy of human history, in a sense one of the earliest of a stream of "general world histories" (*"Allgemeineweltgeschichte"*) which were particularly popular among the Germans in the nineteenth century and which had their analogues elsewhere, as in Condorcet's *Esquisse* of 1795.[32]

---

[31] See Johann Gottfried von Herder, *Über den Ursprung der Sprache*, 1772.

[32] Marquis de Condorcet, *Esquisse d'un tableau historique des progrès de l'esprit humain* (Paris: Masson, 1822). See also, as examples of *Allgemeineweltgeschichte*, C. Meiners, *Grundriss der Geschichte der Menscheit* (Lemgo: Meyerschen Buchhandlung, 1785); Gustav Klemm, *Allgemeine Cultur-Geschichte der Menscheit* (Leipzig: Leubner, 1843); G. W. F. Hegel, *The Philosophy of History*, trans. J. Sibree (New York: Dover, 1956; first published in German, 1837); Friedrich Ratzel, *The History of Mankind*, trans. A. J. Butler (London: Macmillan, 1896; first published in German, 1885–1888); E. B. Tylor, *Anthropology: An Introduction to the Study of Man and Civilization* (New York: D. Appleton, 1899; first published, 1881); A. L. Kroeber, *Anthropology* (New York: Harcourt Brace, rev. ed., 1948; first ed., 1923), which, besides being a world history, is also a treatise on the scope of anthropology; Ralph Linton, *The Tree of Culture* (New York: Alfred A. Knopf, 1955); C. F. Hockett, *Man's Place in Nature* (New York: McGraw-Hill, 1973). The last book covers not only human

This is not the place to discuss G. W. F. Hegel at length—that has been done in a diluvian literature. Important for our purposes is his overwhelmingly evolutionary view of the world, a view fully developed between about 1800 and his death in 1831. It is not to our purpose to decide whether it is a "good" or a "bad" model, a useful or useless one. It *is* important to note, however, that it became one of the world's most influential ones, especially in one of its secularized, materialist versions —Marxism. "Left" and "right" Hegelians, Neo-Hegelians, Marxists, phenomenologists—all are descendants representing a vast spectrum of political and social philosophies, almost all with some evolutionist, developmentalist, or transformationalist viewpoint, and, finally, many to be found in virtually all Western countries, or, in the form of Marxism and its progeny, throughout the world.

Hegel's is a very complex model of sociocultural evolution. Fundamentally idealist—the world is the progressive expression of spirit in self-realization—it nevertheless deals, as one might expect in Herder's Germany, with world history (*Allgemeineweltgeschichte*) which is the material form of the dialectic of self-realization. World history is not merely a flow; it is no mere time-linked, though accidental, sequence but is, in fact, a *necessary* series of transformations, an inherent succession of forms, although the specific expression each form may take may not be totally determined. Thus all observed and observable history is causally linked, has an orderly sequence leading to a final end (total self-realization or consciousness), is progressive both in the sense of linearity and in the sense of improvement, and is inevitable. The model permits one to embrace all human experience (*Allgemeineweltgeschichte!*) and hence, as it were, to preserve the pre-Darwinian conception of species, since each *form* of human experience (by analogy a quasi-species) is part of the fecund unfolding of self-realization by *Geist* (read God-as-Platonic-Aristotelian-philosopher). It permits the retention of finalism and of such concepts as germ (or its equivalent) and stage. It permits a *material* dynamic in addition to the spiritual one, although the latter is causally fundamental (as it continued to be among German philosophers, such as Arthur Schopenhauer, Friedrich Wilhelm Nietzsche, and the rationaliz-

---

history, but primate, mammalian, all evolutionary, and, indeed, galactic and universal history, perhaps like Alexander von Humboldt's *Kosmos* of 1845–1862.

ers of the Third Reich). This view, in turn, conforms with Lamarckian and Cuvierian evolutionary views, socially applied. Finally, the model not only permits the retention of the Christian god or a god-likeness but makes it central to the whole evolutionary process. From a political point of view, thus, immanent in Hegel's thought is its ability to serve all ends: radical, reactionary, and in between;[33] his system, unlike Darwin's, does not inherently threaten all the established orders of political, moral, and philosophical thought, and even, in some versions, greatly strengthens them.

However, from the point of view of later nineteenth-century pre- and non-Darwinian evolutionism and its development, the Hegelian evolutionary dialectic between the material and the spiritual, in which both are part of the developmental process, also allows for a radical departure in evolutionary viewpoint by converting Spirit, as an ontological prime-moving entity, into 'spirit,' as a culture-filled, societally determined projection of human mind whose ontological status is dependent on human existence as biological fact and on human action, especially work. In this view, assertedly, human evolution (*Allgemeineweltgeschichte*) becomes a purely material process, where mind is itself conceived of as part and expression of that process.

I say "assertedly" because there remains, or even is created, a problem, when one removes *Geist*, as to what the generator of the sociocultural evolutionary process is. In *The German Ideology*, for example, Marx and Engels slip in the generation of evolution as one of the five "moments"—or axioms—concerning society and history: the one referred to as the "creation of new needs."[34] It is slipped in again in the corpus of the Marxist literature in the form of "antagonism" and "contradiction" as generators of dialectical motions toward new syntheses, but it is not at all clear why antagonism and contradiction should arise in the first

[33] See Frederick Engels, *Ludwig Feuerbach and the Outcome of Classical German Philosophy* (New York: International Publishers, 1941), pp. 10–11 (first published in German, 1888). Engels makes this point quite strongly.

[34] Karl Marx and Frederick Engels, *The German Ideology*, Parts 1 and 2 (New York: International Publishers, 1947), pp. 16–17 (written, 1844–1846). It is not clear why new needs are created. One easily enough accepts as self-evident, and hence axiomatic, the continuous repetition of needs—e.g., by continuous consumption of food and clothing, but this does *not* automatically lead to *new* needs.

place out of a state of early primitive communism, or why they should synthesize (a teleological view) rather than lead to dissolution of society. This problem of human life and organization is perhaps least discussed by Marx and Engels or their multifarious followers; it seems to me also the most (or only) ambiguous one of the "moments"—and the one that still has a germ of the old *Geist* in it, since it does not give a clear materialist explanation for departure from stasis. It only becomes explicable if one assumes some absolute deprivation, existing from primordial times, which men strive to reduce even in *nonclass* societies.[35] But note that such an assumption reintroduces finalism and teleology, that is, an essentially purposeful universe, even if construed in purely human terms —a non-Darwinian view of evolution, since the processes are not random, not probabilistic, and not individual.

The finalism—sometimes even millennialism—of Marx,[36] paralleling the finalism of most non-Darwinian biological and social evolutionism, carries over till the present and can be found in the writings of the Marxists, neo-Marxists, "leftist deviationists," Trotskyites, etc., etc.[37] But, as I shall show below, since finalism had also been part of the non-Darwinian view of biological and social evolutionism, which derived from the same ultimate roots as Marxist evolutionism, it continued to appear in most evolutionism dealing with the sociocultural universe *even after Darwin*, whose evolutionism, despite forays into that universe,

---

[35] Once class exists, the deprivation theory works in a material and social sense so that the generation of evolution is in society itself by virtue of real and absolute deprivation. For a fascinating glimpse of how "contradiction" and "antagonism," discussed in the text, are introduced as supposedly *empirically* derived dynamics of *all* societies, see Irmgard Sellnow's untitled paper in *VIIme Congrès International des Sciences Anthropologiques et Ethnologiques* (Moscow: Science Publishers, 1967), pp. 474–479. It seems quite evident that they are not empirically derived at all but are axiomatic presuppositions derived from the logic of the dialectic, which happens, at times, to correspond to historical events.

[36] David McLellan in "Introduction" to his translation of Karl Marx's *The Grundrisse* (New York: Harper and Row, Torchbooks, 1971), p. 12, refers to "a Utopian and almost millennial strain" (in the *Grundrisse* and the *Manuscripts*).

[37] See, among others, Georg Lukács, *History and Class Consciousness: Studies in Marxist Dialectics*, trans. Rodney Livingstone (Cambridge: MIT, 1971; first published in German, 1923); Nicos Poulantzas, *Pouvoir politique et classes sociales*, 2 vols. (Paris: Maspero, 1968); Herbert Marcuse, *One-Dimensional Man: Studies in the Ideology of Advanced Industrial Society* (Boston: Beacon, 1964); C. B. Macpherson, *The Real World of Democracy* (Oxford: Clarendon, 1966).

whose models of the evolutionary process, and, above all, whose largest mass of data dealt with the biological world, not with society.

More narrowly, more empirically conceived is the evolutionism of Wilhelm von Humboldt, based in part on his monumental amassment of linguistic and related ethnographic data from various parts of the world, much of it collected by his astonishing brother, Alexander, and on this brother's probable evolutionary views. Particularly important for our purposes are von Humboldt's philological works, represented in a number of writings of wide influence and collectively responsible for creating the field of comparative philology. A three-volume work on the Kawi language of Java, published in 1836, one year after his death, has a separately published introduction entitled, in English translation, *On the Differences in the Construction of Language and Its Influence upon the Intellectual Development of the Human Race*.[38] This work, clearly evolutional in its orientation, sets problems of linguistics and communicational analysis that still have focal relevance in the domain of linguistic science today. The model of evolution, whose data referents include the vast array of "civilized" and "barbarian" languages of the Old and New Worlds with which Wilhelm von Humboldt was conversant, is one of a succession of stages in the evolution of language forms and of communication.

Von Humboldt's work in linguistics and comparative philology influenced all later linguistic discourse, including, first, the work of the brothers Jacob and Wilhelm Grimm, about twenty years von Humboldt's juniors, who formulated the famous Grimm's law of phonetic change—essentially a statement of a nonteleological, transformationalist kind of evolutionism. Second were the philologists, comparative religionists, and comparative mythologists influenced both by von Humboldt and by the Grimms (both of them inveterate mythologists), including Friedrich Max Müller, Andrew Lang, and others, all of them post-Darwinian, all of them non-Darwinian evolutionists, and all of them involved in the maelstrom of discourse concerning sociocultural evolution with such figures as Edward B. Tylor and, later, Sir James George Frazer. Third was Ferdinand de Saussure, one of the founders of modern structuralist lin-

---

[38] *Ueber die Verschiedenheit des menschlichen Sprachbaues und ihrer Einfluss auf die geistige Entwicklung des Menschengeschlechts* (Berlin, 1836).

guistics, some of whose exponents, like Charles Hockett,[39] are deeply concerned with problems of the evolution of language, communication, and culture.

Finally, the fourth center of evolutionary thought was Italy, chiefly in the person of Giovanni Battista Vico (1668–1744). Only brief mention is made of him here, since it is not clear what influences emanated from him until his "rediscovery" in relatively recent times. Vico presents a philosophy of history in which he sees all "nations," i.e., societies, as obeying common principles, and their histories and history in general as being determined by certain and immutable laws such that particular societies and society in general go through cyclical and developmental patterns.[40] Vico displays an interesting combination of idealism and materialism in his historiography, which, like its counterparts and successors in other parts of Europe, was infused with a teleological or finalist view. It is of considerable interest to the general point of my paper—that non-Darwinian, pre-1859 models of evolutionism are still today the predominant ones in social science—that Vico has enjoyed a renewed interest, al-

---

[39] See the following by Hockett: "The Origin of Speech," *Scientific American* (September 1960): 88–96; "Animal 'Languages' and Human Language," in *The Evolution of Man's Capacity for Culture*, ed. J. N. Spuhler (Detroit: Wayne State University Press, 1959), pp. 32–39; (with R. Ascher) "The Human Revolution," *Current Anthropology* 5 (1964): 135–168; *Man's Place in Nature* (New York: McGraw-Hill, 1973). Chapter 23 of this last work, entitled "Intermezzo: Of Time and the River" (pp. 281–313), is a general statement about the nature of evolution, while Part Two (pp. 314–665) traces the evolution of man from the beginning of the universe.

[40] Generically, concern with evolution is a special case of concern with change characterized by a time-linked sequence of forms. All the variant forms of evolutionism—teleological, nonteleological, orthogenetic, random, opportunistic, linear, universal, unilinear, multilinear, transformational, progressivist, etc.—have this in common. Another class of changes also displays time-linked sequences of forms, but in these the sequence is seen to repeat itself cyclically. Cyclicity can be seen as a special form of evolution or as an independent form of change. Furthermore, each can be seen as an epiphenomenon of the other, as in some views of Chinese history or in the Hindu world view of immense cycles of time with evolution through reincarnational stages to reach Nirvana. Cyclical theory, another non-Darwinian variant of evolutionist thought, has reappeared in human thought in the sociocultural evolutionism of Ibn Khaldûn in the Near East in the fourteenth century, Vico in the eighteenth century, and many nineteenth- and twentieth-century writers, such as Arnold Toynbee, A. L. Kroeber, Oswald Spengler, Nikolai Ia. Danilevsky, and Pitrim A. Sorokin.

most a vogue, in the post–World War II resuscitation of evolutionary interests.

J. W. Burrow is wrong about the disappearance of evolutionary interest. In various places in this paper I have shown that it has continued actively till the present in the social sciences and especially in anthropology; this continuity is most marked in the United States, whose anthropological tradition stems from the German Romantic historical tradition in combination with the nineteenth-century British tradition (influenced heavily by the German tradition) represented by E. B. Tylor and Francis Galton.[41]

DOMINANT MODELS OF NINETEENTH-CENTURY SOCIOCULTURAL EVOLUTIONISM

The major outlines of my argument have already been adverted to in the preceding: all significant models of human sociocultural evolution in the later nineteenth century and beyond were non-Darwinian. This included even that of Alfred Russel Wallace, whose publication in 1858 of "On the Tendency of Varieties to Depart Indefinitely from Original Type" (with Darwin's own statement on the same subject) established Darwinian biological evolution.[42] Further, the models of sociocultural evolution, with the possible exception of Marxist ones, were all, if with varying emphases and in different degree, teleological, finalist, immanentist, and essentialist, and displayed various combinations of idealism and materialism—as, I believe, was to be expected, given their placement in the stream of Western intellectual history. The one least con-

[41] What requires explanation is not why there was an overpowering evolutionism in Victorian England—archeological and historical documents demonstrate that some form of evolution has occurred—but rather why twentieth-century British anthropology and other social sciences have been both ahistorical and nonevolutionary. Nevertheless, an evolutionary interest lurks in their thought: Malinowski is explicit in his evolutionary considerations; Radcliffe-Brown speaks of evolution in his more theoretical works; Raymond Firth has recently rediscovered both evolutionism and history—and even Karl Marx! What is interesting about the British is how they repress the endemic problem in favor of an atemporal structuralism, such as has only begun to be popular in the United States since 1960. Perhaps such orientations appear when empire has reached its peak or begun to decline.

[42] See his "The Origin of Human Races and the Antiquity of Man Deduced from the Theory of 'Natural Selection,' " *Journal of the Anthropological Society* 2 (1864): clviii–clxx, and discussion on it, pp. clxx–clxxxvii.

forming to this general pattern, that of Marx and Engels, is also the only one whose formulation was accompanied by a thoroughgoing re-examination of the metaphysics, ontology, epistemology, and axiology of both physical-biological and social sciences.[43] Note in this regard, in connection with his discussion of the effect of Darwin in Russia, Rogers's citation of Marx's attacks on Darwin's theory of evolution as being modeled on the structure of bourgeois entrepreneurial society.

It will be useful to explore each of the terms above and give exemplifications. By 'teleological' I refer to a concept, variously expressed, that the systems ('nations,' 'races,' 'societies,' '*Völker*,' later 'cultures' and 'parts' of 'culture' like 'religion') under discussion move in a direction toward some end usually left undefined. The teleology is closely related to the finalist conception but not identical with it, except when the end is defined as in Hegel or in Christian eschatological evolutionary doctrines. Teleological notions take form in the concept of 'progress,'[44] that of 'improvement,' and the nineteenth-century notion of 'degeneration' (as opposed to Buffon's eighteenth-century usage, which meant the same as 'evolution'—'to generate *from*' rather than 'to generate [badly] *down*' as it became in the later century), which implies an opposite Higher or Better. Even the concept of 'survivals,' so popular in the latter half of the latter century, especially in E. B. Tylor's work and among the folklorists such as Lang, may be related to the teleological notion, since the 'survivals' are consistently thought to be nonfunctional, a residuum in a society that has, so to speak, moved "further"—implicitly along some line of development or improvement.[45] Note the phrase in the subtitle to Lewis Henry Morgan's *Ancient Society*: "*Lines of* Human *Prog-*

[43] See particularly Frederick Engels's *Dialectics of Nature*, ed. and trans. Clemens Dutt (New York: International Publishers, 1940; written, 1872–1882); his *Anti-Dühring* (Moscow: Progress Publishers, 1959; first published, 1878); his and Marx's *German Ideology* (see note 36 above); see also *Reader in Marxist Philosophy from the Writings of Marx, Engels, and Lenin*, ed. Howard Selsam and Harry Martel (New York: International Publishers, 1963), esp. Parts 2, 3, and 4, "Materialism vs. Idealism," "Dialectics and Dialectical Method," "Theory of Knowledge and the Philosophy of Science."

[44] See J. B. Bury, *The Idea of Progress* (London: Macmillan, 1920).

[45] See Burrow, *Evolution and Society* (cited in note 14), esp. pp. 240–241. The notion of 'survivals' may, *in part*, have been modeled on Darwin's residual organs, which are carried along willy-nilly by biological reproduction, unlike 'survivals,' which remain without adequate explanation for their retention.

ress from Savagery through Barbarism to Civilization" (emphasis mine).

'Finalism' refers to a determinate end state to be achieved—a classless society, as in Marxist thought; a state of society whose social conditions, like those of Victorian England, represent the ultimate possible in human achievement, as in Spencer's thought; monogamy and private property, as in Morgan; monotheism, as in a number of nineteenth-century theorists; the Last Judgment or immortality, as in Christian thought, etc.

'Immanentism' involves the conception that the sequence of development inheres in the unit undergoing development from its very beginning. The concept involves the notion of stages, although the two are not linked by logical necessity. It is also always associated with several other related conceptions. First, ideally, if not in actual historical fact, every society—one of the units of development—exists indefinitely backward in time till the beginnings of human life itself and indefinitely forward in time till the final end is reached. It is to be noted that this assumption *must*, by logical necessity, be present if one is to take as axiomatic, or even discuss, the idea that *all* societies must go through all stages of development, as, in principle, Morgan, McLennan, Spencer, Tylor, and others do. They cannot go through *all* stages if (*a*) they had at one time not existed but later sprang from some more advanced state of another society; (*b*) they cease to exist as identifiable societies. We have here a transformed version of the species fixity/extinction problem still universally problematic at the time.

Second, every unit undergoing evolution has an origin—a recognizable First Form which is distinguishable, per se, from all other forms, not gradually—that is, evolutionarily—but at the very moment of origin. For example, a First Form of religion is said to be animism (Tylor), animatism (Marett), or some other form[46]—but there is no discussion

---

[46] See, for example, Waldemar Bogoras, *The Chuckchee*, Part 2, vol. 6 of the Jesup North Pacific Expedition, Memoirs of the American Museum of Natural History (New York, 1907), pp. 277 ff. He says, "I avoid using the term 'animism' because it presupposes the conception of the human soul, which, in my opinion, belongs to a later stage. E. B. Tylor says that animism includes two great dogmas, forming parts of one consistent doctrine. . . . According to my theory, these two dogmas belong to the last stage of development" (p. 277, n. 2). Bogoras's first stage is one in which a people consider material objects alive (p. 277, pp. 280 ff.). He has four other stages before one gets to animism. Religion, incidentally, is another of the units of development very commonly used in the nineteenth century; the reader will note the conceptual realism involved.

of a gradual, piecemeal taking shape of the Form through the fusion of different cultural innovations at different times and in different places—as a culturalized Darwinian model would require. Nor is there a discussion of what selective advantage such a Form had over other Forms. The latter problem *could not* be discussed since there were no competitive Forms around at the beginning—there was no variation—in the evolutionary world view of the nineteenth-century sociocultural evolutionists. It is even the fact that competition does not really figure as a mechanism until writers like Spencer discuss relatively complex, internally divided societies in which *person* and *classes* compete, but not institutions or other units of development. Another way of saying this is that the stages, universal in linear development of culture, *themselves* had an ontological status as Things with characteristics, essences, organization, and boundaries, and, being universal, were also unique. Interesting, in respect of the foregoing discussion, is the fact that they spoke of 'culture' (in general and in the singular), but not of 'cultures.' Speaking of 'cultures' is a twentieth-century development which recognizes the particularity, the organization, and the historicity of individual societies. It is most important to recognize that variation in empirically observed details was recognized but did not affect the argument because these details comprised not varied forms or variants of the Form, but merely exemplars of the Form, hypostasized as something real—and unitary—a Thing which became the object of inquiry.[47]

I deliberately capitalize Form in order to suggest the Platonic ontology and epistemology involved and, further, to suggest how much the whole model of sociocultural evolution which prevailed throughout the nineteenth and most of the first half of the twentieth century, if not still, was yet informed by the Christianized, Neo-Platonist conception of the Great Chain of Being, timelessness and all.

Third, in the germ are contained all the stages through which the unit of development will evolve. This underlying conception seems to me

---

[47] Note R. H. Lowie's comment on "Totemism" on pp. 141–142 of *The History of Ethnological Theory* (New York: Rinehart, 1937): "Totemism [is] an artificial unit . . . applied to diverse phenomena presenting superficial analogies" without common psychological or historical origins—"spurious units" leading to a "sham problem of a generalized origin." "The question how 'Totemism' evolved is recognized as nonsensical." Lowie considers this sort of Thing "an arid conceptual realism based on premature classification."

clearest in Lewis Henry Morgan's *Ancient Society*, where each of a number of institutions is traced from its Original Form through its stages, which are to appear in all societies; as latter-day phrasing has it, they all follow a unilinear evolutionary path.[48]

Fourth, all of these evolutionists have a strong idealist strain perhaps most clearly seen, again, in Morgan. Each of his major sections dealing with a major institution of society—marriage, property, government—is headed "Growth of the *Idea* of . . . ," and the first section deals with "Growth of *Intelligence*" (emphasis mine), albeit in interaction with technology. It is very difficult to assert that he is speaking about an idea in his own head rather than an idea in the mind of Man[49] or guiding force in evolution—however materially evolution may also have worked itself out. The "evolution" of 'mind' is a general concern of the time and beyond; see Wallace's "Origin of Human Races," cited in note 42 above. 'Mind' is an ambiguous term in these usages. It sometimes meant 'culture' in a milieu where that word had not become current, but even in

[48] A few brief quotations from Lewis Henry Morgan's *Ancient Society, or Researches in the Lines of Human Progress from Savagery through Barbarism to Civilization* (ed. Eleanor B. Leacock [Cleveland: World Publishing Company, 1963]; first published, 1877) will illustrate this: (*a*) "The germ of government must be sought in the organization into gentes in the Status of savagery; and followed down, through the advancing forms of this institution, to the establishment of political society" (p. 5). (*b*) "With respect to the family, the stages of its growth are embodied in systems of consanguinity and affinity, . . . by means of which, collectively, the family can be definitely traced through several successive forms" (p. 5). (*c*) "The germs of the principal institutions and arts of life were developed while man was still a savage." From savagery, through barbarism and civilization, each germ has been developed (p. 8). (*d*) "Since mankind were one in origin, their career has been essentially one, running in different but uniform channels upon all continents, and very similarly in all the tribes and nations of mankind down to the same status of advancement" (preface).

[49] The very conception of the idea being in the mind of Man, guiding all men toward improvement, aside from its Platonic Formalism, is tied to, if not identical with, the "psychic unity of mankind," a central axiom uniting all major sociocultural evolutionists, despite contradictory positions they also held regarding racial inferiority, especially of Negroes, Amerindians, and Chinese. The linked notions of psychic unity and monogenesis remain axiomatic in all anthropological schools of 1973, despite occasional suggestions like that of Franz Weidenreich, in *Apes, Giants and Man* (Chicago: University of Chicago Press, 1946), for a quasi-polygenist origin of the different "stocks" of "man." From either a Darwinian or a polygenist perspective, "Man" is a product of conceptual realism, a fact that becomes clearer when even scientists use the term 'Man' to refer to Neanderthals, Pithecanthropus, Zinjanthropus, etc., as if they were all one *species*, 'Man.' This is a pre-Darwinian usage.

that usage it *also* had the sense of intelligence and mental capacity, which today we see as separate. There is virtually no interest, now, in the evolution of the latter—an anthropological gap consonant with the underlying metaphysics of the "psychic unity of mankind." The evolutionary interest in mind may be related to Darwin and to English abstract evolutionary sociocultural theory à la Spencer; the *lack* of interest in mental evolution on the grounds of psychic unity appears to come out of the German Romantic historical and geographical tradition through Tylor and others into anthropology.

The dialectic evolutionary relationship between idealistically and materialistically conceived aspects is characteristic of the century, while the relationship between idealism and materialism referred to just above has its analogue in Hegel and a variety of Hegelians.[50]

The reader will have recognized in the second point above a version of creationism clearly related to both Lamarckian and Cuvierian creationism but far afield from Darwin's evolutionism. In the third point is seen the transference of the models of Robinet, Lamarck, and all their intellectual kin into a full-blown *societal* theory of evolution.

All three aspects are part and parcel of an evolutionism with its origins deep in Plato, deep in the Middle Ages, deep in the Great Chain of Being, but flowering in the eighteenth and throughout the nineteenth and into the mid-twentieth century, largely unaffected by Darwin's *model* of evolution,[51] if not by the *fact* of Darwin's advent on the intellectual scene. Pfeifer describes a renewed vigor and general diffusion of this flowering in the last third of the nineteenth century in the United States, a vigor which, as I shall briefly show below, continues to the present.

'Essentialism,' related to immanentism and the problem of species, again involves a metaphysics deriving from Plato's conception of Forms or Ideals and, with it, a certain kind of epistemology—that of acquiring knowledge by reflection and logic. One can arrive at a knowledge of the

---

[50] See my comments on Hegel and Marx; also see note 27 and Appendix B.

[51] Leacock, in her introduction to Part 1 of Morgan's *Ancient Society* (see note 48 above), p. liv, says: "Morgan . . . was, as yet, reluctant to accept the full implications of Darwin's discovery for social evolution. . . . As he later wrote, he was 'compelled' to stop 'resisting' Darwin, and 'to adopt the conclusion that man commenced at the bottom of the scale from which he worked himself up to his present status.' " To adopt that conclusion does not mean, of course, that one became a Darwinian; the evidence seems to me conclusive against this view.

essence of some postulated Thing, untroubled by variants and variations, not by induction concerning the variation, but by contemplatively disregarding the variation in order to find the essence. It is not that variants are not recognized, but they do not become, as in Darwin, the very cornerstone of the theory itself. Rather, they are dismissed by the devices of classification or typologizing (whereby one brushes over differences, speaks of them as exceptions to some reified type, thinks of them as sports, etc.) or by conceptual devices which diminish the intrinsic significance of the variations either by contrasting them to a generic trend, as in the notion of 'survivals,' or by failure to discover a function for the variants. This is a metaphysics and epistemology that, hydralike, has reared many heads in the social sciences and continues to do so in the twentieth century. For example, it is recognizable in Father Wilhelm Schmidt's and his followers' *Kulturkreis* notion that complexes of culture traits can be diffused, untrammeled by intervening conditions, to geographically disconnected parts of the globe; or in the Lévi-Straussian archetypal rules, or Form-of-Forms, with respect to which all variants of "a" myth, for example, are simply (platonistically imperfect) expressions, knowledge of which is gained by reflection and logical reconstruction alone; rules of correspondence for *any* direct empirical observation or test of the imputed rules are totally lacking.

Connected with essentialism, as I have said above, is conceptual realism, whereby categories or classifications, e.g., religion, totemism, magic, class, the state, irrigation, are taken to be Homogeneous Real Things Out There,[52] that is, ontological entities whose totality is directly knowable, rather than mental products of abstraction of aspects and qualities in specific kinds of language, chiefly scientific metalanguages. As Real Things, they can then have histories and actions all their own and may even be said to act and to do things—to be capable of "real," usually purposive action, rather than mental approximations of what may represent some sort of relationship: societies "act," religion "does," kinship "functions to do something," rituals "produce" solidarity, classes "rule," and so on and on. A characteristic form of discourse connected with conceptual realism, as I remarked above, has to do with the discovery of the essence of the reified Things—to discuss what class "really is" or "what

[52] See Lowie's comment cited in note 47 above.

it is in itself" (Marx's "class *an sich*"), what war "really is," what the "true" basis of law "is," what the "true" city "is" (Weber), what the essential or elementary aspects of religion (Durkheim) or kinship (McLennan, Morgan, Lévi-Strauss, Elman Service) "really are," what the "essence" of man "is" (Teilhard de Chardin).[53] The search for essences asserts a fixed state of being, reflected in the verb "is,"[54] and in this sense is related to the fixity-of-species notion.

One final word on the nineteenth-century models of sociocultural evolutionism, again with the possible exception of Marxism. Almost none are concerned with time as a *variable* of evolution. All that is necessary for the kind of evolution that we have discussed—as in Lamarck's evolutionism—is "sufficient time." But no consideration need be given to *rates*, that is, to degrees of change and units of replacement per unit of time—a consideration which *forces* one to empirical observation, to statements of hypotheses empirically verifiable (for example, by historical documentation), to observed correlations, and to some degree of predictability. Looked at another way, no grid of time (years or convenient multiples thereof) was ever established in sociocultural evolutionism. Nineteenth-century and even many twentieth-century evolutionists dealing with society and culture, except for some of the anthropological archaeologists, do not seem bothered by the question of how *much* time is minimally or maximally necessary for a given kind of change to occur —do not, indeed, fully take cognizance that the changes occur only with some amount of, presumably "sufficient," time. All evolution, with the exception noted, takes place in a mythical time represented by such phrases as ". . . must once have been . . ." (see, in this connection, "once upon a time . . ."!), ". . . would have been likely . . . ," ". . . could then have occurred . . . ," etc., phrases that occur again and again in the nineteenth-century evolutionary literature and its only slightly modified de-

[53] With respect to kinship, in the nineteenth century McLennan proposed exogamy as essential, while Morgan proposed promiscuity and matriliny; in the twentieth century Lévi-Strauss proposed various simplifying rules, and Elman Service proposed patriliny and patrilocality. With respect to essence, Man is said to be *essentially* different from animals in that he is a toolmaker or simply a maker ("Homo faber"), a player of games and roles ("Homo ludens"), a symbol user (presumably "Homo sapiens"), a laugher, a music maker ("Homo musicans"), or that he has culture, has language, etc.

[54] To say an essence "*was*" would already be to limit its universality, to imply its extinction.

scendants of the twentieth, like the works of Leslie A. White, Elman Service, Raoul Narroll, Robert Carneiro, Gertrude Dole, and many others, including myself in earlier works.[55] Some might argue the defense that the conditional mode represents "hypothesis construction." In some cases, of course, this is true, but in most of the instances it can be shown that hypothesis is not properly involved, since no rules of correspondence for testing it are given. In fact none can be given, since the state or condition postulated in the conditional is unknowable because no records remain (if there ever were any)—this lack giving the *excuse* for the conditional. Such rules of correspondence probably *could not exist*, judging by all known empirical data and its demolitional import for the unilinealism underlying the use of the conditional in the first place.

I think this timeless conception of time and the language associated with it are highly diagnostic as to the nature of the evolutionary model I am discussing, and both are impossible without the essentialist conceptual realism. Once one introduces the notion of rates, the fixity of Things dissolves, essences evanesce, conditionals are no longer acceptable, and formalistic analysis gives way to processual analysis—a process happening here and there in the social sciences, as in anthropology, since the turn of the century, but in a systematic way for most spheres of the discipline only after mid-twentieth century and in the most limited way in other social-science disciplines, such as political science.

TWENTIETH-CENTURY EVOLUTIONISM AND THE PROBLEM OF DARWINISM IN SOCIOCULTURAL EVOLUTIONISM

Despite a persistent and even ramifying purification of the metaphysical and methodological foundations of the sciences, one form or another of nineteenth-century evolutionism has persisted in all of the social sciences into the present.[56] The resulting situation is that each of the disciplines has eradicated or diminished the influence of one or several of the features of nineteenth-century evolutionism I have discussed above while more or less unconsciously retaining others.

For example, a modified immanentism resulting in an omnipresent

---

[55] For examples see Appendix C.

[56] This assertion can be heavily documented with ease but would unduly extend these notes; the references in the notes following give a few characteristic examples.

unilinearism permeates the disciplines of economics, political science, sociology, and even to some extent anthropology (though less virulently) in the form of "development theory" concerning "traditional," "underdeveloped," "developing," "newly emergent," and, thank God, for a brief moment only, "expectant" societies. The terms themselves reveal the implicit unilinearity and universality, indicated by progressive stages of removing "underdevelopedness."[57] The assumption is that all such societies will, eventually, evolve to a state of being—so desirable!— just like ours (note, too, the Neo-Victorian finalism here). That the *observed* states of being might be understood as alternative developments, varied possibilities, differing forms of order is an intellectual alternative never considered or at most lightly discussed. We find in this view a version of fixity of species (*each* society will go through all the stages of development after it has finally reached "take-off").[58] That the whole conception of "development," in the sense of development economics or sociology or political science, may simply be false is an idea that breeds instant hostility, ridicule, and rejection.[59]

[57] Two examples suffice: "At a time when no nation had progressed beyond a primitive stage, a politically developed nation was one that was politically unified. In the next stage of advancement a 'developed' political system governed not only a unified nation but also one that was successfully industrializing. Today political development requires national unification, economic modernization, and also a welfare state. Tomorrow it will require all of these functions as well as an automated economy that is in addition politically responsible" (A. F. K. Organski, *The Stages of Political Development* [New York: Alfred A. Knopf, 1965], p. 7). Another: "It is possible and . . . useful to break down the story of each national economy . . . according to this set of stages" (p. 1); "It is possible to identify all societies . . . as [economically] lying within one of five categories: the traditional society, the preconditions for take-off, the take-off, the drive to maturity, and the age of high mass-consumption" (p. 4) (W. W. Rostow, *The Stages of Economic Growth: A Non-Communist Manifesto* [Cambridge: The University Press, 1962]).

[58] The term 'take-off' comes from Rostow's *Stages of Economic Growth* and, as is well known, became a catch phrase in all the ramified spheres of development activity— government, academe, business, etc.—and even in household use. Note: "Particularly in Mesopotamia . . . between the establishment of effective food production and the 'take-off' into urbanism . . ." (Robert M. Adams, *The Evolution of Urban Society* [Chicago: Aldine, 1966], p. 44).

[59] Note the *institutionalization* of the development idea: Institute of Latin American Development Studies (Boston University), Institute of Development Studies (Sussex University), The Interamerican Development Bank, and, of course, such agencies as the International Monetary Fund, the World Bank, the UN's Social Development Fund, etc., all of whose major efforts are directed toward "development."

The germ notion appears still today, in American anthropology, for example, in two forms. One is the notion that one will find a purer, more "aboriginal" state of *a* society by getting the oldest informants (increasingly dying off nowadays) or by "ethnohistory"—i.e., history of a "native" people—in which one reconstructs "how they were 'aboriginally,' " that is, before European conquest when they were still "pure" and "unadulterated" (or, as we say more "technically," when they were still "in their native state" and "unacculturated"). That they had gone through an indefinitely long historical process of shift, change, acculturation, variation, before then—that, in fact, they may not have existed *at all* as a people in some more recent or remote past—is not part of the "ethnohistorical" model.[60] Thus, specific societies conceptually exist indefinitely into the past and presumably went through, as we are so often wont to say in anthropology, "as long an evolution as our own society."[61] The second form is the repetitive and repetitious effort *logically* and *by reflection* to reconstruct a single master-sequence of kinship forms for which it can be claimed either that all societies passed through them or that they represent earliest, earlier, middle, later, and more recent forms developed by society or culture in general in its progressive trajectory, even if specific societies did not go through each of the forms.[62]

Essentialism is miasmic—the State, Warfare, Class, Irrigation, Health, Hunting-and-Gathering, Peasants, Ideology, endlessly on and on, are reified unitary Things with essences. That they vary *empirically* is recognized, but not that they vary *conceptually* or *ontologically*; so one

[60] Contrast "ethnohistory" with the Russian "ethnogenesis." The latter is more Darwinian in approach than the former, in part due to the influence of Marx and Engels. It emphasizes process and relationship rather than essence and state of being, emergence and extinction rather than continuity and fixity.

[61] The phrase is both meaningfully true, in that any observed people must both biologically and socioculturally have derived from somewhere, and meaningless and false, if indeed the observed people are a "breed," "race," or—interestingly enough we have no word to convey a parallel concept in culture—a sociocultural entity which sprouted in some past from an ancestral sociocultural stock. The absence of a term for such an entity reflects the fact that we do not think about "cultures" this way—i.e., with a Darwinian idea of descent, modification, and *speciation*. Note, however, the Russian ethnogenetic model, mentioned in note 60 above, which implies such a model.

[62] See, in this connection, Elman R. Service, *Cultural Evolutionism: Theory and Practice* (New York: Holt, Rinehart and Winston, 1971), and more especially his *Primitive Social Organization: An Evolutionary Perspective* (New York: Random House, 1962) and *The Hunters* (Englewood Cliffs: Prentice-Hall, 1966).

searches for essences and produces endless conceptual and descriptive contradictions, because any asserted essences *will not* encompass the empirically observed variations. The response is to select this or that aspect as being "closer to the essence," to create schools of thought, polemics of partisanship, fads and fancies, rather than to develop a non-essentialist theory of the variants of the phenomena—perhaps (but not necessarily) with stimulus from the Darwinian model.

Through all this, there appears virtually no element of the Darwinian or postgenetic, modified Darwinian ("genetico-Darwinian") models. William Graham Sumner and A. G. Keller, at the beginning of the century, attempted to develop a consistently Darwinian view.[63] Their attempt almost totally disappeared with them, although some of Sumner's concepts, like that of *mores*, remain with us. Only today do we find a scattering of works more or less explicitly Darwinian or genetico-Darwinian in character,[64] or modified from those models, but they have not generated significant response or consistent theoretical and research effort.

These brief comments on the persistence of nineteenth-century models, even if in purified, empiricized, and temporalized forms, and the marginal struggle for survival of Darwinian models in the social sciences lead me to my final, rather speculative remarks. I tend increasingly to believe that sociocultural phenomena do not, *in general*, lend themselves in a simple way to Darwinian or Darwin-derived models, except in certain

[63] See William Graham Sumner, *Folkways: A Study of the Social Importance of Usages, Manners, Customs, Mores, and Morals* (New York: Dover, 1959; first published, 1906); A. G. Keller, *Science of Society* (London: H. Milford, Oxford University Press, 1922); idem, *Societal Evolution: A Study of the Evolutionary Basis of the Science of Society* (New York: Macmillan, 1931); idem, *Man's Rough Road: Backgrounds and Bearings from Mankind's Experience* (New York: F. A. Stokes, 1932).

[64] See the milestone work of Marshall D. Sahlins, *Social Stratification in Polynesia* (Seattle: University of Washington, 1958); A. P. Vayda, "Polynesian Cultural Distributions in New Perspective," *American Anthropologist* 61, no. 5 (1959): 817–828, underlying whose argument is a notion of drift, borrowed from evolutionary genetics; Sahlins's "Segmentary Lineage: An Organization of Predatory Expansion," *American Anthropologist* 63, no. 2 (1961): 322–345; Anthony Leeds, "Microinvention as an Evolutionary Process," *Transactions of the New York Academy of Sciences* 24, no. 8 (1962): 930–943; Alexander Alland, Jr., *Evolution and Human Behavior* (New York: Natural History Press, 1967); idem, *Adaptation in Cultural Evolution: An Approach to Medical Anthropology* (New York: Columbia, 1970); Eugene E. Ruyle, "Genetic-Cultural Pools: Some Suggestions for a Unified Theory of Bio-Cultural Evolution," *Human Ecology* 1, no. 3 (1973): 201–215.

limited spheres and in certain perspectives.[65] The significant units of analysis are, only for some problems, individuals, but much more importantly various types of superorganic, supraindividual orders, such as groups, associations, agencies, bureaucracies, institutions, polities, economies, ecosystems, societies or clusters of societies, and units of multisociety organization,[66] which do not lend themselves to reductionist analysis or Darwinian models. All of these, whether or not they can in themselves be considered teleological, minister to and are influenced by a pre-eminently teleological species, when its members are seen as individuals or in its group-organizational aspect—human beings. The teleological orientation of human individual and group living cannot be denied by anyone but need not be conceived of as "idealist," since it can also be understood in "materialist" frameworks as part of the biology of the species. The teleology of individuals and groups does not logically imply or necessitate a teleology (in a finalist or goal-oriented sense) of the larger system into which they fit,[67] and indeed the latter may be con-

[65] The two major perspectives are those of adaption, short- or long-term, by individuals or organized groups, on one hand, and, on the other, the distribution and ordering of culture traits. Cultural and social *systems* do not seem easily amenable to Darwinian models, despite efforts to introduce "laws" of a "Darwinian" sort, such as the Law of Cultural Dominance, the Law of Evolutionary Potential, and the Principle of Phylogenetic Discontinuity of Progress (Marshall D. Sahlins and Elman R. Service, eds., *Evolution and Culture* [Ann Arbor: University of Michigan, 1961], p. 98)—none of which are testable and all of which are built upon pyramids of deductive hypotheses such as are presented in Appendix C. Further, it is rarely the case that a rationale for choosing the Darwinian model over some other is given: because it has been "successful" in biology, it must therefore ipso facto be a Good for the study of society. The evolution of structured social systems may eventually look more like astronomical evolution, which is clearly non-Darwinian and non-Lamarckian, than it does like any evolution derived from biological experience.

[66] Note, with respect to the last two, such notions as the 'ecumene,' the 'co-tradition,' the 'culture area,' 'civilization,' 'the West,' 'the East,' 'the international labor system,' 'the international division of labor,' 'the multinational corporation,' 'the communist bloc,' 'the world order'—and even the agencies, bureaucracies, and institutions associated with them.

[67] Note C. F. Hockett's remark in *Man's Place in Nature* (see note 32 above), p. 289: "Despite all the above [loci of purpose among humans], the role of purpose in history is very limited. . . . to say the universe contains purpose does not imply that the universe is purposive. . . . There was no purposive behavior in the universe . . . until Earth spawned life and until genetic variation and selection, operating quite purposelessly, had led to life forms that contain homeostatic mechanisms. . . . such mechanisms necessarily have survival value—for if an organism's inbuilt purpose is

trary to the ends and goals of the former (e.g., the incidence of warfare).

In short, the ordering of life characteristics of the human species means that the Darwinian model will not work for sociocultural phenomena, even though the evolutionary result of sociocultural evolution is "descent with modification." The ordering is at least quadruple and arranged in a hierarchy of systems—individuals, groups, society, and culture. The last two, constituting a distinction not clearly recognized until late in the nineteenth century, operate at supraindividual and supragroup levels of organization in large-scale units which are not teleological (in the sense of finalism) and are interlocked across all the varieties of human experience ("cultures") by diffusion and acculturation processes which are specifically human, with virtually no counterparts among non-human animal species, much less plants. The large-scale orders, unlike the biological species with which Darwinian and genetico-Darwinian models are concerned, are structured, but not by random processes (though some probabilistic models may be applicable to certain features), in part because of the teleological orientations of the lower-level constituents, in part because of the supraindividual structure of norms and roles. In view of these characteristics, the Darwinian and especially the genetico-Darwinian model of *random* mutations and variation, of essentially a statistical process of selection (operating with, but not comprehensible in terms of, individuals alone) for or against some *trait* (as opposed to a relationship, which, as a unit, is so important in sociocultural life), and of the importance of sexual selection as the mechanism is of very limited usefulness. The analogous results of biological and sociocultural evolution are the products of different processes operating on different realities; ultimately only obfuscation and theoretical confusion can result from treating them as identical in process because of the similarity of results. The "descent with modification" of sociocultural systems is not of *a* "species"—i.e., *a* sociocultural variety—but of several

---

self-destruction, it is not likely to reproduce. Thereafter, the behavior of *individuals* was in part goal-directed. *Collective* purposes appeared only as selection led to certain types of community-formation. . . . not even human history can usefully be characterized as composed of purposes. . . . the consequence of an action is rarely in exact accord with the intent of the actor, and it is the actions and their consequences, not the purposes, that have so far formed the fabric of human history."

and/or all of them, since sociocultural varieties do not "observe," as it were, "species" boundaries, as in the biological world, but pass over into each other relatively freely by that major process of sociocultural evolution, diffusion.

In sum, the sociocultural evolutionary models beginning in the sixteenth, seventeenth, and eighteenth centuries and reaching a peak in nineteenth- and twentieth-century evolutionism had grasped some fundamental truths about the sociocultural evolutionary process which Darwin missed and which Darwinism could not solve.[68] This seems to me to explain the extraordinary situation Rogers recounts for the Russian reaction to Darwin—*not one* of the models of sociocultural evolution (or even of biological evolution) he describes for them is Darwinian. It explains why there was such a successful "social Lamarckism," as Pfeifer calls it, in the United States. It explains why, in the sociocultural domain, there is no Darwinian evolutionism in France even among the archeologists and possibly even among the paleontologists like Teilhard de Chardin.

Divested of the metaphysics of essentialism, of fixity of species (Darwin's ultimate contribution to sociocultural evolutionism), of immanentism, and of finalism; with a clear specification of the treatment of time; with the retention of some of the empirical insights of earlier evolutionists; and with a modified view of the teleological aspects of human life and the systematic institutional and superorganic orientation that characterizes that life at certain levels, nineteenth–twentieth-century evolutionism can generate viable, productive, focal models for the social sciences. Whether these directions are possible in societies around the world—

---

[68] This has been clearly understood by nonsocial scientists. B. F. Skinner, in *Beyond Freedom and Dignity* (New York: Alfred A. Knopf, 1971), pp. 130–131, remarks: "The parallel between biological and cultural evolution breaks down at the point of transmission. There is nothing like chromosome-gene mechanism in the transmission of a cultural practice. Cultural evolution is Lamarckian in the sense that acquired practices are transmitted. . . . a culture which develops a practice . . . can transmit that practice not only to new members but to contemporaries or to surviving members of an earlier generation [or] through 'diffusion' to other cultures—as if antelopes, observing the usefulness of the long neck in giraffes, were to grow longer necks. Species are isolated from each other by the nontransmissibility of genetic traits, but there is no comparable isolation of cultures. A culture is a set of practices, but it is not a set which cannot be mixed with other sets."

deeply influenced by eschatological Christianity, by growth-oriented capitalism, by future-oriented Marxism, and generally by the desire to create and to believe in the possibility of creating a better world in the future—remains to be seen. But, meanwhile, these same influences, operative in societies immersed in mortal conflict, tend to re-emphasize the elements of evolutionism that need eradication and to recreate evolutionist viewpoints in the social sciences that are destructive of human life and values—for example, developmentalism which destroys "traditional" ways and values, that is, viable extant ones that have stood the practical tests of time, in the name of Progress.

APPENDIX A

Evolutionism in America: The Persistence of the
War between Religion and Science

The dogmas of creation are still being defended, not only by fundamentalist spokesmen in general, but also by representatives of their numbers who claim scientific credentials. Thus, for example, in the *Bible Science Newsletter* 11, no. 11 (November 1973), we find an article by Walter Peters, "geologist" at Round Lake Beach, Illinois, "proving" that "Geology Reinforces the Bible" (pp. 3, 5, 6), while Charles Roessger, "educator" at the Milwaukee Museum, in an article called "Biology and Theology" (pp. 1–3), continually cites Prof. Theodore L. Jahn of U.C.L.A., writing on protozoa, to "prove" that the protozoa are "not simple" in order to deny the "leap" from inorganic absence of life to life because of "the total absence of simple life even on a single-cell level." He then argues that evolutionists accept "spontaneous generation"—the idea that "life spontaneously generated from non-living things"—at a new level, the molecular. In fact, "an evolutionist must believe" in it. Then he argues that this belief should have died with Pasteur's development of the law of biogenesis—"life comes only from life." Spontaneous generation at the molecular level, he says, has been shown to be an erroneous idea by a Dr. Duane Gish, "a biochemist." Since one cannot believe in spontaneous generation, and since life is so complex, one must therefore believe in creation. Mr. Peters gives a long argument against evolutionary time scales, including an attack against various dating methods like the use of Carbon 14 because of their "discrepancies" and faulty assumptions; against "uniformitarian geologists"; and against various geological interpretations. The aim is to collapse geological time scales into Biblical time scales. Of the resulting period correspondences, I give the first few, abbreviating his notes:

ARCHAEOZOIC ERA: Genesis 1:1 to 1:8 (about 4898 B.C.)

PROTEROZOIC ERA: Genesis 1:9 to 6:14 (about 4898 to 3518 B.C.) First orogeny: separation of the waters and appearance of the dry land (earth). . . .

LIPALLIAN INTERVAL: Genesis 6:14 to 7:5 (about 3518 to 3398 B.C.) The last 120 years before the Flood.

PALEOZOIC ERA: Genesis 7:6 to 8:5 (about 3398 to 3397 B.C.) The Noachian Flood, second . . . orogeny: Cambrian to Mississippian periods. . . .

PENNSYLVANIAN PERIOD: Genesis 8:6 to 8:13 (about 3397 B.C.) The waning of the Flood waters, cyclical deposition of vast coal deposits of the world.

And so on to the Pleistocene (through Joshua 6:20, about 2000–1447 B.C.). The *Newsletter* is "DEDICATED TO: Special Creation, Literal Bible Interpretation, Divine Design and Purpose in Nature, A Young Earth, A Universal Noachian Flood, . . ."

These explicit dogmatic expressions comprise only one of the forms that antievolutionism has taken in the United States. In America there has been a very widespread resistance not to be underestimated as a generating milieu for antagonism to science, intellectualism, and financing of scientific endeavor. It may even provide a generating milieu for a generalized ideology which, in attenuated form, affects the axiomatic structures used and choices of models made by scientists. This would be entirely consistent with the linkage of the frontier, do-goodism, Lamarckism, etc., that Pfeifer so lucidly shows in this volume. It is congruent, too, with the mutual interaction found by Richard Hofstadter, in *Social Darwinism in American Thought* (Philadelphia: University of Pennsylvania Press, 1944), pp. 9–12, among social Darwinism, conservatism, a repressive work ethic, and savings. It will be noted in the passages that follow that the attack is on *Darwin* or *Darwinism* in particular, but not necessarily, thereby, on developmentalism, for example. It would be an interesting research project to discover if the kinds of attitudes reflected in the quotations below, put into practice in the education system, selectively affect American developmentalist views and hence the policies that flow from them.

In connection with the resistance, one usually mentions the Tennessee

Monkey Trial of the 1920's. Unhappily that was not the end. Observe, for example, the biology-textbook controversies in Texas and California in the 1960's and 1970's. Texas at least allows high-school biology textbooks to speak about evolution "as a hypothesis" (but not as a theory). Note the letter from John E. Summers (*Science*, November 9, 1973, p. 535):

It's fantastic! It's unbelievable! It's out of sight! I'm talking about the State of California science textbooks for the elementary grades. The textbooks on science, on life, and on heredity do not mention, . . . anywhere in the text, the vocabulary, or the index, the word "evolution" or any form of the word "evolve." . . . They tell about and show pictures of many great scientists. But how about Charles Darwin. . . . No picture. Only in grade 5 is he mentioned and in this back-handed way: "George Darwin, son of the famous English scientist, Charles Darwin." . . . But for what is Charles Darwin famous? You won't find it in the California elementary science textbooks. As an extra class activity, text 5 suggests: "*On Your Own.* 1. Who was George Darwin's father, and for what famous work is he known?"

Note also the following (from Letters to the Editor, *Bulletin of the American Association for the Advancement of Science* 18, no. 3 [1973]: 6–7). This (unsigned) "letter" reports that on March 17, 1973, the radio program "Life Line," originating in Dallas, Texas, and heard throughout the United States, broadcast these thoughts:

A group of scientists adopted a sad resolution. They want the Genesis Creation omitted from public schools because they say there is nothing scientific about it. It is strange how men of science can be so unscientific. . . . On October 13, 1972, [the Commission on Science Education of the AAAS] held its regular fall meeting in Washington, D.C. In that meeting the Commission . . . adopted a resolution. . . . There is some essential background to [their] unanimous resolution. Some liberal-minded scientists move heaven and earth to clear the way for the compulsory study of the theory of human evolution as the true, scientific record of human creation. Our textbooks today are soaked through with this altogether unproven, untested, and unscientific explanation of one of the great secular mysteries of all time. . . . Once the Darwinian cult had won the legal battle to allow the teaching of the Darwin theory of animal-to-man human evolution as a scientific theory, humanist scientists thought they had it made. But then came an impressive array of religious scientists—and there are truly such people—calling for equal time.

Their argument is that it is just as valid to teach the Genesis version of human creation as an acceptable theory as it is to teach the Darwin theory of creation. The anti-Genesis crowd of scientists is amazingly inconsistent. They create a set of human, secular, and scientific rules which they say must be applied to research into any scientific fact. They then claim that the Bible story of creation meets none of these major tests of science and, therefore, is not a valid theory.

Let it here be clearly stated that a vastly greater percentage of factors involved in the Bible story of man's origin have been proved by secular and scientific discoveries than in the Darwin theory of animal-to-man evolution.

The "Life Line" program then gave the entire resolution, which did *not* in fact say some of the things that "Life Line" said it did (it *did* say that the scientists were opposed to teaching religious accounts of creation in *science* classes; that, from a *scientific* point of view, there was no acceptable alternative theory that explained all the kinds of data gleaned from geologists, paleontologists, biologists, geneticists, etc.; they *did* say that religious statements of creation may be believed but are not subject to scientific tests). "Life Line" goes on to further attacks on the scientists, "who owe much to the Bible for the factual material they have at hand." Note, in this connection, the *Bible Science Newsletter* cited above. It is published in Idaho and makes extensive reference to the Lutheran Research Forum in Wisconsin. In other words, creationism, with its antiscientism and its use of science to bolster its worldview, is widespread in the United States.

In connection with the above, note the following books, which attempt reconciliations between science and Christian dogmas; they are written, in part, by persons involved in scientific activity. Russell L. Mixter, ed., *Evolution and Christian Thought Today* ("a symposium by thirteen Christian scientists and theologians") (Grand Rapids, Mich.: Eerdmans, 1959) ; and P. G. Fothergill, *Evolution and Christians* (London: Longmans, 1961), the latter under imprimatur. Note, also, the papal encyclical of 1950, *Humani Generis*, which maintains the creationist position with respect to soul, though the physical body is allowed to have evolved.

APPENDIX B

Evolutionism in the Soviet Union and America:
The Persistence of the War between Materialism and Idealism

Lewis Henry Morgan's *Ancient Society, or Researches in the Lines of Human Progress from Savagery through Barbarism to Civilization* (ed. Eleanor B. Leacock [Cleveland: World Publishing Company, 1963]; first published, 1877) was discovered by Marx and Engels and was used by the latter in writing *The Origin of the Family, Private Property, and the State* (New York: International Publishers, 1942; first published, 1884), both in terms of its scheme of evolution, which generally confirmed various postulations about early society and its evolution that Marx and Engels had made almost from their earliest writings, and in terms of its data, which gave substantive illustration of the evolutionary sequences and of actual cases of different types of society. Engels stated that "the key [to the history of primitive society] . . . was provided by Morgan only in 1877" ("Preface to the Second Edition" [1885], *Anti-Dühring* [Moscow: Progress Publishers, 1959], p. 15). Morgan's strongly stage-conscious schema is reflected also in Engels's stages and periods of history. From the point of view of this paper, Morgan's "confirmation" of Marx's and Engels's postulates is not surprising, since some of the underlying axioms both of Morgan and of Marx and Engels—and some of their logic, derived from the same eighteenth-century roots—are alike. Even the fact that both display mixtures of idealism and materialism is to be expected (see discussion of Morgan's and Marx-Engels's idealism in the text).

The Morgan-Engels viewpoint became part of the world view and ideology of socialism and communism and forms part of the basis for the prediction of evolutionary outcomes of advanced forms of communist society. Even Morgan said, "It seems probable that democracy, once universal in a rudimentary form and repressed in many civilized states, is destined to become again universal and supreme" (p. 351 in the edition cited above). However, the specific formulation of stages or periods disintegrated under the cumulation of contrary empirical data (see ch. 6, "Lewis H. Morgan," in Robert H. Lowie's *The History of Ethnological Theory* [New York: Rinehart, 1937], pp. 54–67). Since, however, the axiomatic basis, the logic, and the schema of Morgan's theory are *ideologically* necessary, and perhaps even *logically* necessary, to current communist theory, there is still great concern in communist circles with the "periodization" problem. Irmgard Sellnow's *Grundprinzipien einer Periodisierung der Urgeschichte* (Berlin: Akademie-Verlag, 1961) is a thoroughgoing investigation of the philosophical and empirical foundations for periodization and a study of the outcomes of various efforts at periodizing from Hesiod through the Middle Ages, the Enlightenment, and the nineteenth century, to Leslie White, Julian Steward, and others in the United States, and an array of scholars in the Soviet Union. More interesting, perhaps, for our present purpose, is the three-session symposium on the "reperiodization" of Morgan held in Moscow in 1964 and published in volume 4 of the *VIIme Congrès International des Sciences Anthropologiques et Ethnologiques* (Moscow: Science Publishers, 1967), pp. 439–511, ed. A. I. Perchitz. The symposium was organized by the Russians, who felt reperiodization to be an ur-

gent anthropological problem. In fact, the events of the meeting made it evident that it was indeed critical and central to the Russians—it was the key to
a series of ideological battles in the political sphere. All the major speakers,
except Sellnow, presented their papers in the first morning session. About
fifteen minutes from the end, one speaker, E. Konstas (her name, like those
of the other participants, is printed in Cyrillic, though their papers are published in their native languages), set out to defend Julian Steward's multi-
lineal evolutionism, while degrading Morgan and making a sharp attack on
Marx and "his" use of Morgan (ibid., pp. 455–461) because Marx could
not cope with Oriental despotism and Morgan never mentioned it. D. A.
Olderogge, the chairman, had a copy of the paper, saw the drift, and after a
few minutes cut her off. Sellnow's paper (ibid., pp. 474–479—the papers
were not published in the order read), quite critical of Morgan's immanentism
and idealism while recognizing his major anthropological achievement and
his importance for Marxist historiography, was in German. She was told that
there were no concurrent translators for German. If she wanted to give her
paper she would have to translate it into English or French in the lunch period—which she did (with Frederick Rose's help). At the end of the first session all audience participants who wished to comment were asked to inscribe
their names—and *this list was read aloud.* In the afternoon, after Sellnow's
paper, the comments were given (ibid., pp. 479–488, all published in Russian, including those, like Margaret Mead's, David Maybury-Lewis's, and
mine, made in English!). The last two—Maybury-Lewis's and mine—were
very strong attacks on the periodization. After questioning why one should
periodize at all, I commented that the idealist strain necessarily showed in
Morgan, given his era. The effect of Konstas's and Sellnow's papers (both
of which Olderogge had obviously tried to suppress) and the drift of the comments plainly dismayed him. After the inscribed list had all commented, there
suddenly appeared other commentators (ibid., pp. 488 ff.) who had not been
read as on the list, all Russian, all senior anthropologists of the scientific
establishment, who each made pronouncements of the order: "We *know* that
patrilineality succeeded matrilineality"—utterances of dogmatic faith in the
Morganian periods. In short, they exemplified ideological and scientific forces
of a non-Darwinian evolutionism whose beginnings lie long before Darwin
and one of whose chief expressions emerged in 1877.

After the Moscow Congress I asked Professor Frederick Rose in East Berlin
what he thought lay behind the events of the periodization symposium. He
pulled off his shelf two editions of Engels's *Origin of the Family, Private
Property and the State*, both published in the U.S.S.R., the first about 1952
(just before Stalin's death), the second some years later, during the opener

Krushchev years. A footnote appears in the earlier edition which says, in effect, "Here, Engels falls into a dualism," meaning an idealist-materialist dualism. Rose said the note had been attributed to Stalin. The later edition no longer has the footnote. Rose felt that the nature of the hidden jockeying going on at, or through, the symposium was that Stalin, in his acquisition of power carried to excess, had tried to discredit Engels (and, indirectly, Morgan); therefore, to *re-establish* Morgan and Engels as good materialists was an appropriate *political* act in the mid-sixties, a statement of loyalty and commitment to the (new) Establishment. The effort backfired because of the strong statements (especially, perhaps, Sellnow's, Konstas's, and my own) arguing that there were marked strains of idealism and/or ideology in Morgan and Engels. These remarks were of course anathema to the Soviet colleagues, who had to reaffirm the materialist basis by suppressing Konstas's paper, unsuccessfully trying to stave off Sellnow's, and dragging out their more venerable anthropologists—who had survived the Stalin years—like B. O. Dolgikh and D. A. Olderogge himself, to argue that their archaeology and ethnology proved the evolutionary order in a materialist way, rather than that it had been established by imposition of an idealist-derived immanent principle.

APPENDIX C

A Nineteenth–Twentieth-Century Evolutionary Methodology: The Conditional Mode as Historic "Proof"

Some examples: "There were innumerable occasions upon which mutual aid would be a decided asset in the business of daily living, and we may assume that, given the capacity and the means of cooperation, this mutual aid would be forthcoming. Did men and women have some recognition of the value of cooperation, some appreciation of the advantages of mutual aid, conscious or unconscious, at the very outset of their career as human beings? It seems reasonable to suppose that they did; an appreciation of cooperation would go hand in hand with the capacity for it" (Leslie A. White, *The Evolution of Culture: The Development of Civilization to the Fall of Rome* [New York: McGraw-Hill, 1959], pp. 80–81). One appreciates the piling of deductive hypothesis upon deductive hypothesis, the lack of any empirical proof whatever, and the single-time origin of the state of being human in this quotation. Again, "This stage is hypothetical and idealized. No known society is completely isolated, but relative isolation is a significant factor and to place a hypothetical familial group in a social vacuum helps make clear the significance of complexity. Only in isolation would a family be altogether without sociocentric terms. If one group of 'other' people were present then there would undoubtedly be names for both 'we' and 'they'—and these names, of course, would be sociocentric. And they would be used as status terms to the extent that social relations occurred between the two groups" (Elman R. Service, *Cultural Evolutionism: Theory and Practice* [New York: Holt, Rinehart and Winston, 1971], p. 105). Another: "In a circumscribed, densely settled area a defeated group could not make a strategic withdrawal. There would be no place for it to go; all of the arable land would be occupied. Instead it would have to remain where it was and suffer the consequences. And the consequences of defeat under these conditions would generally be, first, the payment of tribute, and, at a later stage, outright incorporation into the territory of the victor. Under the necessity of having to pay tribute in kind, the vanquished group would have to work their lands even more intensively than before"—and thus arise the State, internal stratification, slavery, specialization, and so on (Robert L. Carneiro, "Slash-and-Burn Cultivation among the Kuikuru and Its Implications for Cultural Development in the Amazon Basin," in *The Evolution of Horticultural Systems in Native South America: Causes and Consequences*, ed. J. Wilbert, *Antropológica* [Caracas], Supplemental Publication, no. 2 [1961], pp. 60–61). One more: "It is the amount [of energy] so trapped . . . and the degree to which it is raised to a higher state that would seem to be the evolutionary measure of life; that would seem to be the way that a crab is superior to an amoeba, a goldfish to a crab, a mouse to a goldfish, a man to a mouse. We put all this in quite qualified form because we lack any competence in physical biology, and do not know how to specify the operations required to ascertain this measure" (Marshall D. Sahlins, "Evolution: Specific and General," in *Evolution and Culture*, ed. idem and Elman R. Service [Ann Arbor: University of Michigan, 1960], p. 21). Finally: "It may be assumed, therefore, that primal human social organization was more similar to that of anthropoid apes than to that of any other living sub-human form. Moreover it can be assumed that primal human organization differed from ape societies in the direction of the forms exhibited by contemporary human societies. Hence we may interpolate from these two forms and converge on the conditions which constitute the limits within which primal human society could have developed. . . . The social behavior of primitive peoples resembles that of apes first because of the fundamental similarities of ape and human biological organisms and second because primitive human societies have not undergone the vast changes involved in the rise of civilization" (Gertrude E. Dole, "A Preliminary Consideration of the Origin of Incest Prohibitions, Based on the Comparative Study of Human and Sub-human Primate Societies," mimeographed, ca. 1961). Note the unilineal principle on which the whole last argument is built. This kind of literature is not always written in the conditional but often appears in (or readily shifts into) simple declarative statements ("it was the case that . . .") for which no evidence is adduced. Most of White's book and most of the Sahlins-Service book are phrased that way. See, also, Anthony Leeds, "The Evolution of Horticultural Systems in Native South America: Causes and Consequences—A Symposium; Introduction," in *The Evolution of Horticultural Systems* (cited above), pp. 1–12.

NOTES ON CONTRIBUTORS

MICHELE L. ALDRICH (b. 1942) is assistant editor of the Joseph Henry Papers at the Smithsonian Institution. She is preparing a dissertation for The University of Texas on the New York Natural History Survey, 1836–1845.

NAJM A. BEZIRGAN (b. 1931) is an associate professor in the Department of Oriental and African Languages and Literatures and the Center for Middle Eastern Studies at The University of Texas at Austin. Educated at the Universities of Baghdad and Manchester, he is a specialist in Islamic philosophy.

ILSE N. BULHOF (b. 1932), assistant professor of history at The University of Texas at Austin, was educated at the Universities of Groningen and Utrecht, the Netherlands. She is the author of *Apollo's Wiederkehr*, a book in German on Nietzsche's philosophy of history.

FREDERICK BURKHARDT (b. 1912), the author of *Science and the Humanities*, taught philosophy at the University of Wisconsin from 1937 to 1943 and in 1946–1947. He was president of Bennington College from 1947 to 1957 and president of the American Council of Learned Societies from 1957 to 1974.

THOMAS F. GLICK (b. 1939) is an associate professor of history and geography at Boston University. He is the author of *Irrigation and Society in Medieval Valencia* and is presently pursuing research on medieval Spanish history as well as on the reception of scientific ideas in modern Spain.

M. J. S. HODGE (b. 1940) is a specialist in post-Newtonian accounts of origins and species (including Darwinian theories). He is a member of the History and Philosophy of Science division of the Philosophy Department at the University of Leeds.

DAVID L. HULL (b. 1935), professor of philosophy at the University of Wisconsin—Milwaukee, has written on the history and philosophy of science in the nineteenth century and is the author of *Darwin and His Critics* and *Philosophy of Biological Science.*

ANTHONY LEEDS (b. 1925) is a professor of anthropology at Boston University. He has long been interested in evolutionary theory in anthropology and is the author of "Microinvention as a Cultural Evolutionary Process" and "Capitalism, Colonialism, and War—An Evolutionary Perspective."

WILLIAM M. MONTGOMERY (b. 1942) is pursuing research in Germany on the reception of evolutionary biology. He is the author of "The Origins of the Spiral Theory of Phyllotaxis."

ROBERTO MORENO (b. 1943) holds a research appointment at the Instituto de Investigaciones Bibliográficas in Mexico City and teaches History of Science at the Universidad Nacional Autónoma de México. He has written extensively on Mexican science.

HARRY W. PAUL (b. 1933), professor of history at the University of Florida, is the author of books and articles on Catholicism and on science in modern France. He is especially interested in the relationship between science and Catholicism.

EDWARD J. PFEIFER (b. 1920) is a professor of history at St. Michael's College, Winooski, Vermont. His research interests center upon the impact of evolutionary ideas in the United States.

JAMES ALLEN ROGERS (b. 1929), a specialist in Russian history and the history of Darwinism, is a professor of history at Claremont Men's College and Claremont Graduate School. He is the author of a forthcoming book, *Darwinism and Russian Thought*, and the editor of *Peter Kropotkin's Memoirs of a Revolutionist.*

ROBERT E. STEBBINS (b. 1931) is a professor of history at Eastern Kentucky University. He is a social historian whose research extends into the impact of scientific ideas on French society and culture.

ALEXANDER VUCINICH (b. 1914), professor of history at The University of Texas at Austin, is the author of *The Soviet Academy of Sciences* and *Science in Russian Culture.* His basic interest is in the interaction of modern science and society.

INDEX